McGraw-Hill

Illustrative Mathematics™

Course 2

Mc
Graw
Hill

Cover: (l, br)McGraw-Hill Education, (tr)Vijay kumar/DigitalVision Vectors/Getty Images, (cr)Michael Tatman/Shutterstock

mheducation.com/prek-12

Send all inquiries to:
McGraw-Hill Education
STEM Learning Solutions Center
8787 Orion Place
Columbus, OH 43240

ISBN: 978-0-07-689387-4
MHID: 0-07-689387-1

Illustrative Mathematics, Course 2
Student Edition, Volume 1

Printed in the United States of America.

10 11 12 LMN 28 27 26 25 24 23 22

Contents in Brief

(t to b, l to r) dibrova/Shutterstock, Collins93/Shutterstock, Blend Images/Image Source, Andrey Armyagov/123RF, Creative Travel Projects/Shutterstock, Daniel Dempster Photography/Alamy Stock Photo, HDRExposed - Dave DiCello Photography/Flickr RF/Getty Images, View Stock/View Stock RF/Getty Images, Corey Jenkins/Image Source

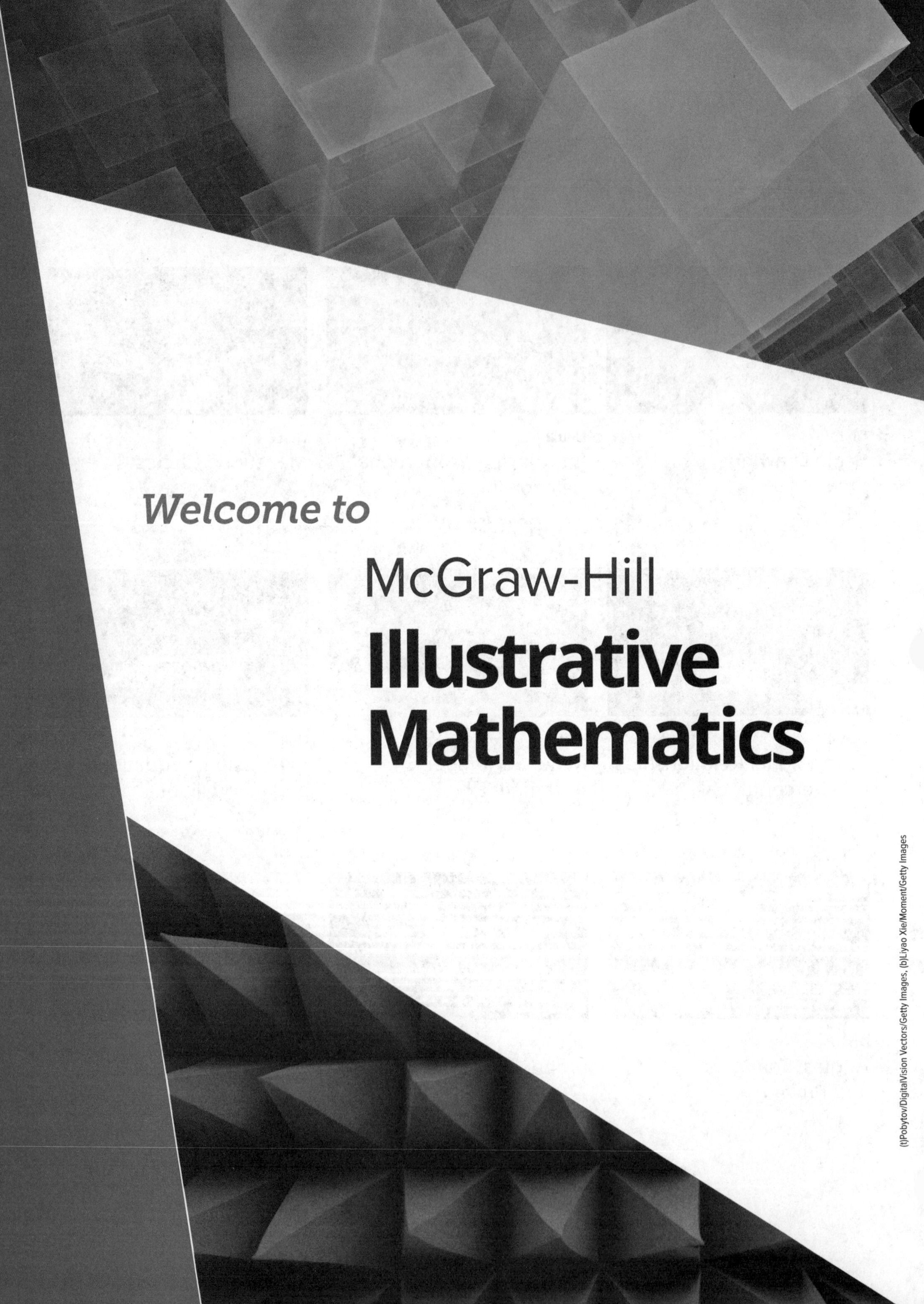
Welcome to
McGraw-Hill
Illustrative
Mathematics

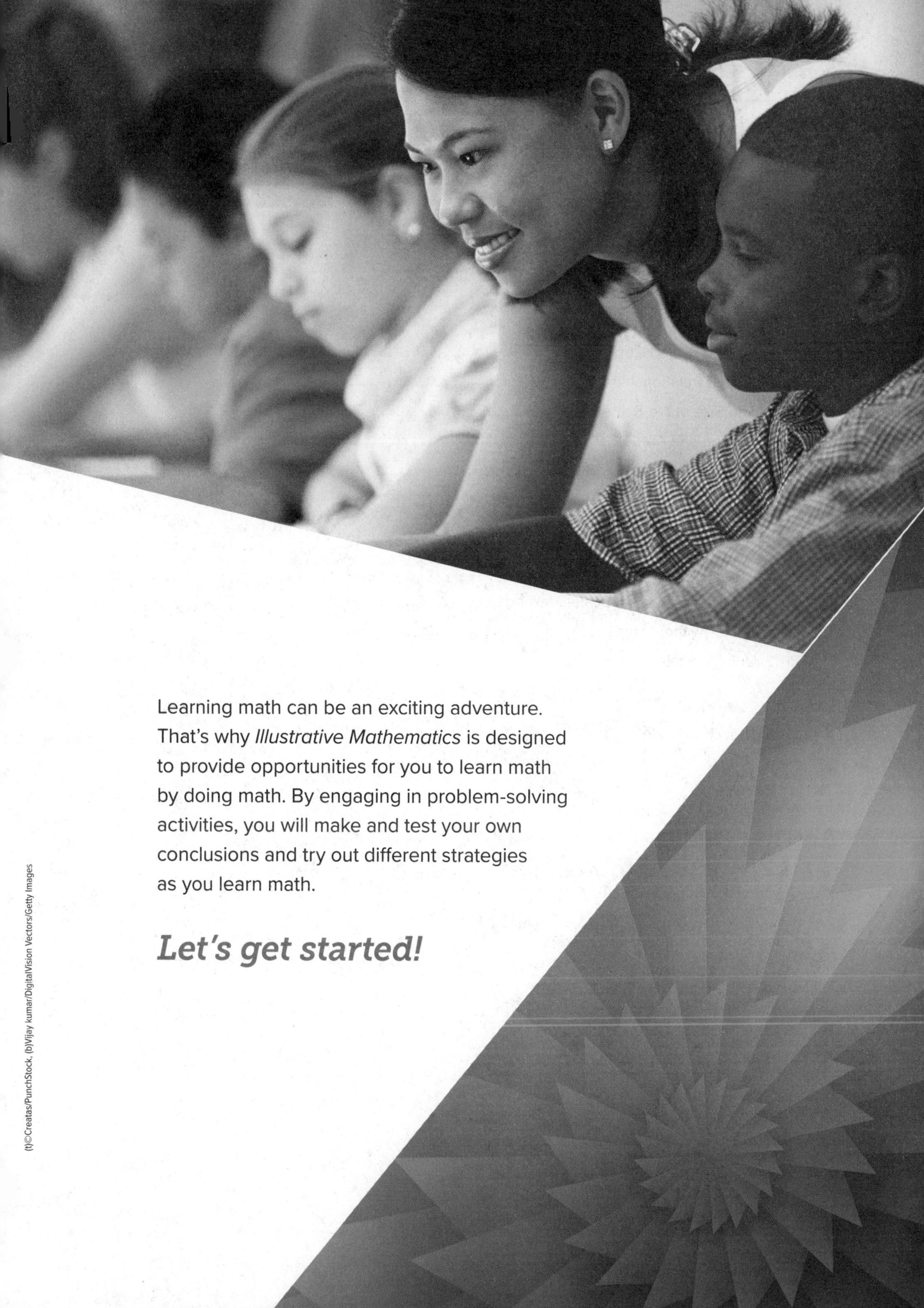

Learning math can be an exciting adventure. That's why *Illustrative Mathematics* is designed to provide opportunities for you to learn math by doing math. By engaging in problem-solving activities, you will make and test your own conclusions and try out different strategies as you learn math.

Let's get started!

Unit 1

Scale Drawings

Scaled Copies

Scale Drawings

Let's Put It to Work

dibrova/Shutterstock

Unit 2

Introducing Proportional Relationships

Collins93/Shutterstock

Unit 3

Measuring Circles

Andrey Armyagov/123RF

Unit 4

Proportional Relationships and Percentages

Proportional Relationships with Fractions

Percent Increase and Decrease

Applying Percentages

Let's Put It to Work

Unit 5

Rational Number Arithmetic

Creative Travel Projects/Shutterstock

Daniel Dempster Photography/Alamy Stock Photo

Unit 6

Expressions, Equations, and Inequalities

Representing Situations of the Form $px + q = r$ and $p(x + q) = r$

Solving Equations of the Form $px + q = r$ and $p(x + q) = r$ and Problems that Lead to Those Equations

Inequalities

Writing Equivalent Expressions

Let's Put It to Work

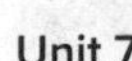

Unit 7

Angles, Triangles, and Prisms

Angle Relationships

Drawing Polygons with Given Conditions

Solid Geometry

Let's Put It to Work

Unit 8

Probability and Sampling

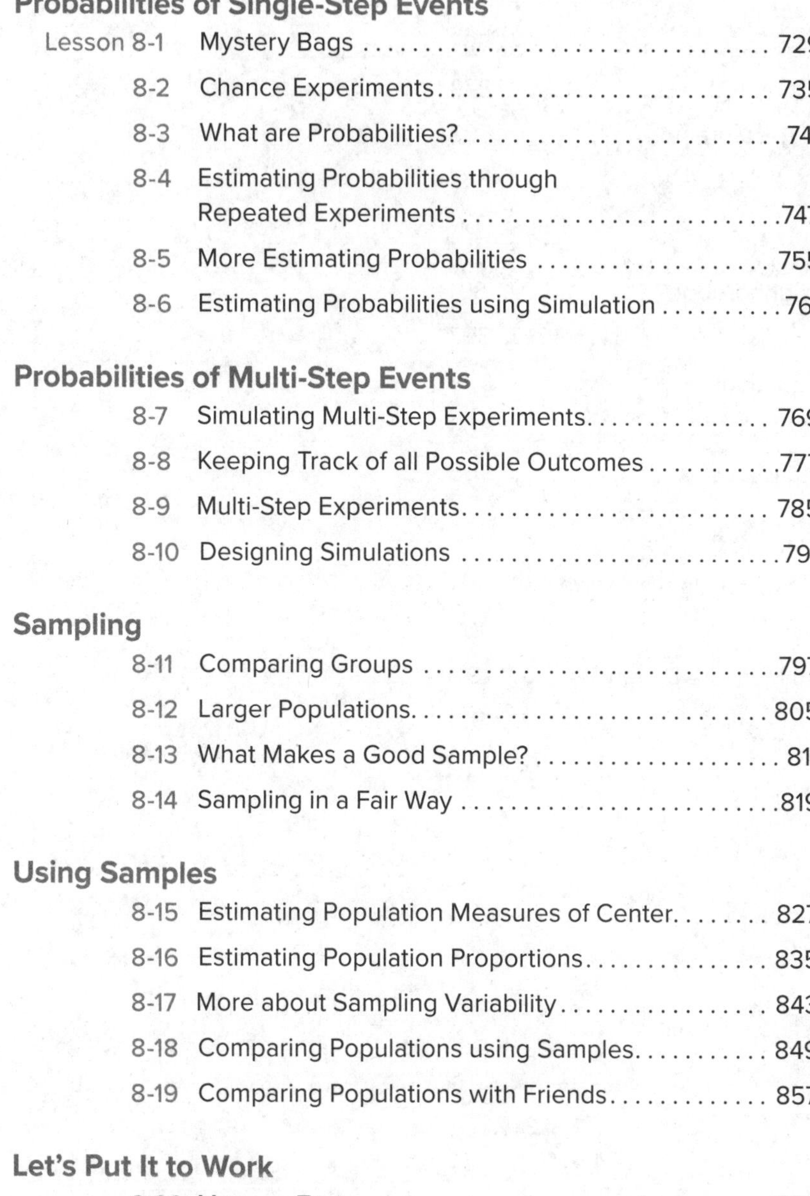

Probabilities of Single-Step Events

Probabilities of Multi-Step Events

Sampling

Using Samples

Let's Put It to Work

Corey Jenkins/Image Source

Unit 9

Putting It All Together

Running a Restaurant

Making Connections

Designing a Course

Unit 1

Scale Drawings

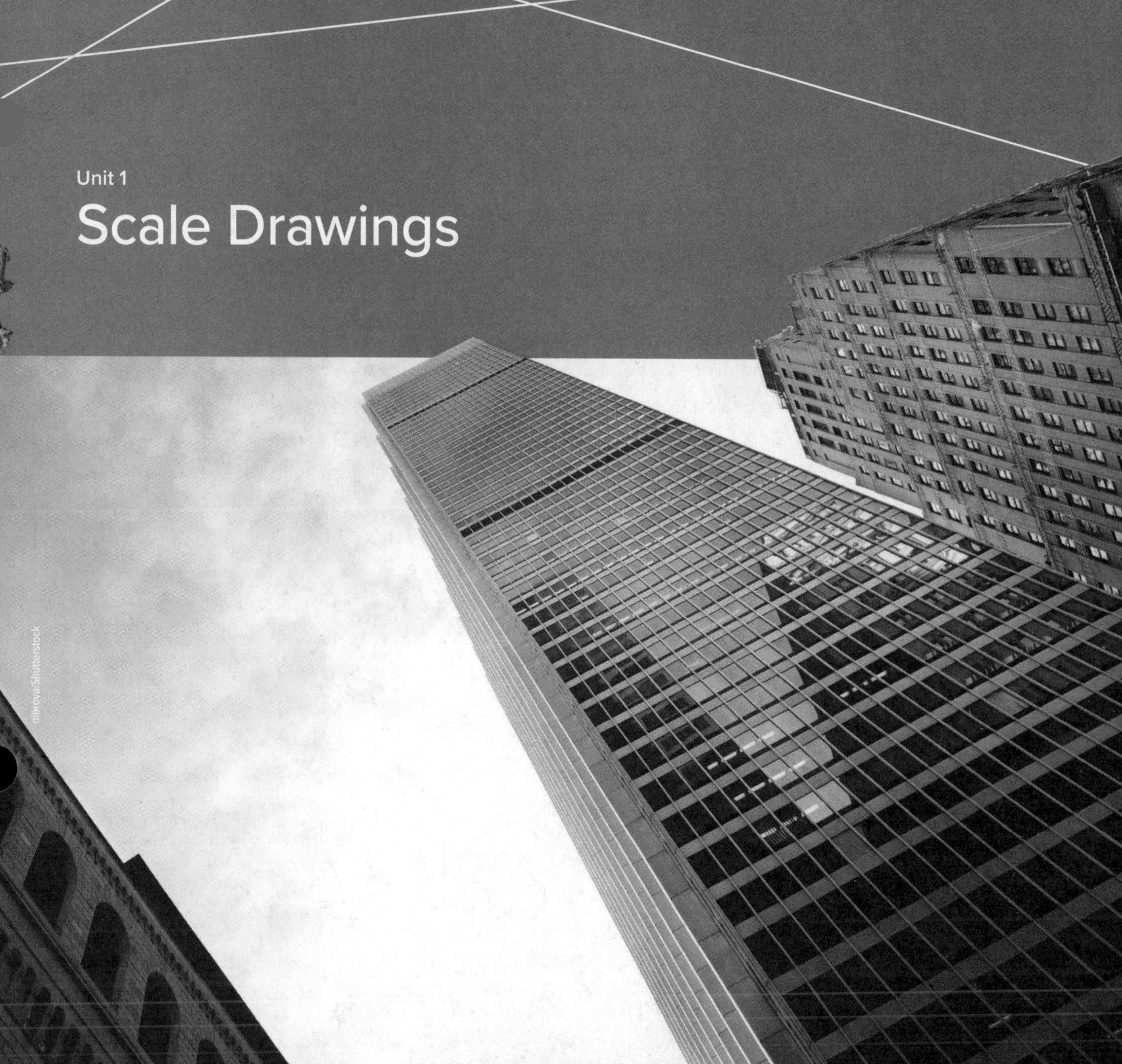

dibrova/Shutterstock

Architects use scale drawings to design buildings. You'll learn more about scale drawings in this unit.

Topics

- Scaled Copies
- Scale Drawings
- Let's Put It to Work

Unit 1

Scale Drawings

Scaled Copies

Scale Drawings

Let's Put It to Work

Lesson 1-1

What Are Scaled Copies?

NAME ______________________ DATE ____________ PERIOD ____________

Learning Goal Let's explore scaled copies.

Warm Up

1.1 Printing Portraits

Here is a portrait of a student.

1. Look at Portraits A–E. How is each one the same as or different from the original portrait of the student?

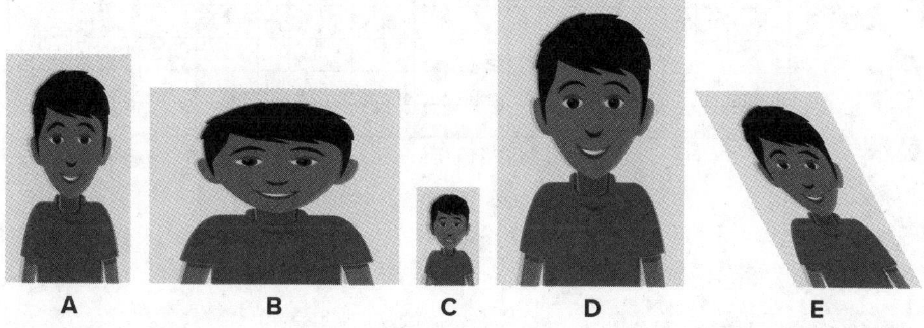

2. Some of the Portraits A–E are **scaled copies** of the original portrait. Which ones do you think are scaled copies? Explain your reasoning.

3. What do you think "scaled copy" means?

Activity
1.2 Scaling F

Here is an original drawing of the letter F and some other drawings.

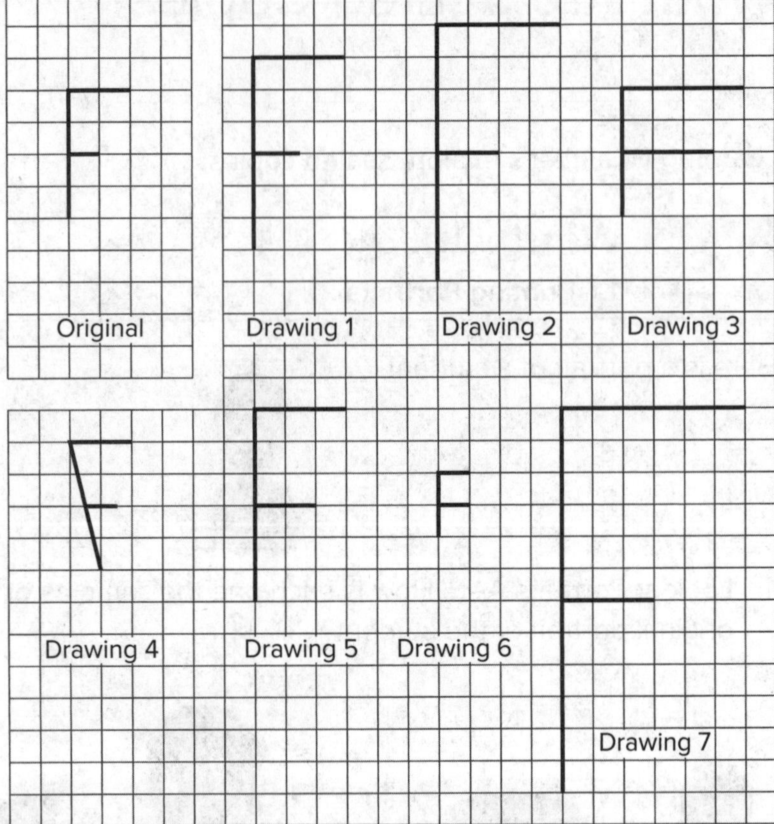

1. Identify **all** the drawings that are scaled copies of the original letter F. Explain how you know.

2. Examine all the scaled copies more closely, specifically the lengths of each part of the letter F. How do they compare to the original? What do you notice?

3. On the grid, draw a different scaled copy of the original letter F.

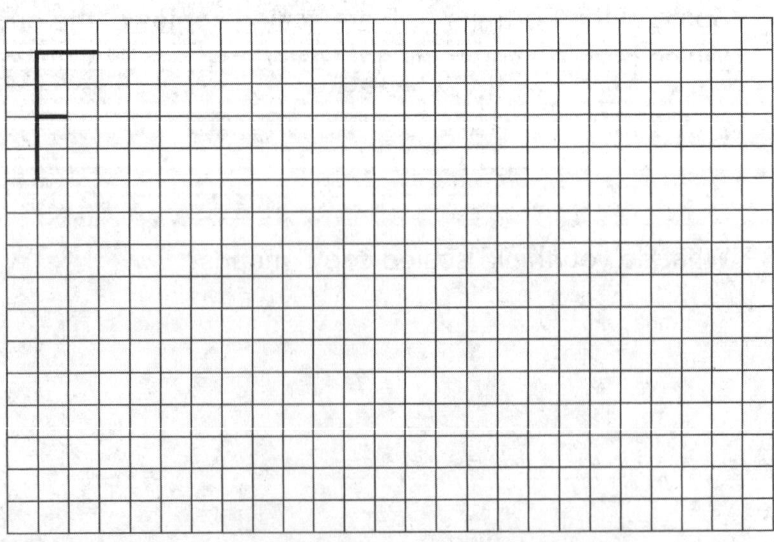

NAME ______________________ DATE ____________ PERIOD ________

Activity

1.3 Pairs of Scaled Polygons

Your teacher will give you a set of cards that have polygons drawn on a grid. Mix up the cards and place them all face up.

1. Take turns with your partner to match a pair of polygons that are scaled copies of one another.
   a. For each match you find, explain to your partner how you know it's a match.
   b. For each match your partner finds, listen carefully to their explanation, and if you disagree, explain your thinking.
2. When you agree on all of the matches, check your answers with the answer key. If there are any errors, discuss why and revise your matches.
3. Select one pair of polygons to examine further. Draw both polygons on the grid. Explain or show how you know that one polygon is a scaled copy of the other.

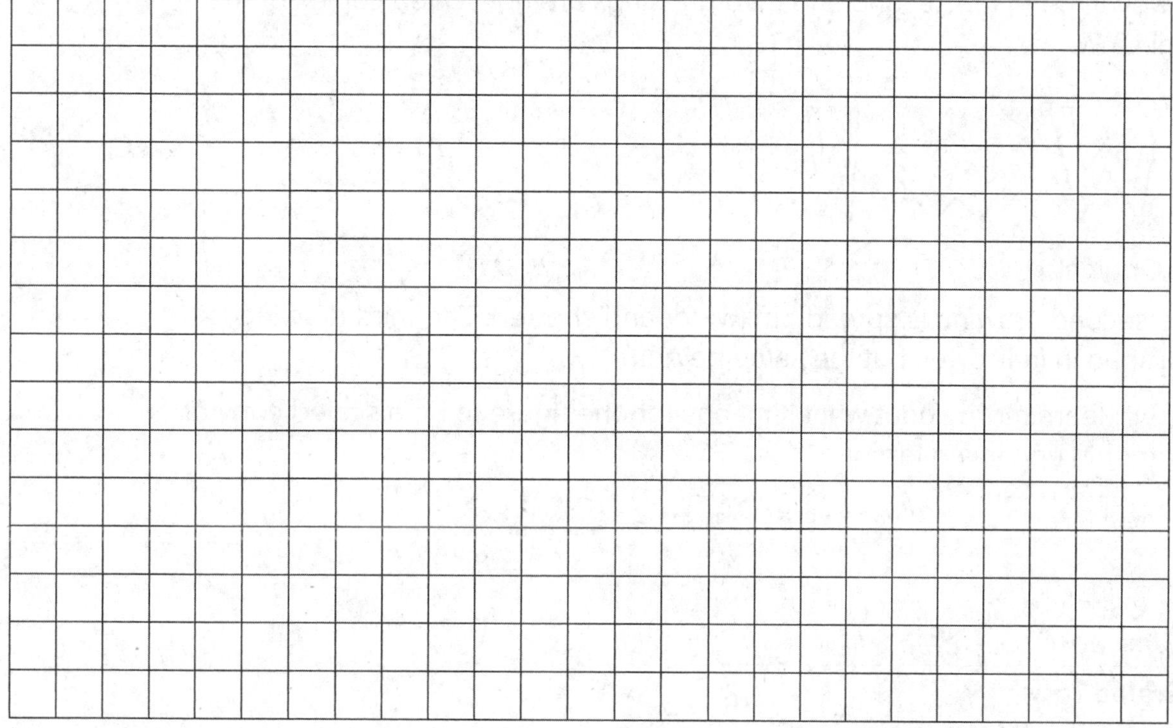

Are you ready for more?

Is it possible to draw a polygon that is a scaled copy of both Polygon A and Polygon B? Either draw such a polygon, or explain how you know this is impossible.

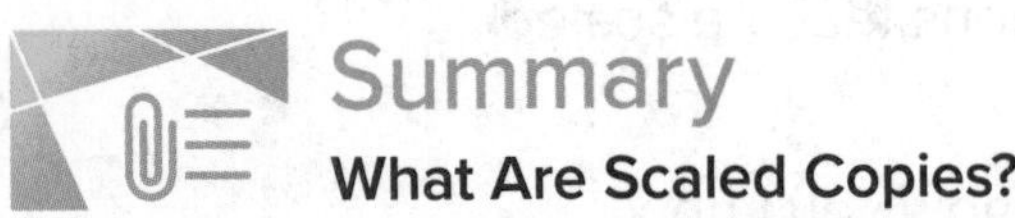

Summary

What Are Scaled Copies?

What is a **scaled copy** of a figure? Let's look at some examples.

The second and third drawings are both scaled copies of the original Y.

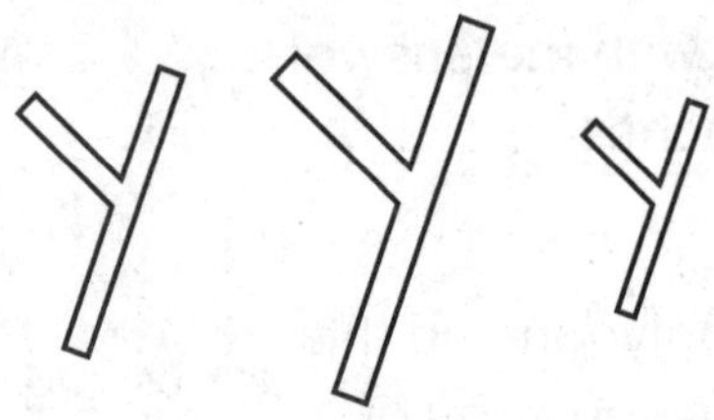

Original

However, here, the second and third drawings are *not* scaled copies of the original W.

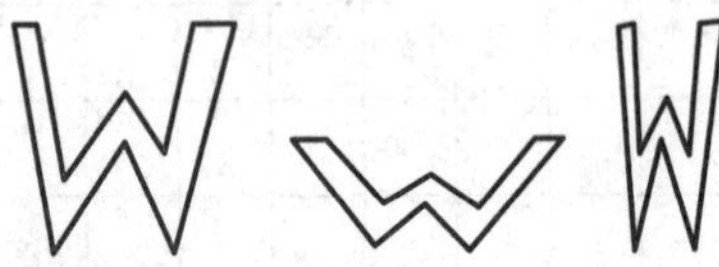

Original

The second drawing is spread out (wider and shorter). The third drawing is squished in (narrower, but the same height).

We will learn more about what it means for one figure to be a scaled copy of another in upcoming lessons.

Glossary

scaled copy

NAME ______________________ DATE ______________ PERIOD ____________

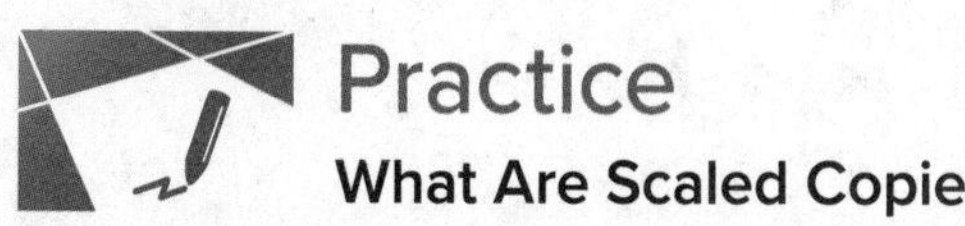

Practice

What Are Scaled Copies?

1. Here is a figure that looks like the letter A, along with several other figures. Which figures are scaled copies of the original A? Explain how you know.

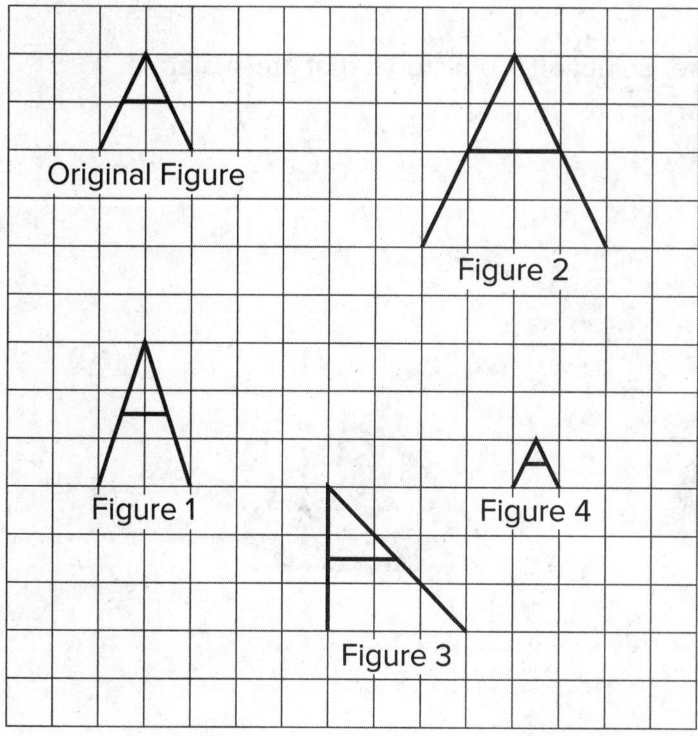

2. Tyler says that Figure B is a scaled copy of Figure A because all of the peaks are half as tall.

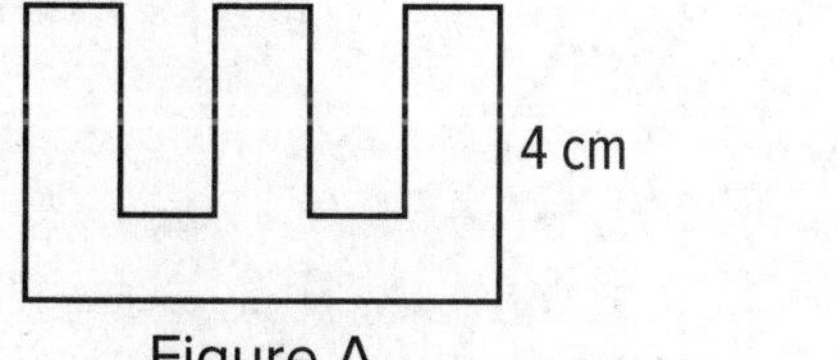

Figure A

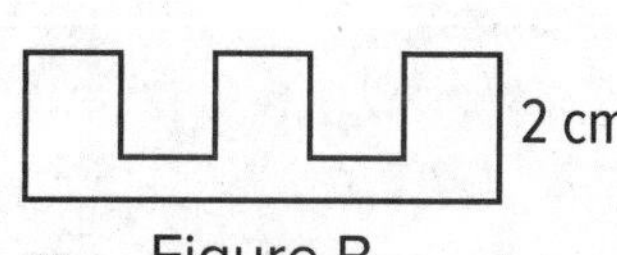

Figure B

Do you agree with Tyler? Explain your reasoning.

3. Here is a picture of a set of billiard balls in a rack.

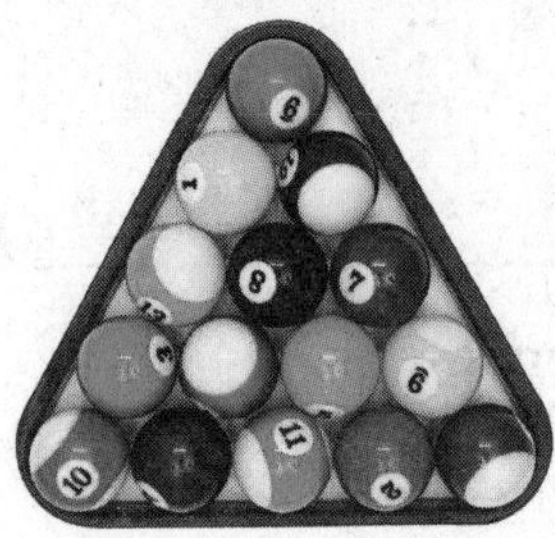

Here are some copies of the picture. Select **all** the pictures that are scaled copies of the original picture.

A. 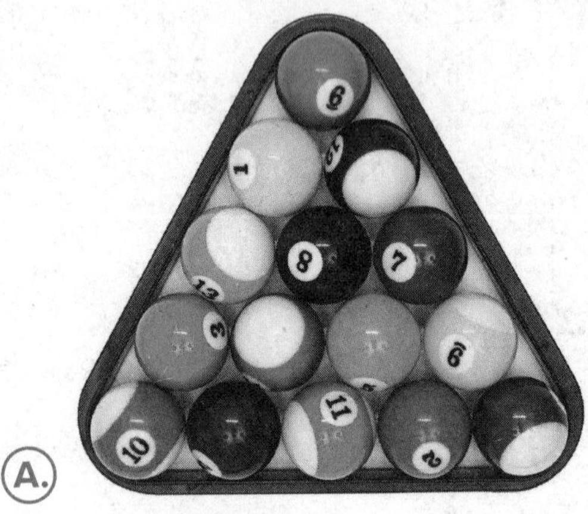

C. 

B. 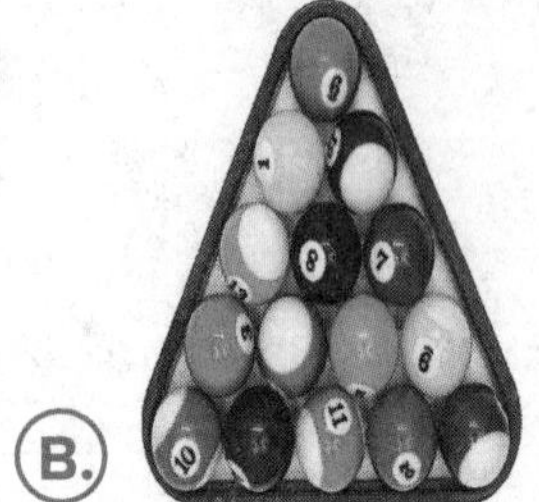

D. 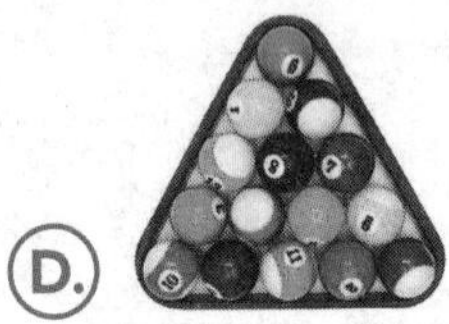

4. Complete each equation with a number that makes it true.

a. 5 • _______ = 15

b. 4 • _______ = 32

c. 6 • _______ = 9

d. 12 • _______ = 3

Lesson 1-2

Corresponding Parts and Scale Factors

NAME ______________________ DATE ______________ PERIOD ____________

Learning Goal Let's describe features of scaled copies.

Warm Up

2.1 Number Talk: Multiplying by a Unit Fraction

Find each product mentally.

1. $\frac{1}{4} \cdot 32$

2. $(7.2) \cdot \frac{1}{9}$

3. $\frac{1}{4} \cdot (5.6)$

Activity

2.2 Corresponding Parts

Here is a figure and two copies, each with some points labeled.

Original

Copy 1

Copy 2

1. Complete this table to show **corresponding parts** in the three figures.

Original	Copy 1	Copy 2
point P		
segment LM		
	segment EF	
		point W
angle KLM		
		angle XYZ

2. Is either copy a scaled copy of the original figure? Explain your reasoning.

3. Use tracing paper to compare angle *KLM* with its corresponding angles in Copy 1 and Copy 2. What do you notice?

4. Use tracing paper to compare angle *NOP* with its corresponding angles in Copy 1 and Copy 2. What do you notice?

NAME ______________________ DATE ______________ PERIOD ____________

Activity
2.3 Scaled Triangles

Here is Triangle O, followed by a number of other triangles.

Triangle O

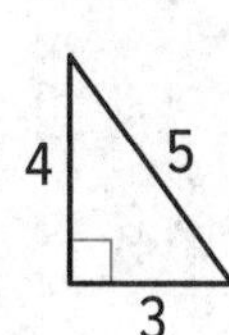

Triangle A

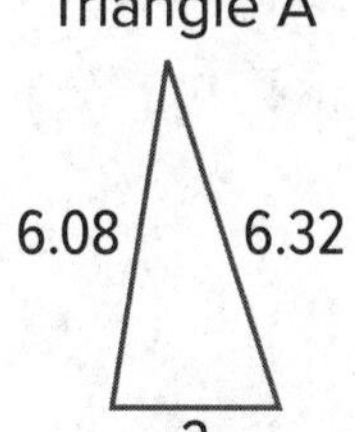

Triangle B

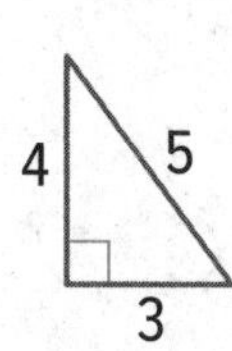

Triangle C

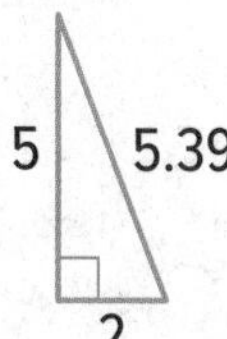

Triangle D

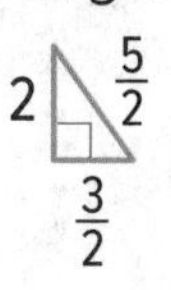

Triangle E

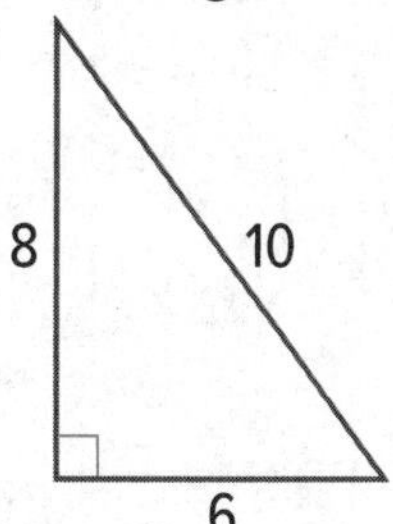

Triangle F

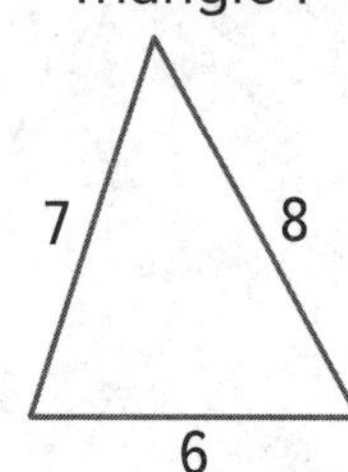

Triangle G

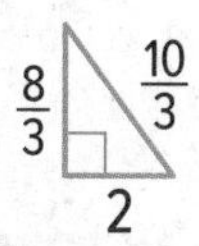

Triangle H

2 2.83

2

Your teacher will assign you two of the triangles to look at.

1. For each of your assigned triangles, is it a scaled copy of Triangle O? Be prepared to explain your reasoning.

2. As a group, identify *all* the scaled copies of Triangle O in the collection. Discuss your thinking. If you disagree, work to reach an agreement.

3. List all the triangles that are scaled copies in the table. Record the side lengths that correspond to the side lengths of Triangle O listed in each column.

Triangle O	3	4	5

4. Explain or show how each copy has been scaled from the original (Triangle O).

Are you ready for more?

Choose one of the triangles that is not a scaled copy of Triangle O. Describe how you could change at least one side to make a scaled copy, while leaving at least one side unchanged.

NAME ______________________ DATE ______________ PERIOD ____________

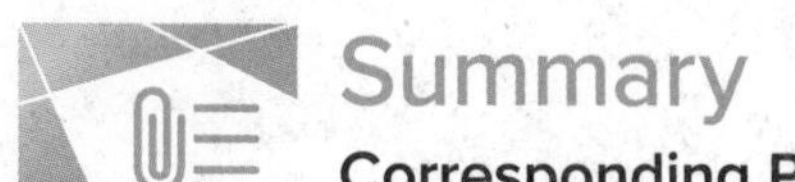

Summary

Corresponding Parts and Scale Factors

A figure and its scaled copy have **corresponding parts**, or parts that are in the same position in relation to the rest of each figure. These parts could be points, segments, or angles. For example, Polygon 2 is a scaled copy of Polygon 1.

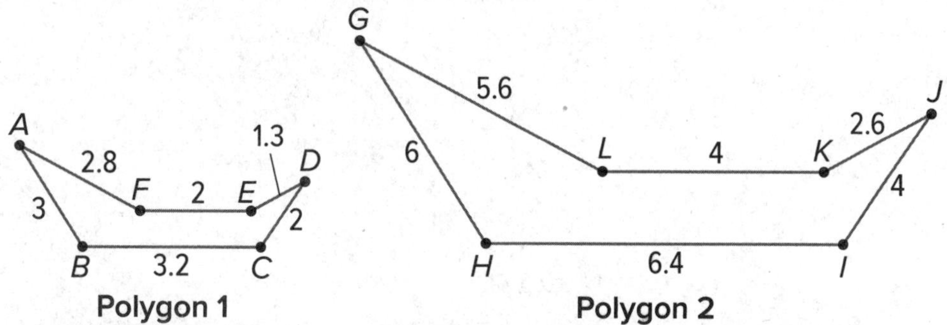

- Each point in Polygon 1 has a *corresponding point* in Polygon 2. For example, point *B* corresponds to point *H* and point *C* corresponds to point *I*.
- Each segment in Polygon 1 has a *corresponding segment* in Polygon 2. For example, segment *AF* corresponds to segment *GL*.
- Each angle in Polygon 1 also has a *corresponding angle* in Polygon 2. For example, angle *DEF* corresponds to angle *JKL*.

The **scale factor** between Polygon 1 and Polygon 2 is 2, because all of the lengths in Polygon 2 are 2 times the corresponding lengths in Polygon 1. The angle measures in Polygon 2 are the same as the corresponding angle measures in Polygon 1.

For example, the measure of angle *JKL* is the same as the measure of angle *DEF*.

Glossary

corresponding
scale factor

Practice

Corresponding Parts and Scale Factors

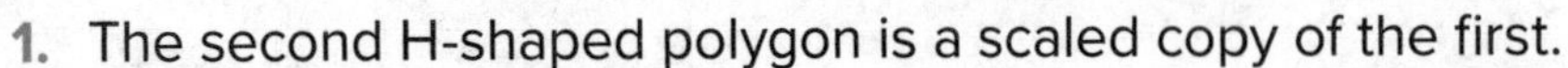

1. The second H-shaped polygon is a scaled copy of the first.

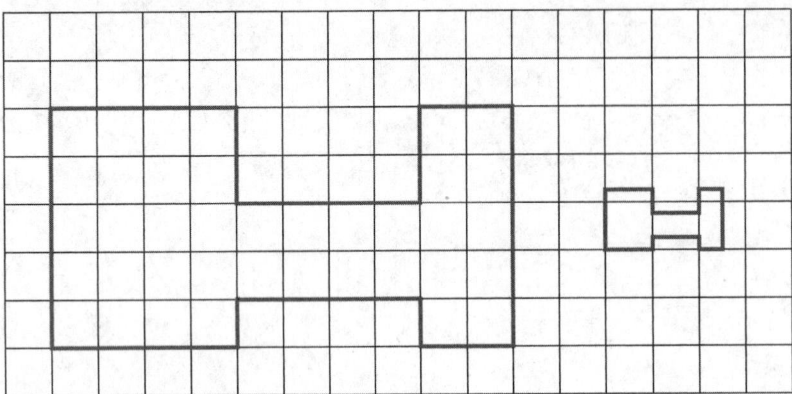

   a. Show one pair of corresponding points and two pairs of corresponding sides in the original polygon and its copy. Consider using colored pencils to highlight corresponding parts or labeling some of the vertices.

   b. What scale factor takes the original polygon to its smaller copy? Explain or show your reasoning.

NAME ______________________ DATE ____________ PERIOD ____________

2. Figure B is a scaled copy of Figure A. Select **all** of the statements that must be true.

(A.) Figure B is larger than Figure A.

(B.) Figure B has the same number of edges as Figure A.

(C.) Figure B has the same perimeter as Figure A.

(D.) Figure B has the same number of angles as Figure A.

(E.) Figure B has angles with the same measures as Figure A.

3. Polygon B is a scaled copy of Polygon A.

Polygon A

2.5

1.5

2.5

?

?

a. What is the scale factor from Polygon A to Polygon B? Explain your reasoning.

b. Find the missing length of each side marked with ? in Polygon B.

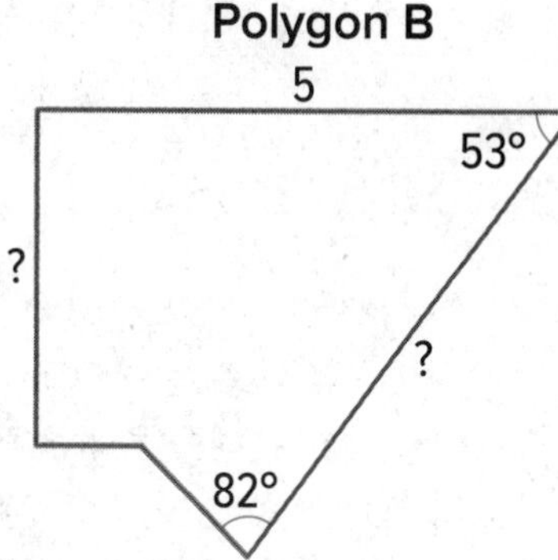

c. Determine the measure of each angle marked with ? in Polygon A.

4. Complete each equation with a number that makes it true.

a. $8 \cdot$ ______ $= 40$

b. $8 +$ ______ $= 40$

c. $21 \div$ ______ $= 7$

d. $21 -$ ______ $= 7$

e. $21 \cdot$ ______ $= 7$

Lesson 1-3

Making Scaled Copies

NAME ______________________ DATE ____________ PERIOD ____________

Learning Goal Let's draw scaled copies.

Warm Up

3.1 More or Less?

For each problem, select the answer from the two choices.

1. The value of 25 • (8.5) is:

   (A.) more than 205

   (B.) less than 205

2. The value of (9.93) • (0.984) is:

   (A.) more than 10

   (B.) less than 10

3. The value of (0.24) • (0.67) is:

   (A.) more than 0.2

   (B.) less than 0.2

Activity

3.2 Drawing Scaled Copies

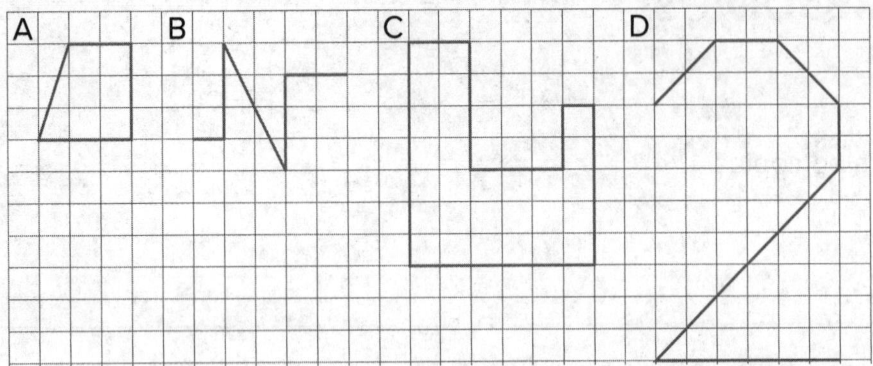

1. Draw a scaled copy of either Figure A or B using a scale factor of 3.
2. Draw a scaled copy of either Figure C or D using a scale factor of $\frac{1}{2}$.

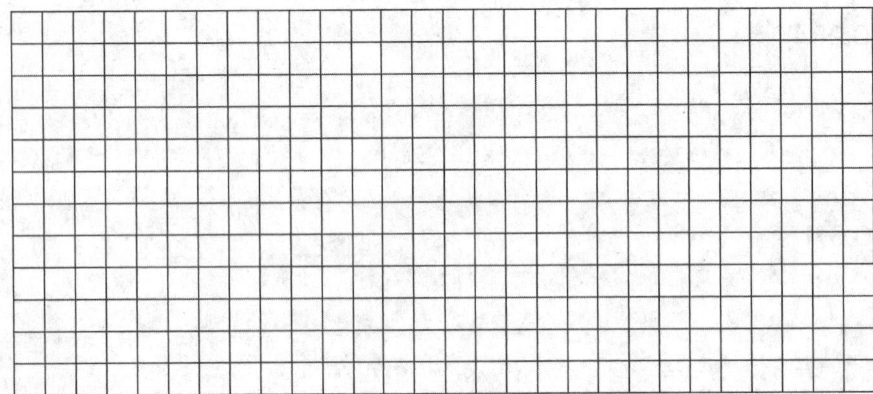

Activity

3.3 Which Operations? (Part 1)

Diego and Jada want to scale this polygon so the side that corresponds to 15 units in the original is 5 units in the scaled copy.

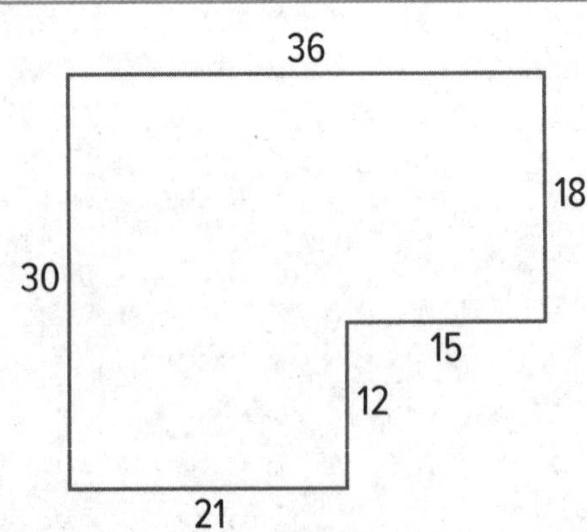

NAME ______________________ DATE ____________ PERIOD ________

Diego and Jada each use a different operation to find the new side lengths. Here are their finished drawings.

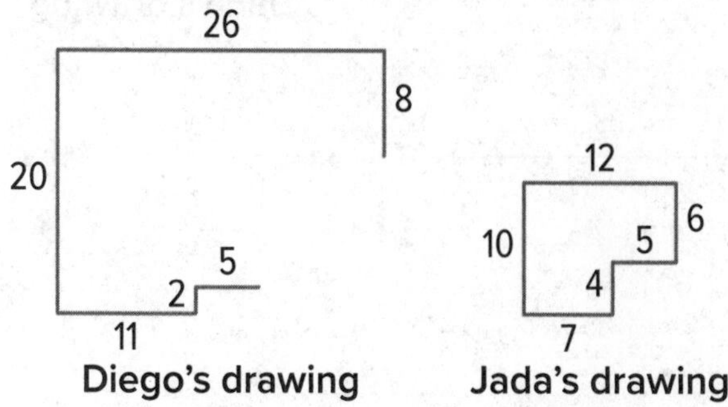

Diego's drawing **Jada's drawing**

1. What operation do you think Diego used to calculate the lengths for his drawing?

2. What operation do you think Jada used to calculate the lengths for her drawing?

3. Did each method produce a scaled copy of the polygon? Explain your reasoning.

Activity

3.4 Which Operations? (Part 2)

Andre wants to make a scaled copy of Jada's drawing so the side that corresponds to 4 units in Jada's polygon is 8 units in his scaled copy.

1. Andre says "I wonder if I should add 4 units to the lengths of all of the segments?" What would you say in response to Andre? Explain or show your reasoning.

2. Create the scaled copy that Andre wants. If you get stuck, consider using the edge of an index card or paper to measure the lengths needed to draw the copy.

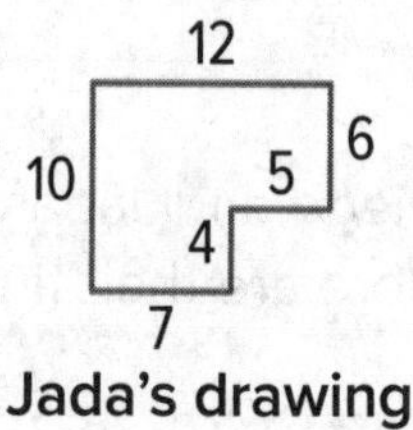

Jada's drawing

The side lengths of Triangle B are all 5 more than the side lengths of Triangle A. Can Triangle B be a scaled copy of Triangle A? Explain your reasoning.

Summary

Making Scaled Copies

Creating a scaled copy involves *multiplying* the lengths in the original figure by a scale factor.

For example, to make a scaled copy of triangle *ABC* where the base is 8 units, we would use a scale factor of 4. This means multiplying all the side lengths by 4, so in triangle *DEF*, each side is 4 times as long as the corresponding side in triangle *ABC*.

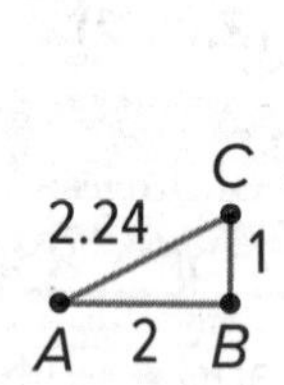

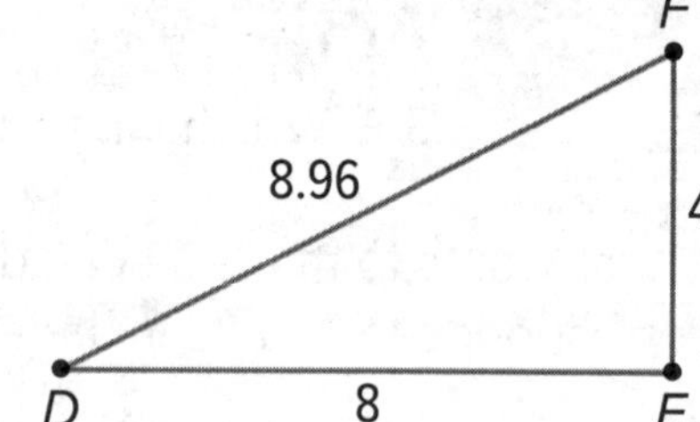

NAME ______________________ DATE ____________ PERIOD ________

Practice

Making Scaled Copies

1. Here are 3 polygons.

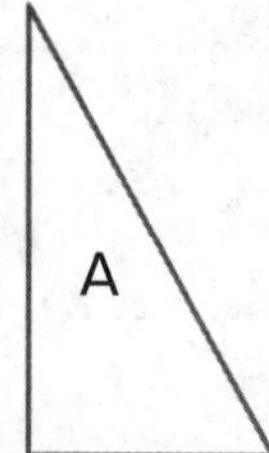

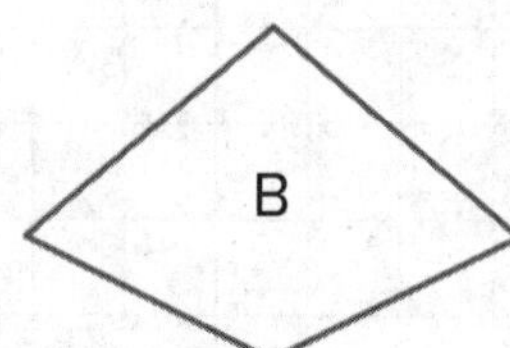

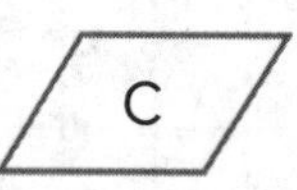

a. Draw a scaled copy of Polygon A using a scale factor of 2.

b. Draw a scaled copy of Polygon B using a scale factor of $\frac{1}{2}$.

c. Draw a scaled copy of Polygon C using a scale factor of $\frac{3}{2}$.

2. Quadrilateral A has side lengths 6, 9, 9, and 12. Quadrilateral B is a scaled copy of Quadrilateral A, with its shortest side of length 2. What is the perimeter of Quadrilateral B?

3. Here is a polygon on a grid.

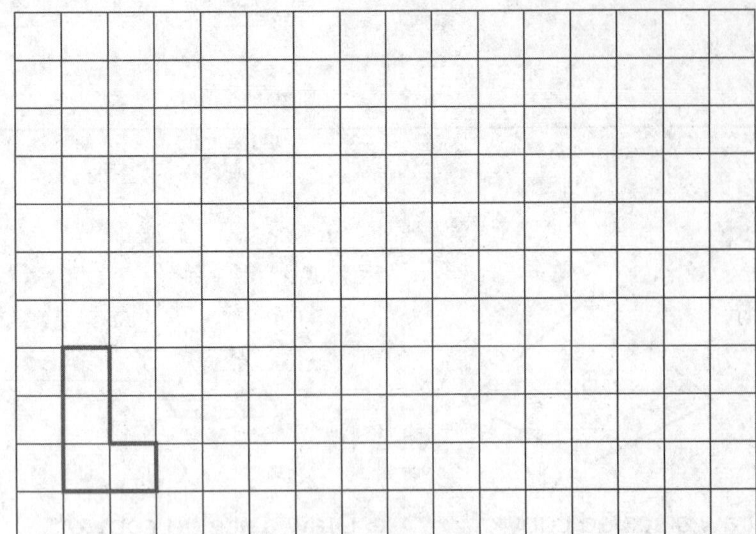

Draw a scaled copy of this polygon that has a perimeter of 30 units. What is the scale factor? Explain how you know.

4. Priya and Tyler are discussing the figures shown. Priya thinks that B, C, and D are scaled copies of A. Tyler says B and D are scaled copies of A. Do you agree with either of them? Explain your reasoning. **(Lesson 1-1)**

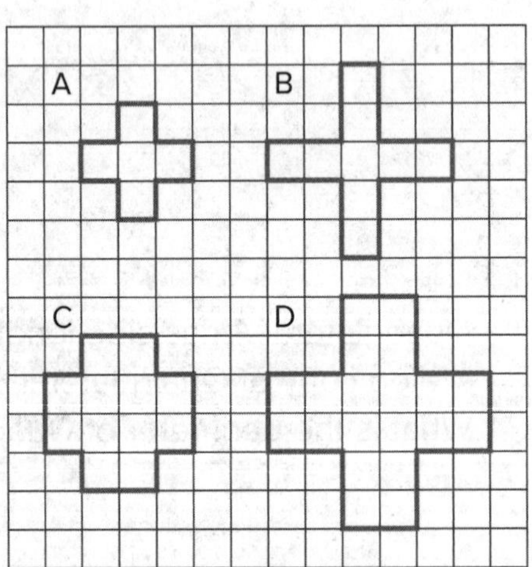

Lesson 1-4

Scaled Relationships

NAME ____________________ DATE ____________ PERIOD ____________

Learning Goal Let's find relationships between scaled copies.

Warm Up

4.1 Three Quadrilaterals (Part 1)

Each of these polygons is a scaled copy of the others.

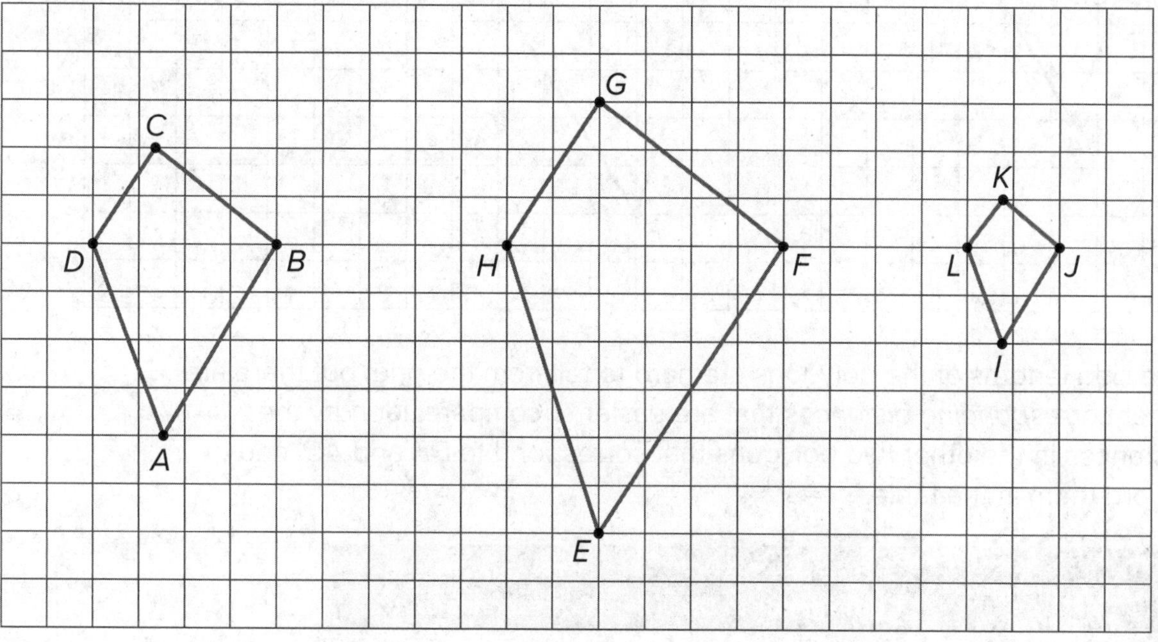

1. Name two pairs of corresponding angles. What can you say about the sizes of these angles?

2. Check your prediction by measuring at least one pair of corresponding angles using a protractor. Record your measurements to the nearest 5°.

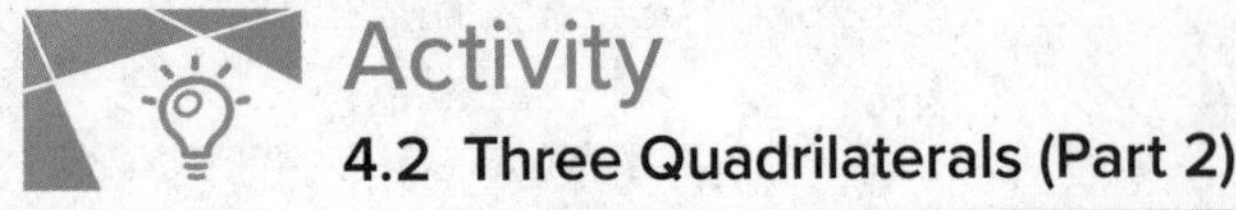

Activity

4.2 Three Quadrilaterals (Part 2)

Each of these polygons is a scaled copy of the others. You already checked their corresponding angles.

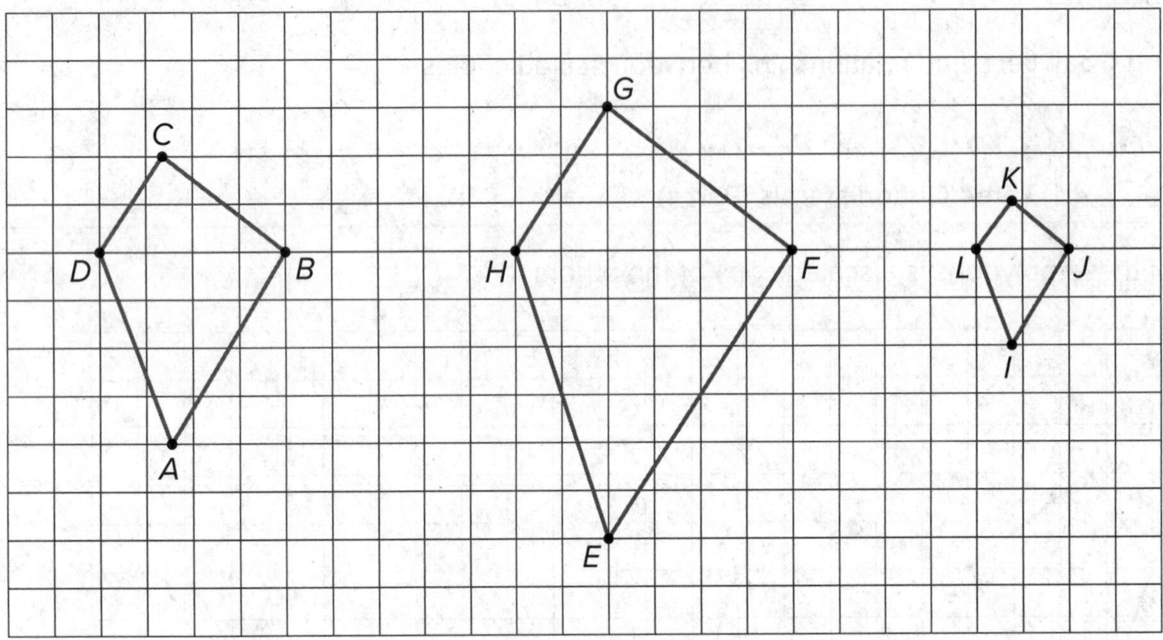

1. The side lengths of the polygons are hard to tell from the grid, but there are other *corresponding distances* that are easier to compare. Identify the distances in the other two polygons that correspond to *DB* and *AC*, and record them in the table.

Quadrilateral	Distance That Corresponds to *DB*	Distance That Corresponds to *AC*
ABCD	$DB = 4$	$AC = 6$
EFGH		
IJKL		

2. Look at the values in the table. What do you notice?

Pause here so your teacher can review your work.

NAME ______________________ DATE ______________ PERIOD __________

3. The larger figure is a scaled copy of the smaller figure.

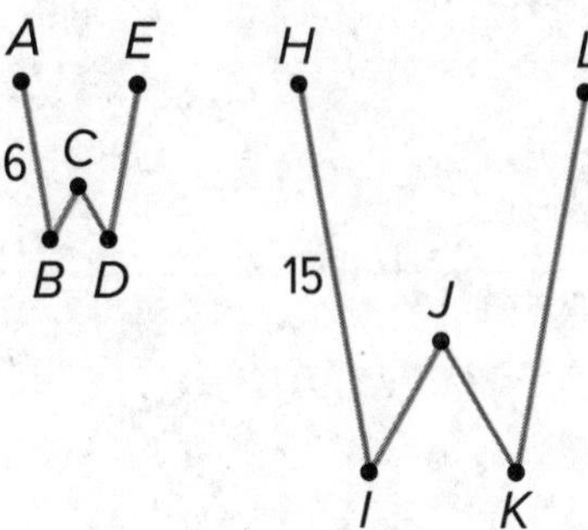

a. If $AE = 4$, how long is the corresponding distance in the second figure? Explain or show your reasoning.

b. If $IK = 5$, how long is the corresponding distance in the first figure? Explain or show your reasoning.

Activity
4.3 Scaled or Not Scaled?

Here are two quadrilaterals.

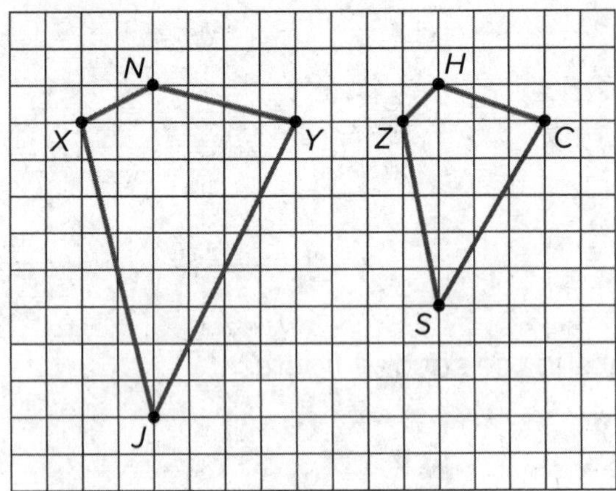

1. Mai says that Polygon *ZSCH* is a scaled copy of Polygon *XJYN*, but Noah disagrees. Do you agree with either of them? Explain or show your reasoning.

2. Record the corresponding distances in the table. What do you notice?

Quadrilateral	Horizontal Distance	Vertical Distance
XJYN	*XY* =	*JN* =
ZSCH	*ZC* =	*SH* =

3. Measure at least three pairs of corresponding angles in *XJYN* and *ZSCH* using a protractor. Record your measurements to the nearest 5°. What do you notice?

4. Do these results change your answer to the first question? Explain.

NAME ______________________ DATE ____________ PERIOD ____________

5. Here are two more quadrilaterals.

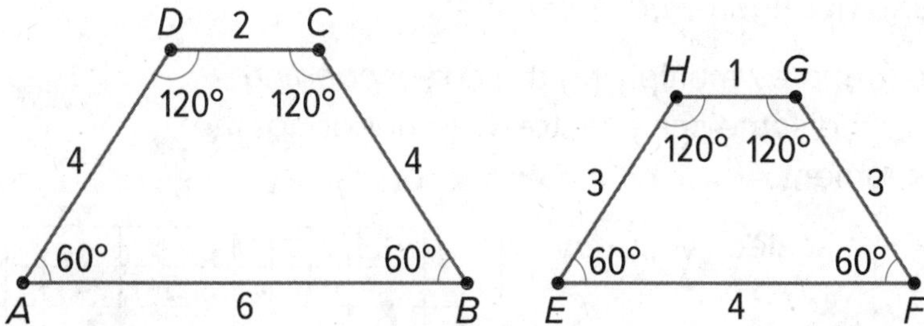

Kiran says that Polygon *EFGH* is a scaled copy of *ABCD*, but Lin disagrees. Do you agree with either of them? Explain or show your reasoning.

Are you ready for more?

All side lengths of quadrilateral *MNOP* are 2, and all side lengths of quadrilateral *QRST* are 3. Does *MNOP* have to be a scaled copy of *QRST*? Explain your reasoning.

Activity

4.4 Comparing Pictures of Birds

Here are two pictures of a bird. Find evidence that one picture is not a scaled copy of the other. Be prepared to explain your reasoning.

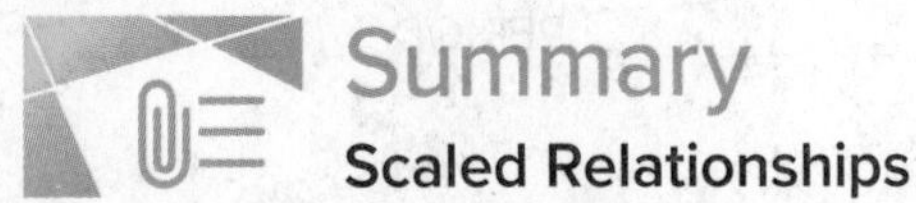

Summary

Scaled Relationships

When a figure is a scaled copy of another figure, we know that:

- All distances in the copy can be found by multiplying the *corresponding distances* in the original figure by the same scale factor, whether or not the endpoints are connected by a segment.

  For example, Polygon *STUVWX* is a scaled copy of Polygon *ABCDEF*.

  The scale factor is 3.

  The distance from *T* to *X* is 6, which is three times the distance from *B* to *F*.

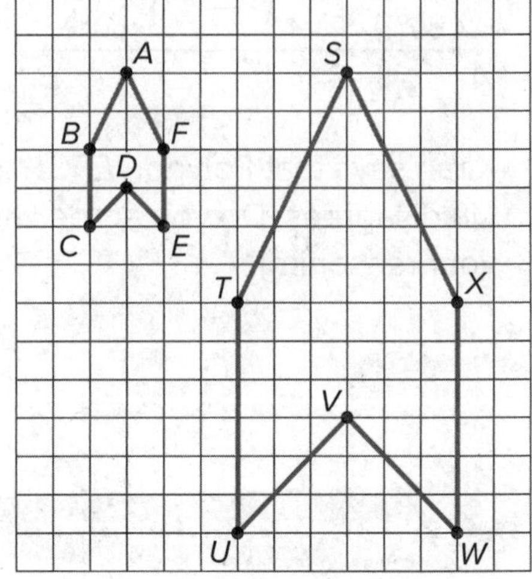

- All angles in the copy have the same measure as the corresponding angles in the original figure, as in these triangles.

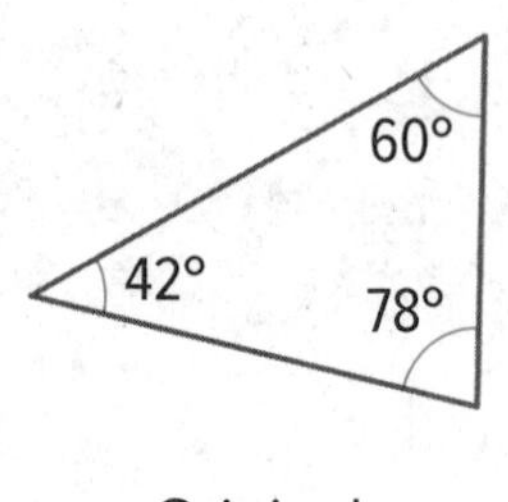

Original

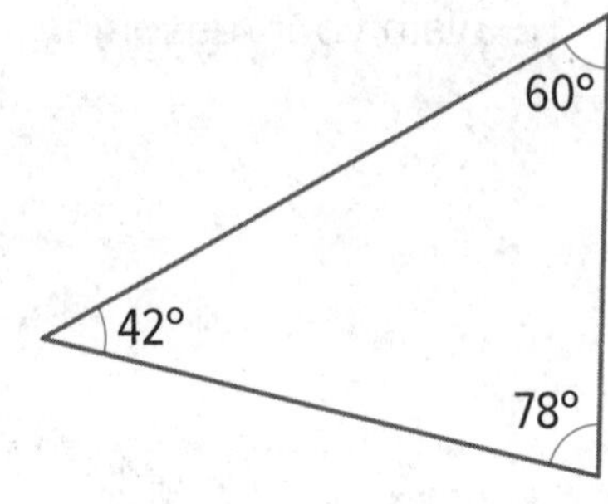

These observations can help explain why one figure is *not* a scaled copy of another. For example, even though their corresponding angles have the same measure, the second rectangle is not a scaled copy of the first rectangle, because different pairs of corresponding lengths have different scale factors, $2 \cdot \frac{1}{2} = 1$ but $3 \cdot \frac{2}{3} = 2$.

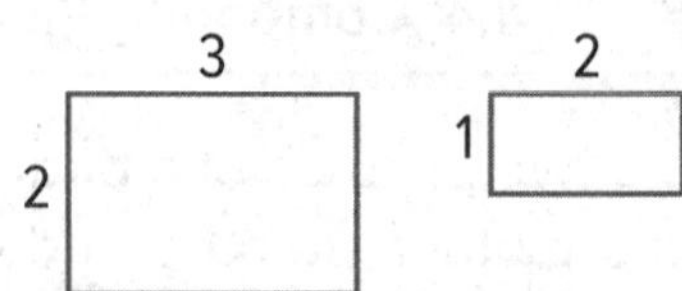

NAME ______________________ DATE ______________ PERIOD ____________

Practice
Scaled Relationships

1. Select **all** the statements that must be true for *any* scaled copy *Q* of Polygon *P*.

   A. The side lengths are all whole numbers.

   B. The angle measures are all whole numbers.

   C. *Q* has exactly 1 right angle.

   D. If the scale factor between *P* and *Q* is $\frac{1}{5}$, then each side length of *P* is multiplied by $\frac{1}{5}$ to get the corresponding side length of *Q*.

   E. If the scale factor is 2, each angle in *P* is multiplied by 2 to get the corresponding angle in *Q*.

   F. *Q* has 2 acute angles and 3 obtuse angles.

Polygon *P*

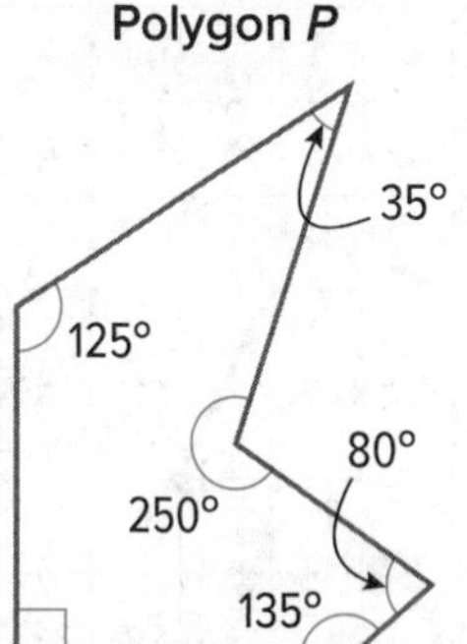

2. Here is Quadrilateral *ABCD*.

   Quadrilateral *PQRS* is a scaled copy of Quadrilateral *ABCD*. Point *P* corresponds to *A*, *Q* to *B*, *R* to *C*, and *S* to *D*. If the distance from *P* to *R* is 3 units, what is the distance from *Q* to *S*? Explain your reasoning.

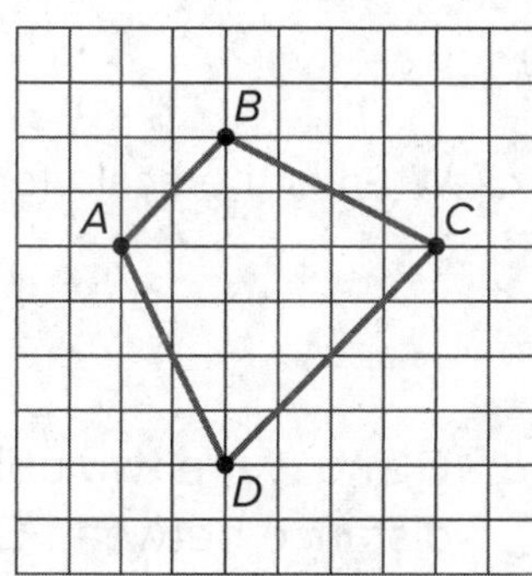

3. Figure 2 is a scaled copy of Figure 1.

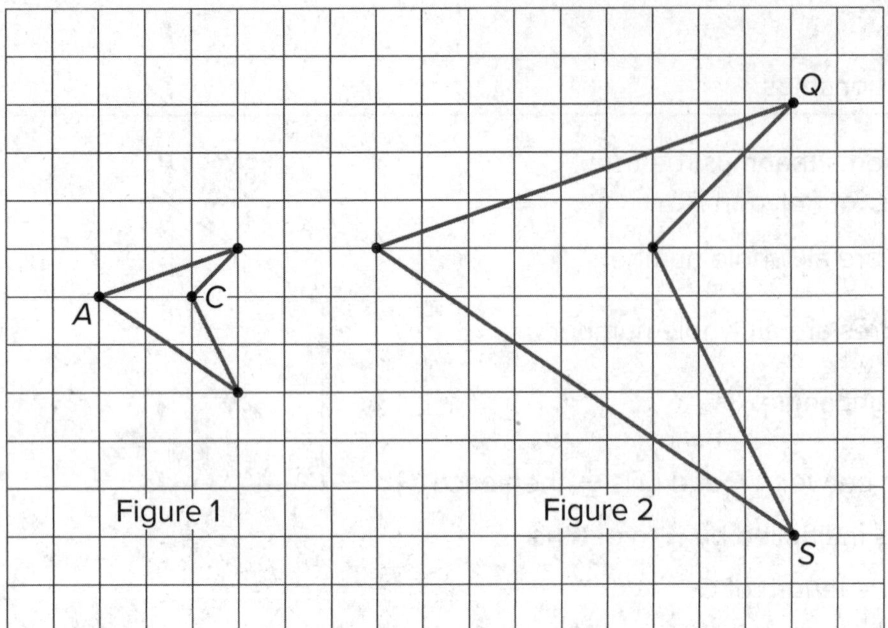

a. Identify the points in Figure 2 that correspond to the points *A* and *C* in Figure 1. Label them *P* and *R*. What is the distance between *P* and *R*?

b. Identify the points in Figure 1 that correspond to the points *Q* and *S* in Figure 2. Label them *B* and *D*. What is the distance between *B* and *D*?

c. What is the scale factor that takes Figure 1 to Figure 2?

d. *G* and *H* are two points on Figure 1, but they are not shown. The distance between *G* and *H* is 1. What is the distance between the corresponding points on Figure 2?

4. To make 1 batch of lavender paint, the ratio of cups of pink paint to cups of blue paint is 6 to 5. Find two more ratios of cups of pink paint to cups of blue paint that are equivalent to this ratio.

Lesson 1-5

The Size of the Scale Factor

NAME ______________________ DATE ______________ PERIOD ____________

Learning Goal Let's look at the effects of different scale factors.

Warm Up

5.1 Number Talk: Missing Factor

Solve each equation mentally.

1. $16x = 176$
2. $16x = 8$
3. $16x = 1$
4. $\frac{1}{5}x = 1$
5. $\frac{2}{5}x = 1$

Activity

5.2 Card Sort: Scaled Copies

Your teacher will give you a set of cards. On each card, Figure A is the original and Figure B is a scaled copy.

1. Sort the cards based on their scale factors. Be prepared to explain your reasoning.
2. Examine cards 10 and 13 more closely. What do you notice about the shapes and sizes of the figures? What do you notice about the scale factors?
3. Examine cards 8 and 12 more closely. What do you notice about the figures? What do you notice about the scale factors?

Are you ready for more?

Triangle B is a scaled copy of Triangle A with scale factor $\frac{1}{2}$.

1. How many times bigger are the side lengths of Triangle B when compared with Triangle A?
2. Imagine you scale Triangle B by a scale factor of $\frac{1}{2}$ to get Triangle C. How many times bigger will the side lengths of Triangle C be when compared with Triangle A?
3. Triangle B has been scaled once. Triangle C has been scaled twice. Imagine you scale Triangle A n times to get Triangle N, always using a scale factor of $\frac{1}{2}$. How many times bigger will the side lengths of Triangle N be when compared with Triangle A?

Activity

5.3 Scaling A Puzzle

Your teacher will give you 2 pieces of a 6-piece puzzle.

1. If you drew scaled copies of your puzzle pieces using a scale factor of $\frac{1}{2}$, would they be larger or smaller than the original pieces? How do you know?
2. Create a scaled copy of each puzzle piece on a blank square, with a scale factor of $\frac{1}{2}$.
3. When everyone in your group is finished, put all 6 of the original puzzle pieces together like this.

1	2	3
4	5	6

Next, put all 6 of your scaled copies together. Compare your scaled puzzle with the original puzzle. Which parts seem to be scaled correctly and which seem off? What might have caused those parts to be off?

NAME ______________________ DATE ____________ PERIOD ________

4. Revise any of the scaled copies that may have been drawn incorrectly.

5. If you were to lose one of the pieces of the original puzzle, but still had the scaled copy, how could you recreate the lost piece?

Activity

5.4 Missing Figure, Factor, or Copy

1. What is the scale factor from the original triangle to its copy? Explain or show your reasoning.

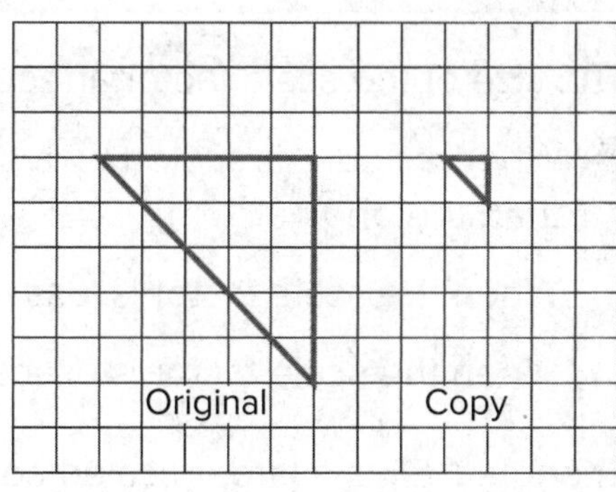

2. The scale factor from the original trapezoid to its copy is 2. Draw the scaled copy.

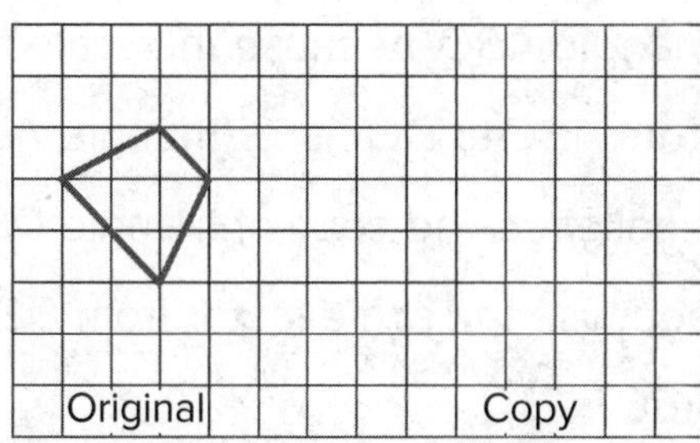

3. The scale factor from the original figure to its copy is $\frac{3}{2}$. Draw the original figure.

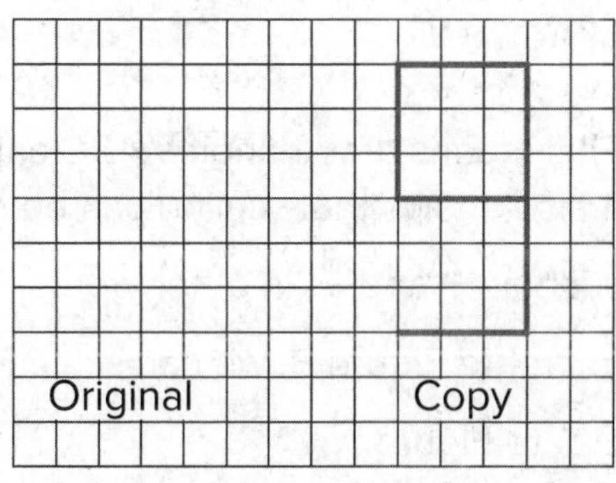

4. What is the scale factor from the original figure to the copy? Explain how you know.

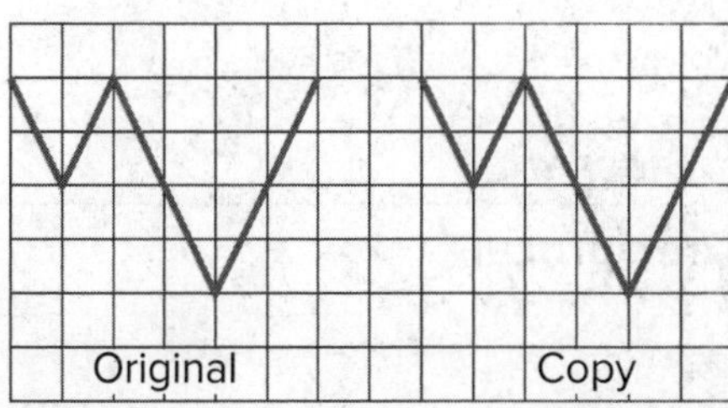

5. The scale factor from the original figure to its scaled copy is 3. Draw the scaled copy.

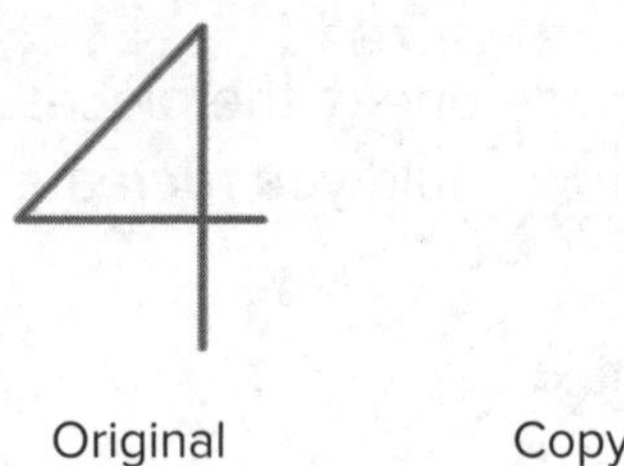

Original Copy

Summary

The Size of the Scale Factor

The size of the scale factor affects the size of the copy.

- When a figure is scaled by a scale factor greater than 1, the copy is larger than the original.
- When the scale factor is less than 1, the copy is smaller.
- When the scale factor is exactly 1, the copy is the same size as the original.

Triangle *DEF* is a larger scaled copy of triangle *ABC*, because the scale factor from *ABC* to *DEF* is $\frac{3}{2}$. Triangle *ABC* is a smaller scaled copy of triangle *DEF*, because the scale factor from *DEF* to *ABC* is $\frac{2}{3}$.

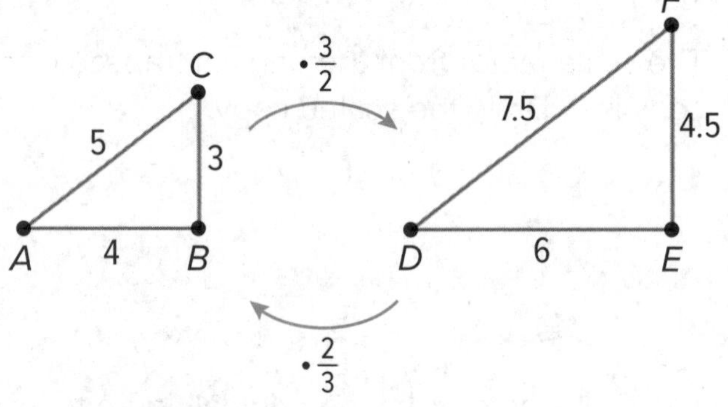

This means that triangles *ABC* and *DEF* are scaled copies of each other. It also shows that scaling can be reversed using **reciprocal** scale factors, such as $\frac{2}{3}$ and $\frac{3}{2}$. In other words, if we scale Figure A using a scale factor of 4 to create Figure B, we can scale Figure B using the reciprocal scale factor, $\frac{1}{4}$, to create Figure A.

Glossary

reciprocal

NAME _______________ DATE _______________ PERIOD _______________

Practice
The Size of the Scale Factor

1. Rectangles P, Q, R, and S are scaled copies of one another. For each pair, decide if the scale factor from one to the other is greater than 1, equal to 1, or less than 1.

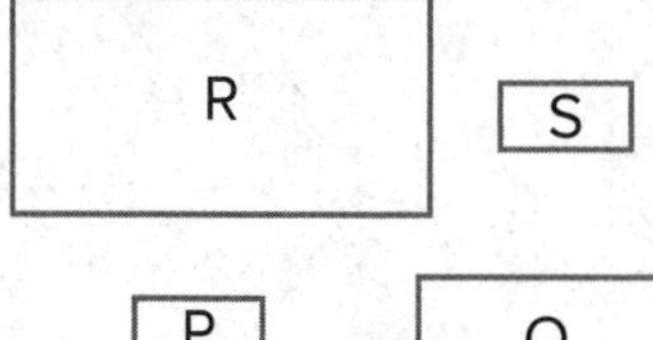

   a. from P to Q

   b. from P to R

   c. from Q to S

   d. from Q to R

   e. from S to P

   f. from R to P

   g. from P to S

2. Triangle S and Triangle L are scaled copies of one another.

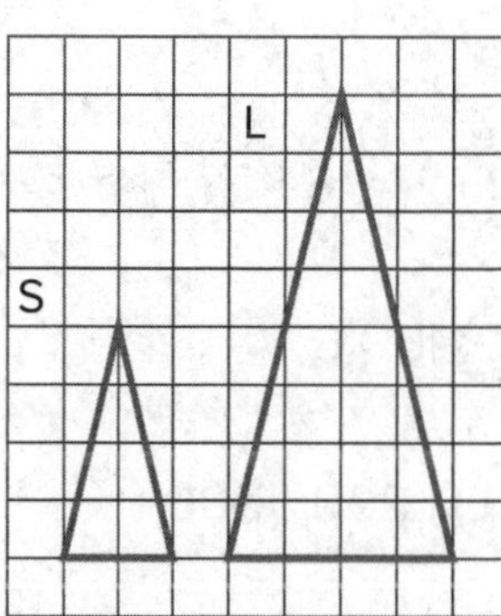

   a. What is the scale factor from S to L?

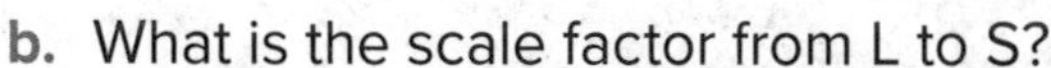

   b. What is the scale factor from L to S?

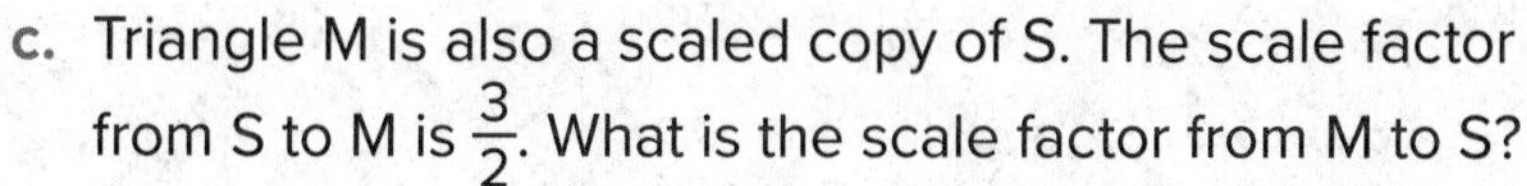

   c. Triangle M is also a scaled copy of S. The scale factor from S to M is $\frac{3}{2}$. What is the scale factor from M to S?

3. Are two squares with the same side lengths scaled copies of one another? Explain your reasoning.

4. Quadrilateral A has side lengths 2, 3, 5, and 6. Quadrilateral B has side lengths 4, 5, 8, and 10. Could one of the quadrilaterals be a scaled copy of the other? Explain. **(Lesson 1-2)**

5. Select **all** the ratios that are equivalent to the ratio 12 : 3.

A. 6 : 1

B. 1 : 4

C. 4 : 1

D. 24 : 6

E. 15 : 6

F. 1,200 : 300

G. 112 : 13

Lesson 1-6

Scaling and Area

NAME ________________ DATE ________ PERIOD ________

Learning Goal Let's build scaled shapes and investigate their areas.

Warm Up

6.1 Scaling a Pattern Block

Your teacher will give you some pattern blocks. Work with your group to build the scaled copies described in each question.

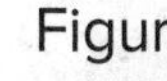

Figure A Figure B Figure C

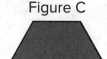

1. How many blue rhombus blocks does it take to build a scaled copy of Figure A:
   a. where each side is twice as long?
   b. where each side is 3 times as long?
   c. where each side is 4 times as long?

2. How many green triangle blocks does it take to build a scaled copy of Figure B:
   a. where each side is twice as long?
   b. where each side is 3 times as long?
   c. using a scale factor of 4?

3. How many red trapezoid blocks does it take to build a scaled copy of Figure C:
   a. using a scale factor of 2?
   b. using a scale factor of 3?
   c. using a scale factor of 4?

Activity

6.2 Scaling More Pattern Blocks

Your teacher will assign your group one of these figures.

Figure D

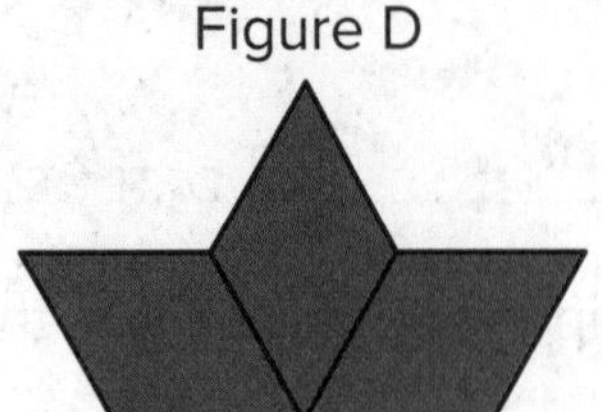

Figure E

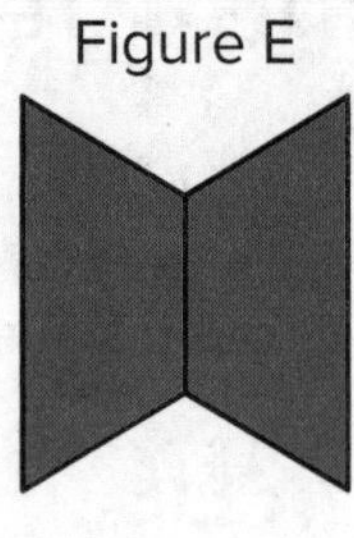

Figure F

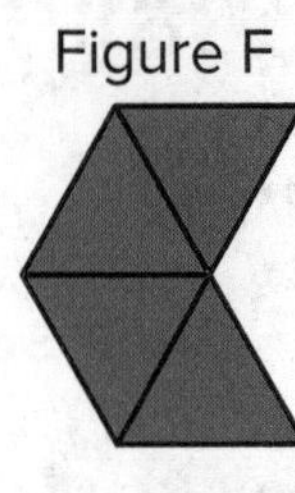

1. Build a scaled copy of your assigned shape using a scale factor of 2. Use the same shape blocks as in the original figure. How many blocks did it take?
2. Your classmate thinks that the scaled copies in the previous problem will each take 4 blocks to build. Do you agree or disagree? Explain your reasoning.
3. Start building a scaled copy of your assigned figure using a scale factor of 3. Stop when you can tell for sure how many blocks it would take. Record your answer.
4. How many blocks would it take to build scaled copies of your figure using scale factors 4, 5, and 6? Explain or show your reasoning.
5. How is the pattern in this activity the same as the pattern you saw in the previous activity? How is it different?

Are you ready for more?

1. How many blocks do you think it would take to build a scaled copy of one yellow hexagon where each side is twice as long? Three times as long?
2. Figure out a way to build these scaled copies.
3. Do you see a pattern for the number of blocks used to build these scaled copies? Explain your reasoning.

NAME ______________________ DATE ____________ PERIOD ____________

Activity

6.3 Area of Scaled Parallelograms and Triangles

1. Your teacher will give you a figure with measurements in centimeters. What is the area of your figure? How do you know?

2. Work with your partner to draw scaled copies of your figure on a separate piece of paper, using each scale factor in the table. Complete the table with the measurements of your scaled copies.

Scale Factor	Base (cm)	Height (cm)	Area (cm^2)
1			
2			
3			
$\frac{1}{2}$			
$\frac{1}{3}$			

3. Compare your results with a group that worked with a different figure. What is the same about your answers? What is different?

4. If you drew scaled copies of your figure with the following scale factors, what would their areas be? Discuss your thinking. If you disagree, work to reach an agreement. Be prepared to explain your reasoning.

Scale Factor	Area (cm^2)
5	
$\frac{3}{5}$	

Summary

Scaling and Area

Scaling affects lengths and areas differently. When we make a scaled copy, all original lengths are multiplied by the scale factor.

If we make a copy of a rectangle with side lengths 2 units and 4 units using a scale factor of 3, the side lengths of the copy will be 6 units and 12 units, because $2 \cdot 3 = 6$ and $4 \cdot 3 = 12$.

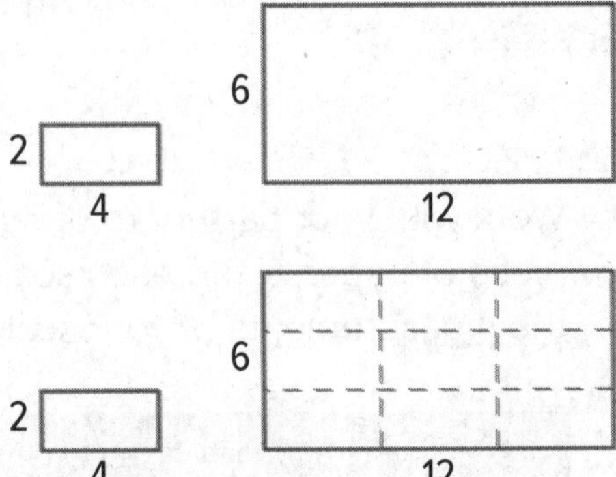

The area of the copy, however, changes by a factor of (scale factor)^2. If each side length of the copy is 3 times longer than the original side length, then the area of the copy will be 9 times the area of the original, because $3 \cdot 3$, or 3^2, equals 9.

In this example, the area of the original rectangle is 8 units^2 and the area of the scaled copy is 72 units^2, because $9 \cdot 8 = 72$.

- We can see that the large rectangle is covered by 9 copies of the small rectangle, without gaps or overlaps.
- We can also verify this by multiplying the side lengths of the large rectangle: $6 \cdot 12 = 72$.

Lengths are one-dimensional, so in a scaled copy, they change by the scale factor. Area is two-dimensional, so it changes by the *square* of the scale factor.

We can see this is true for a rectangle with length l and width w.

If we scale the rectangle by a scale factor of s, we get a rectangle with length $s \cdot l$ and width $s \cdot w$.

The area of the scaled rectangle is $A = (s \cdot l) \cdot (s \cdot w)$, so $A = (s^2) \cdot (l \cdot w)$.

The fact that the area is multiplied by the square of the scale factor is true for scaled copies of other two-dimensional figures too, not just for rectangles.

Glossary

area

NAME ______________________ DATE ____________ PERIOD ________

Practice
Scaling and Area

1. On the grid, draw a scaled copy of Polygon Q using a scale factor of 2. Compare the perimeter and area of the new polygon to those of Q.

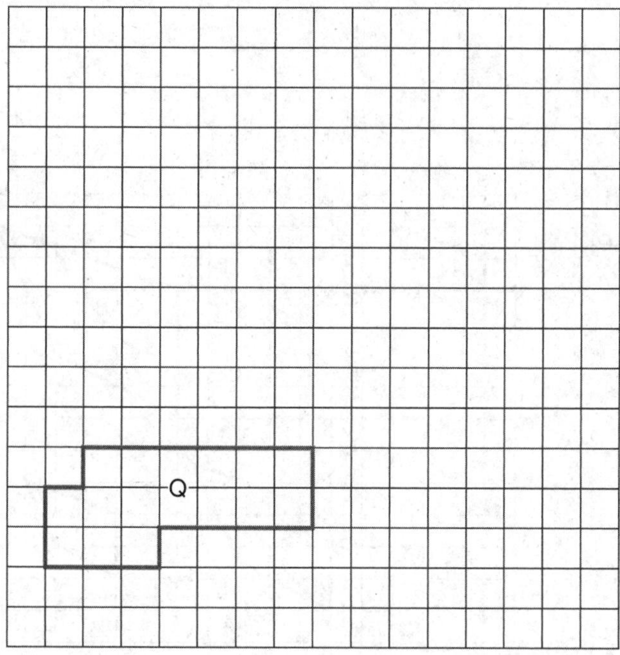

2. A right triangle has an area of 36 square units. If you draw scaled copies of this triangle using the scale factors in the table, what will the areas of these scaled copies be? Explain or show your reasoning.

Scale Factor	Area (units^2)
1	36
2	
3	
5	
$\frac{1}{2}$	
$\frac{2}{3}$	

3. Diego drew a scaled version of a Polygon P and labeled it Q.

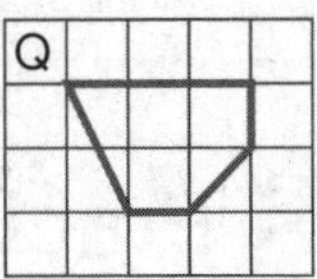

If the area of Polygon P is 72 square units, what scale factor did Diego use to go from P to Q? Explain your reasoning.

4. Here is an unlabeled polygon, along with its scaled copies Polygons A–D. For each copy, determine the scale factor. Explain how you know. **(Lesson 1-2)**

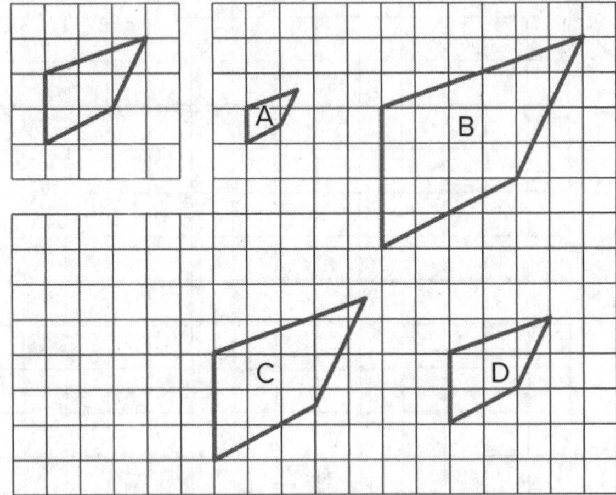

5. Solve each equation mentally. **(Lesson 1-5)**

a. $\frac{1}{7} \cdot x = 1$

b. $x \cdot \frac{1}{11} = 1$

c. $1 \div \frac{1}{5} = x$

Lesson 1-7

Scale Drawings

NAME ______________________ DATE ______________ PERIOD ______________

Learning Goal Let's explore scale drawings.

Warm Up

7.1 What Is a Scale Drawing?

Here are some drawings of a school bus, a quarter, and the subway lines around Boston, Massachusetts. The first three drawings are **scale drawings** of these objects.

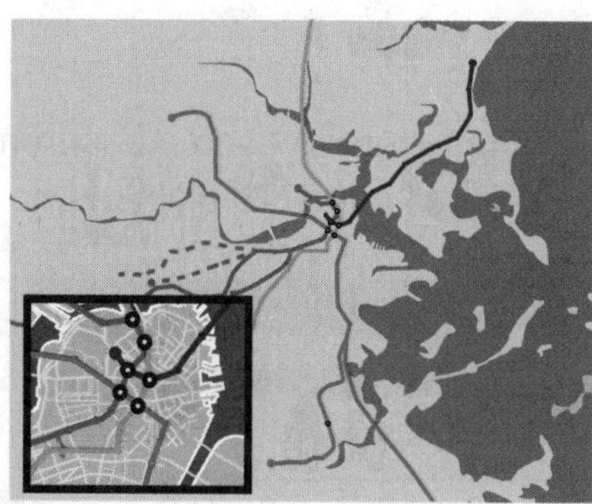

The next three drawings are *not* scale drawings of these objects.

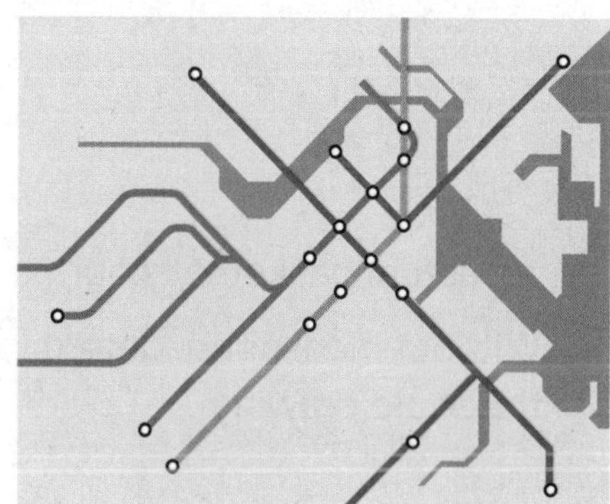

Discuss with your partner what a scale drawing is.

Activity

7.2 Sizing Up a Basketball Court

Your teacher will give you a scale drawing of a basketball court. The drawing does not have any measurements labeled, but it says that 1 centimeter represents 2 meters.

1. Measure the distances on the scale drawing that are labeled a–d to the nearest tenth of a centimeter. Record your results in the first row of the table.

2. The statement "1 cm represents 2 m" is the **scale** of the drawing. It can also be expressed as "1 cm to 2 m," or "1 cm for every 2 m." What do you think the scale tells us?

3. How long would each measurement from the first question be on an actual basketball court? Explain or show your reasoning.

Measurement	(a) Length of Court	(b) Width of Court	(c) Hoop to Hoop	(d) 3 Point Line to Sideline
Scale Drawing				
Actual Court				

4. On an actual basketball court, the bench area is typically 9 meters long.
   a. Without measuring, determine how long the bench area should be on the scale drawing.
   b. Check your answer by measuring the bench area on the scale drawing. Did your prediction match your measurement?

NAME ______________________ DATE ____________ PERIOD ________

Activity
7.3 Tall Structures

Here is a scale drawing of some of the world's tallest structures.

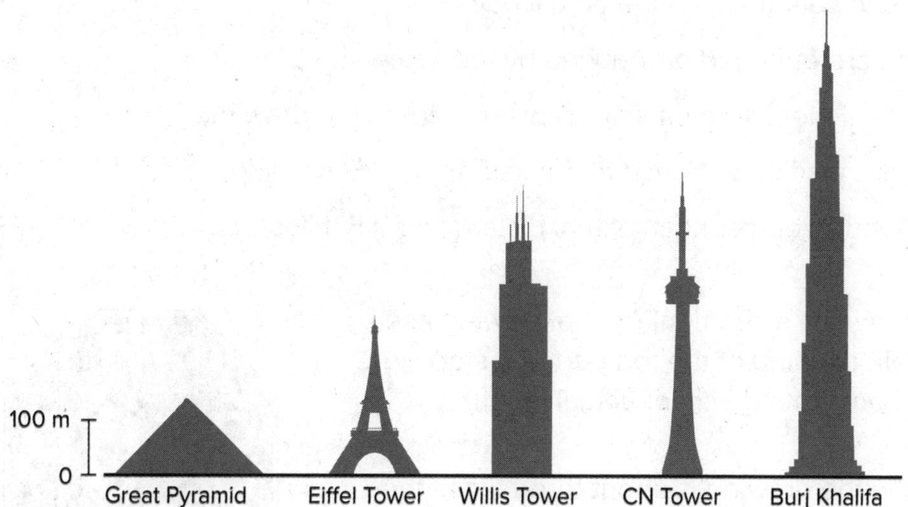

1. About how tall is the actual Willis Tower? About how tall is the actual Great Pyramid? Be prepared to explain your reasoning.

2. About how much taller is the Burj Khalifa than the Eiffel Tower? Explain or show your reasoning.

3. Measure the line segment that shows the scale to the nearest tenth of a centimeter. Express the scale of the drawing using numbers and words.

Are you ready for more?

The tallest mountain in the United States, Mount Denali in Alaska, is about 6,190 m tall. If this mountain were shown on the scale drawing, how would its height compare to the heights of the structures? Explain or show your reasoning.

Summary

Scale Drawings

Scale drawings are two-dimensional representations of actual objects or places. Floor plans and maps are some examples of scale drawings. On a scale drawing:

- Every part corresponds to something in the actual object.
- Lengths on the drawing are enlarged or reduced by the same scale factor.
- A **scale** tells us how actual measurements are represented on the drawing. For example, if a map has a scale of "1 inch to 5 miles" then a $\frac{1}{2}$-inch line segment on that map would represent an actual distance of 2.5 miles.

Sometimes the scale is shown as a segment on the drawing itself. For example, here is a scale drawing of the top part of a stop sign with a line segment that represents 25 cm of actual length.

The width of the octagon in the drawing is about three times the length of this segment, so the actual width of the sign is about $3 \cdot 25$, or 75 cm.

Because a scale drawing is two-dimensional, some aspects of the three-dimensional object are not represented. For example, this scale drawing does not show the thickness of the stop sign.

A scale drawing may not show every detail of the actual object. However, the features that are shown correspond to the actual object and follow the specified scale.

Glossary

scale
scale drawing

NAME ______________________ DATE ____________ PERIOD ________

Practice

Scale Drawings

1. The Westland Lysander was an aircraft used by the Royal Air Force in the 1930s. Here are some scale drawings that show the top, side, and front views of the Lysander.

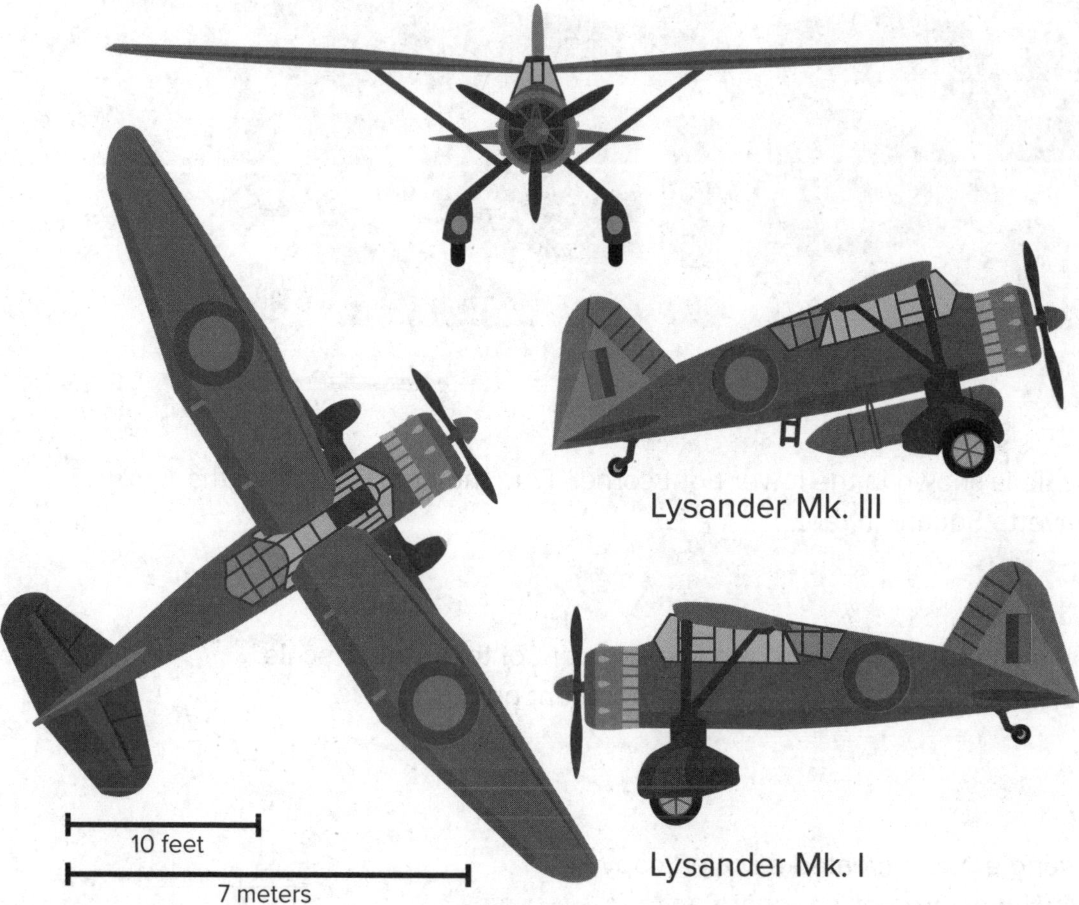

Use the scales and scale drawings to approximate the actual lengths of:

a. the wingspan of the plane, to the nearest foot

b. the height of the plane, to the nearest foot

c. the length of the Lysander Mk. I, to the nearest meter

2. A blueprint for a building includes a rectangular room that measures 3 inches long and 5.5 inches wide. The scale for the blueprint says that 1 inch on the blueprint is equivalent to 10 feet in the actual building. What are the dimensions of this rectangular room in the actual building?

3. Here is a scale map of Lafayette Square, a rectangular garden north of the White House.

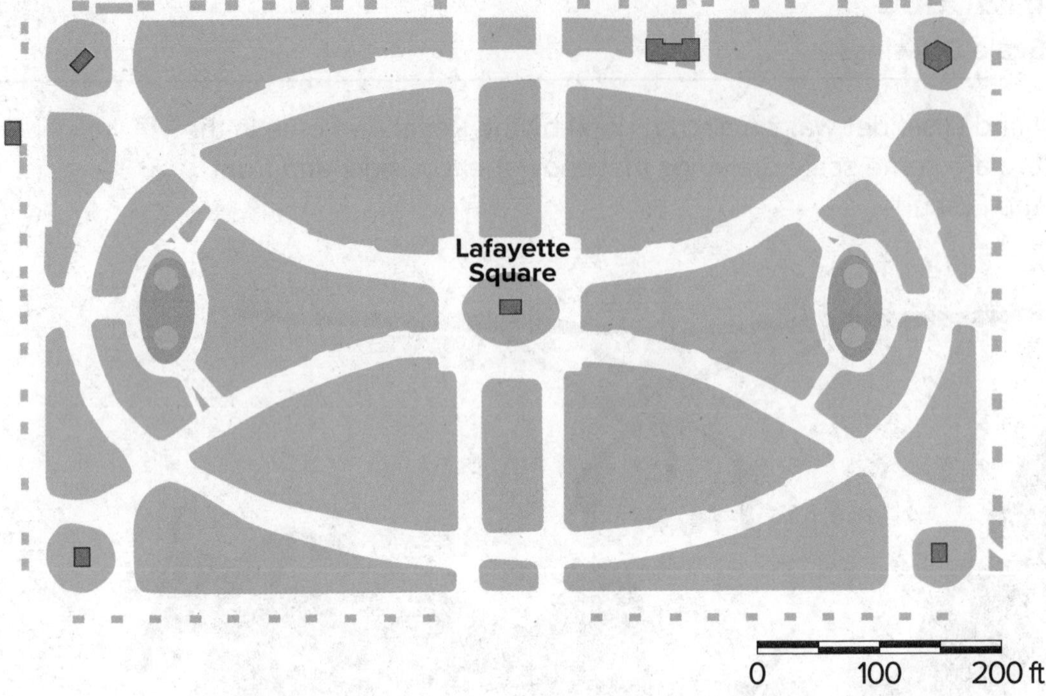

a. The scale is shown in the lower right corner. Find the actual side lengths of Lafayette Square in feet.

b. Use an inch ruler to measure the line segment of the graphic scale. About how many feet does one inch represent on this map?

4. Here is Triangle A. Lin created a scaled copy of Triangle A with an area of 72 square units. **(Lesson 1-6)**

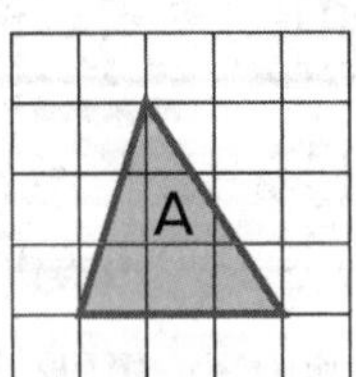

a. How many times larger is the area of the scaled copy compared to that of Triangle A?

b. What scale factor did Lin apply to the Triangle A to create the copy?

c. What is the length of the bottom side of the scaled copy?

Lesson 1-8

Scale Drawings and Maps

NAME ______________________________ DATE ______________ PERIOD ______________

Learning Goal Let's use scale drawings to solve problems.

Warm Up

8.1 A Train and a Car

Two cities are 243 miles apart.

- It takes a train 4 hours to travel between the two cities at a constant speed.
- A car travels between the two cities at a constant speed of 65 miles per hour.

Which is traveling faster, the car or the train? Be prepared to explain your reasoning.

Activity

8.2 Driving on I-90

1. A driver is traveling at a constant speed on Interstate 90 outside Chicago. If she traveled from Point A to Point B in 8 minutes, did she obey the speed limit of 55 miles per hour? Explain your reasoning.

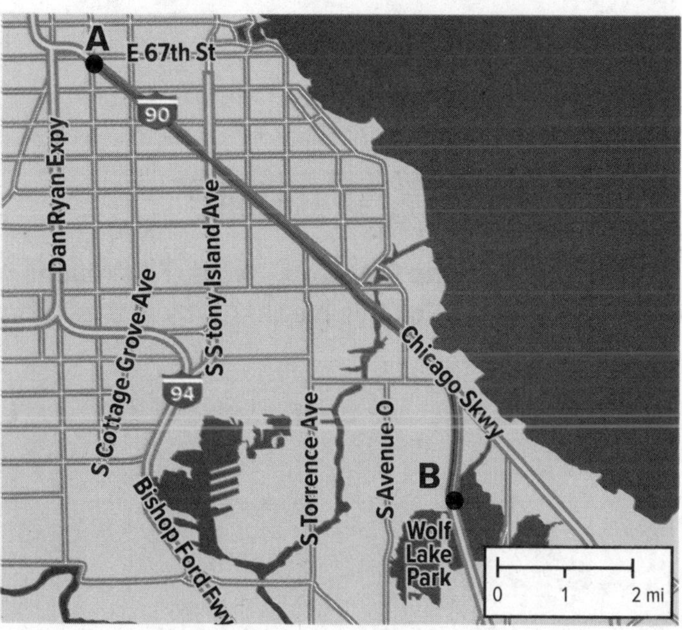

2. A traffic helicopter flew directly from Point A to Point B in 8 minutes. Did the helicopter travel faster or slower than the driver? Explain or show your reasoning.

Activity

8.3 Biking through Kansas

A cyclist rides at a constant speed of 15 miles per hour. At this speed, about how long would it take the cyclist to ride from Garden City to Dodge City, Kansas?

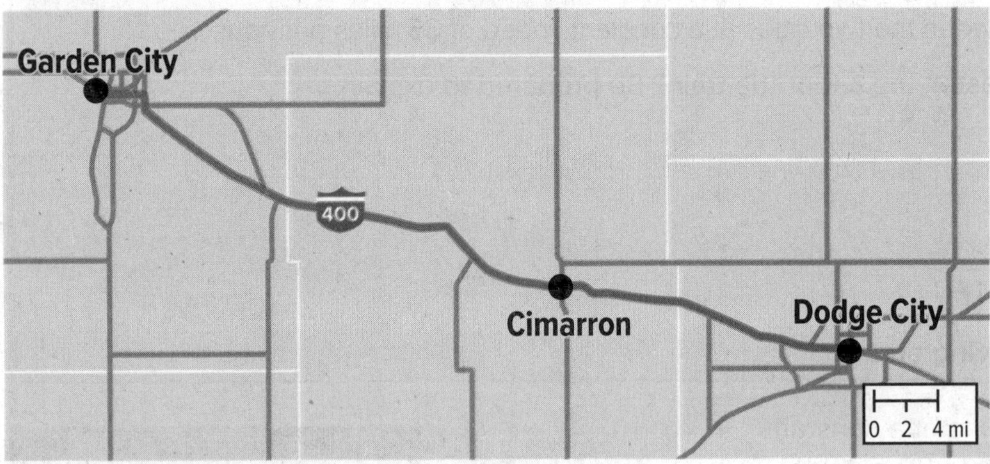

Are you ready for more?

Jada finds a map that says, "Note: This map is not to scale." What do you think this means? Why is this information important?

NAME ______________________ DATE ______________ PERIOD ______________

Summary

Scale Drawings and Maps

Maps with scales are useful for making calculations involving speed, time, and distance.

Here is a map of part of Alabama.

Suppose it takes a car 1 hour and 30 minutes to travel at constant speed from Birmingham to Montgomery. How fast is the car traveling?

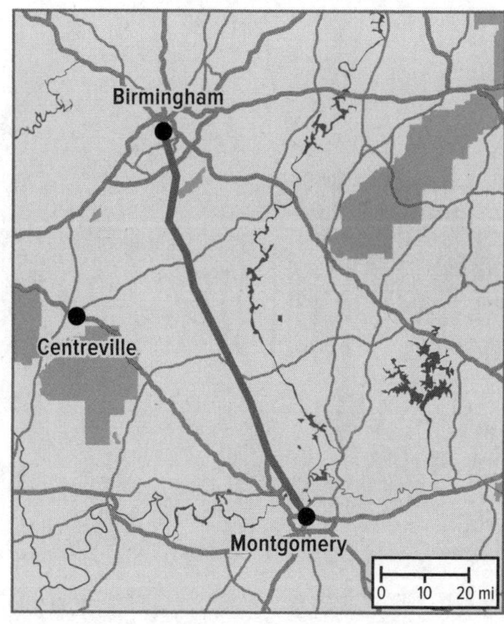

To make an estimate, we need to know about how far it is from Birmingham to Montgomery. The scale of the map represents 20 miles, so we can estimate the distance between these cities is about 90 miles.

Since 90 miles in 1.5 hours is the same speed as 180 miles in 3 hours, the car is traveling about 60 miles per hour.

Time (hours)	Distance (miles)
1.5	90
3	180
1	60

$\cdot 2$ (1.5 to 3; 90 to 180)

$\cdot \frac{1}{3}$ (3 to 1; 180 to 60)

Suppose a car is traveling at a constant speed of 60 miles per hour from Montgomery to Centreville. How long will it take the car to make the trip? Using the scale, we can estimate that it is about 70 miles. Since 60 miles per hour is the same as 1 mile per minute, it will take the car about 70 minutes (or 1 hour and 10 minutes) to make this trip.

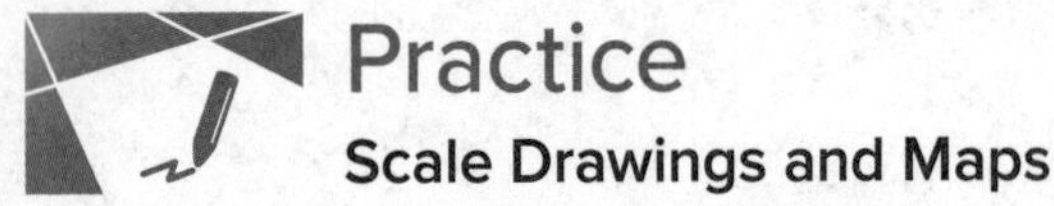

Practice

Scale Drawings and Maps

1. Here is a map that shows parts of Texas and Oklahoma.

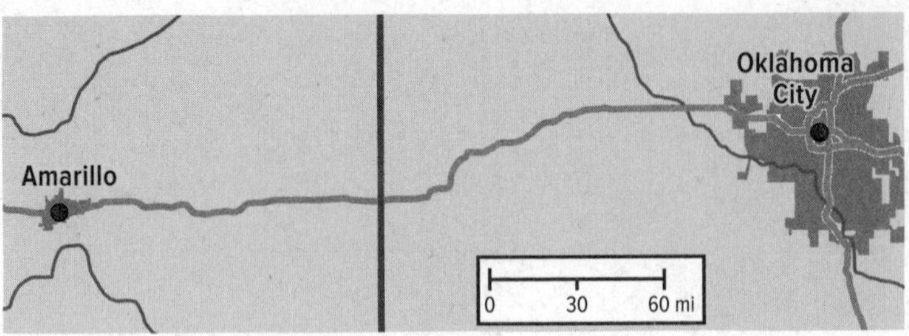

   a. About how far is it from Amarillo to Oklahoma City? Explain your reasoning.

   b. Driving at a constant speed of 70 miles per hour, will it be possible to make this trip in 3 hours? Explain how you know.

2. A local park is in the shape of a square. A map of the local park is made with the scale 1 inch to 200 feet.

   a. If the park is shown as a square on the map, each side of which is one foot long, how long is each side of the square park?

   b. If a straight path in the park is 900 feet long, how long would the path be when represented on the map?

Lesson 1-9

Creating Scale Drawings

NAME ______________________ DATE ____________ PERIOD ________

Learning Goal Let's create our own scale drawings.

Warm Up

9.1 Number Talk: Which is Greater?

Without calculating, decide which quotient is larger.

1. $11 \div 23$ or $7 \div 13$
2. $0.63 \div 2$ or $0.55 \div 3$
3. $15 \div \frac{1}{3}$ or $15 \div \frac{1}{4}$

Activity

9.2 Bedroom Floor Plan

Here is a rough sketch of Noah's bedroom (not a scale drawing).

Noah wants to create a floor plan that is a scale drawing.

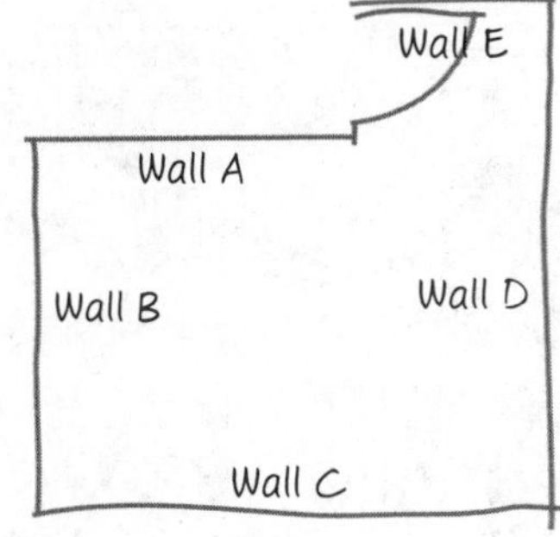

1. The actual length of Wall C is 4 m. To represent Wall C, Noah draws a segment 16 cm long. What scale is he using? Explain or show your reasoning.

2. Find another way to express the scale.

3. Discuss your thinking with your partner. How do your scales compare?

4. The actual lengths of Wall A and Wall D are 2.5 m and 3.75 m. Determine how long these walls will be on Noah's scale floor plan. Explain or show your reasoning.

Are you ready for more?

If Noah wanted to draw another floor plan on which Wall C was 20 cm, would 1 cm to 5 m be the right scale to use? Explain your reasoning.

Activity

9.3 Two Maps of Utah

A rectangle around Utah is about 270 miles wide and about 350 miles tall. The upper right corner that is missing is about 110 miles wide and about 70 miles tall.

1. Make a scale drawing of Utah where 1 centimeter represents 50 miles.

   Make a scale drawing of Utah where 1 centimeter represents 75 miles.

2. How do the two drawings compare? How does the choice of scale influence the drawing?

NAME ________________ DATE ________ PERIOD ________

Summary
Creating Scale Drawings

If we want to create a scale drawing of a room's floor plan that has the scale "1 inch to 4 feet," we can divide the actual lengths in the room (in feet) by 4 to find the corresponding lengths (in inches) for our drawing.

Suppose the longest wall is 15 feet long. We should draw a line 3.75 inches long to represent this wall, because $15 \div 4 = 3.75$.

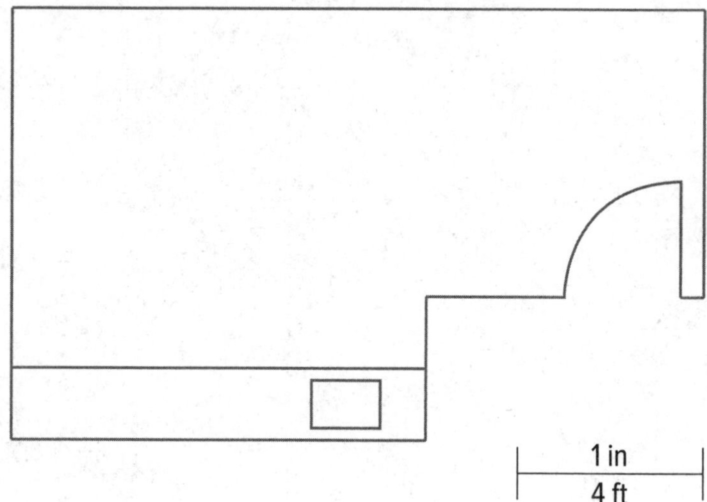

There is more than one way to express this scale.

These three scales are all equivalent, since they represent the same relationship between lengths on a drawing and actual lengths.

- 1 inch to 4 feet
- $\frac{1}{2}$ inch to 2 feet
- $\frac{1}{4}$ inch to 1 foot

Any of these scales can be used to find actual lengths and scaled lengths (lengths on a drawing). For instance, we can tell that, at this scale, an 8-foot long wall should be 2 inches long on the drawing because $\frac{1}{4} \cdot 8 = 2$.

The size of a scale drawing is influenced by the choice of scale.

For example, here is another scale drawing of the same room using the scale 1 inch to 8 feet. Notice this drawing is smaller than the previous one. Since one inch on this drawing represents twice as much actual distance, each side length only needs to be half as long as it was in the first scale drawing.

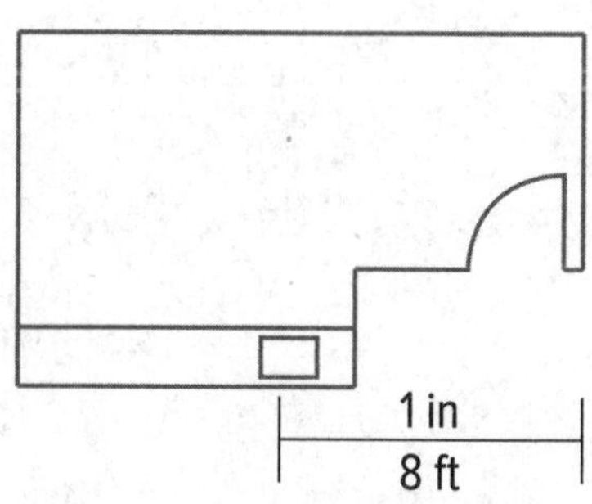

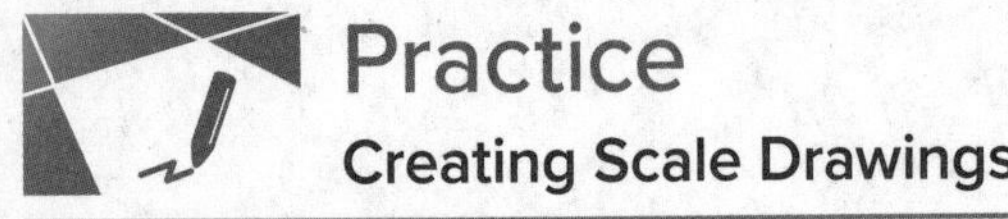

Practice
Creating Scale Drawings

1. An image of a book shown on a website is 1.5 inches wide and 3 inches tall on a computer monitor. The actual book is 9 inches wide.

   a. What scale is being used for the image?

   b. How tall is the actual book?

2. The flag of Colombia is a rectangle that is 6 ft long with three horizontal stripes.

   - The top stripe is 2 ft tall and is yellow.
   - The middle stripe is 1 ft tall and is blue.
   - The bottom stripe is also 1 ft tall and is red.

   a. Create a scale drawing of the Colombian flag with a scale of 1 cm to 2 ft.

   b. Create a scale drawing of the Colombian flag with a scale of 2 cm to 1 ft.

NAME ______________________ DATE ____________ PERIOD ____________

3. These triangles are scaled copies of each other.

Triangle F Triangle B Triangle G Triangle H

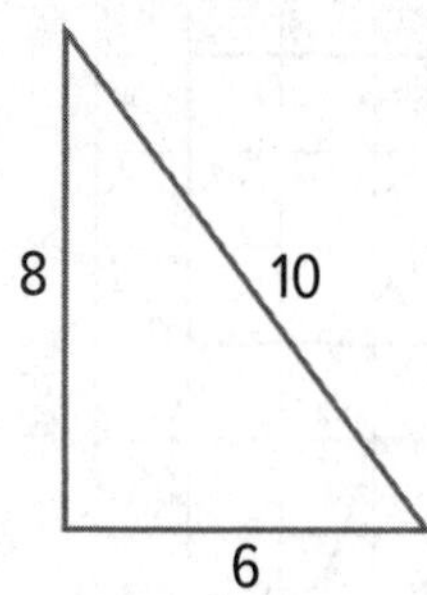

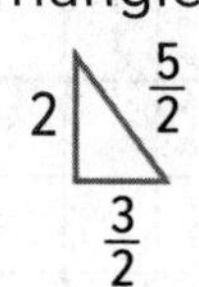

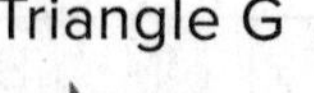

4 5 3

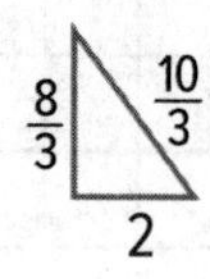

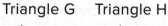

For each pair of triangles listed, the area of the second triangle is how many times larger than the area of the first? **(Lesson 1-6)**

a. Triangle G and Triangle F

b. Triangle G and Triangle B

c. Triangle B and Triangle F

d. Triangle F and Triangle H

e. Triangle G and Triangle H

f. Triangle H and Triangle B

4. Here is an unlabeled rectangle, followed by other quadrilaterals that are labeled.

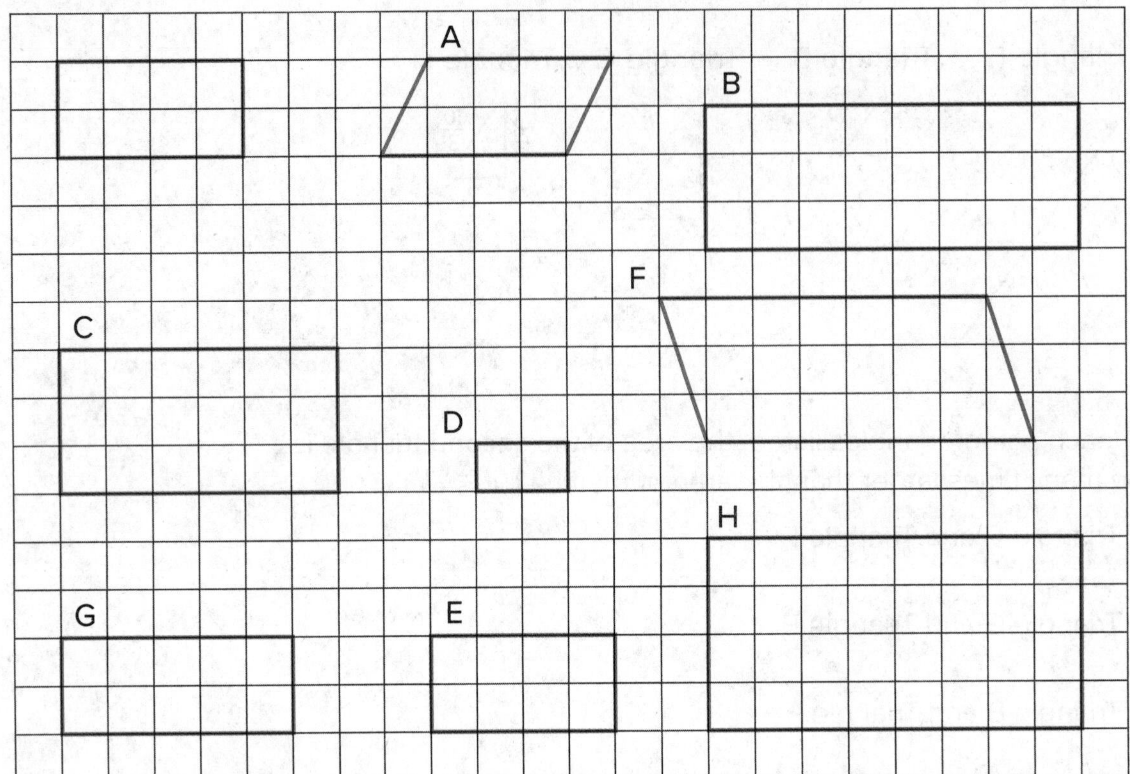

a. Select **all** quadrilaterals that are scaled copies of the unlabeled rectangle. Explain how you know.

b. On graph paper, draw a different scaled version of the original rectangle. **(Lesson 1-3)**

Lesson 1-10

Changing Scales in Scale Drawings

NAME ______________________ DATE ____________ PERIOD ____________

Learning Goal Let's explore different scale drawings of the same actual thing.

Warm Up

10.1 Appropriate Measurements

1. If a student uses a ruler like this to measure the length of their foot, which choices would be appropriate measurements? Select **all** that apply. Be prepared to explain your reasoning.

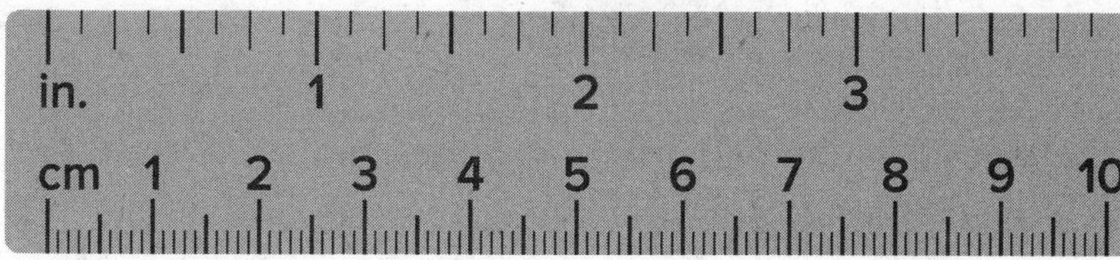

A. $9\frac{1}{4}$ inches

B. $9\frac{5}{64}$ inches

C. 23.47659 centimeters

D. 23.5 centimeters

E. 23.48 centimeters

2. Here is a scale drawing of an average seventh-grade student's foot next to a scale drawing of a foot belonging to the person with the largest feet in the world. Estimate the length of the larger foot.

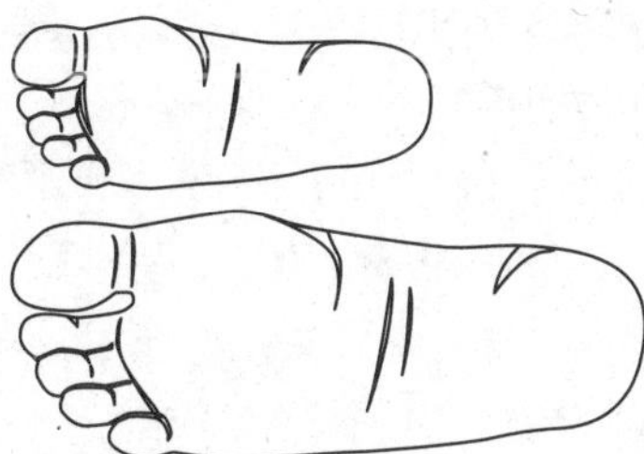

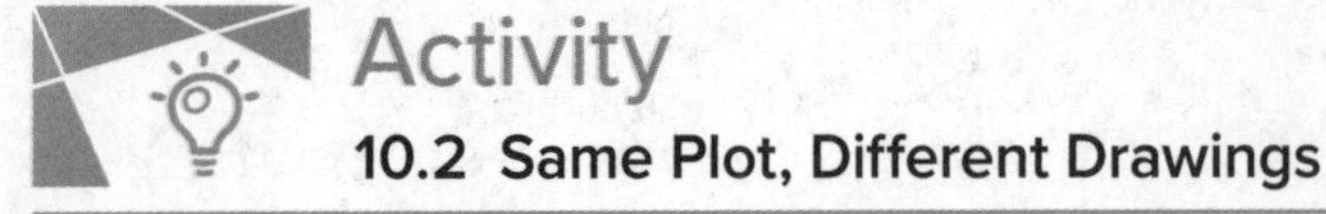

Activity

10.2 Same Plot, Different Drawings

Here is a map showing a plot of land in the shape of a right triangle.

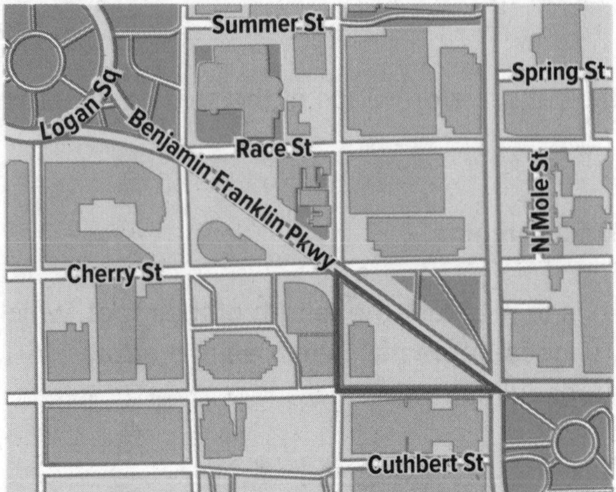

1. Your teacher will assign you a scale to use. On centimeter graph paper, make a scale drawing of the plot of land. Make sure to write your scale on your drawing.
2. What is the area of the triangle you drew? Explain or show your reasoning.

3. How many square meters are represented by 1 square centimeter in your drawing?

4. After everyone in your group is finished, order the scale drawings from largest to smallest. What do you notice about the scales when your drawings are placed in this order?

NAME ______________________ DATE ______________ PERIOD ____________

Are you ready for more?

Noah and Elena each make a scale drawing of the same triangular plot of land, using the following scales. Make a prediction about the size of each drawing. How would they compare to the scale drawings made by your group?

1. Noah uses the scale 1 cm to 200 m.
2. Elena uses the scale 2 cm to 25 m.

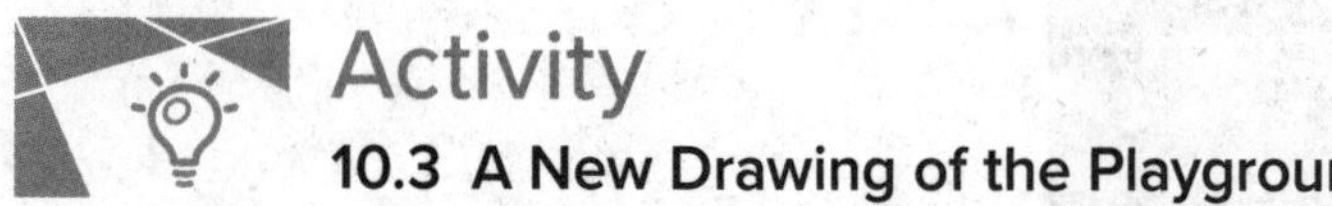

Activity
10.3 A New Drawing of the Playground

Here is a scale drawing of a playground. The scale is 1 centimeter to 30 meters.

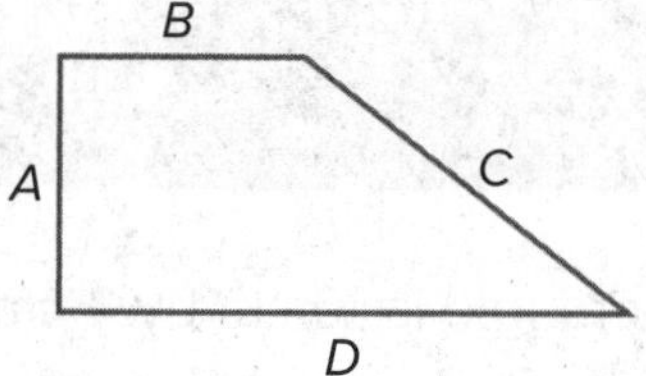

1. Make another scale drawing of the same playground at a scale of 1 centimeter to 20 meters.

2. How do the two scale drawings compare?

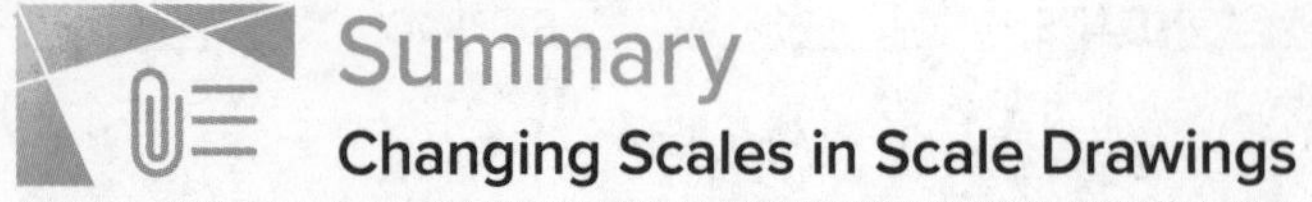

Summary

Changing Scales in Scale Drawings

Sometimes we have a scale drawing of something, and we want to create another scale drawing of it that uses a different scale.

- We can use the original scale drawing to find the size of the actual object.
- Then we can use the size of the actual object to figure out the size of our new scale drawing.

For example, here is a scale drawing of a park where the scale is 1 cm to 90 m.

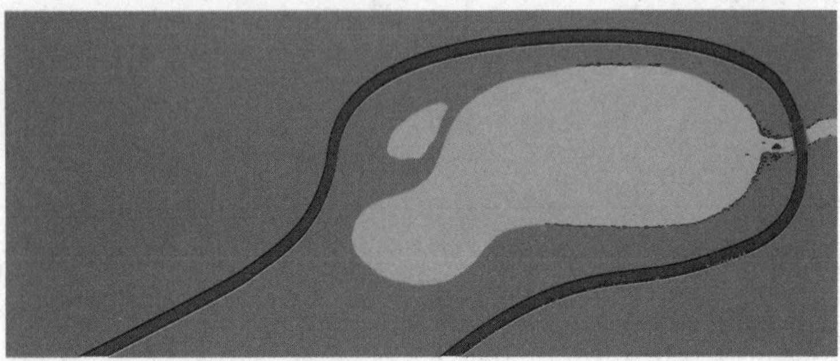

The rectangle is 10 cm by 4 cm, so the actual dimensions of the park are 900 m by 360 m, because $10 \cdot 90 = 900$ and $4 \cdot 90 = 360$.

Suppose we want to make another scale drawing of the park where the scale is 1 cm to 30 meters. This new scale drawing should be 30 cm by 12 cm, because $900 \div 30 = 30$ and $360 \div 30 = 12$.

Another way to find this answer is to think about how the two different scales are related to each other.

- In the first scale drawing, 1 cm represented 90 m.
- In the new drawing, we would need 3 cm to represent 90 m.

That means each length in the new scale drawing should be 3 times as long as it was in the original drawing. The new scale drawing should be 30 cm by 12 cm, because $3 \cdot 10 = 30$ and $3 \cdot 4 = 12$.

Since the length and width are 3 times as long, the area of the new scale drawing will be 9 times as large as the area of the original scale drawing, because $3^2 = 9$.

NAME ______________________ DATE ____________ PERIOD ____________

Practice

Changing Scales in Scale Drawings

1. Here is a scale drawing of a swimming pool where 1 cm represents 1 m.

   a. How long and how wide is the actual swimming pool?

   b. Will a scale drawing where 1 cm represents 2 m be larger or smaller than this drawing?

   c. Make a scale drawing of the swimming pool where 1 cm represents 2 m.

2. A map of a park has a scale of 1 inch to 1,000 feet. Another map of the same park has a scale of 1 inch to 500 feet. Which map is larger? Explain or show your reasoning.

3. On a map with a scale of 1 inch to 12 feet, the area of a restaurant is 60 in^2. Han says that the actual area of the restaurant is 720 ft^2. Do you agree or disagree? Explain your reasoning.

4. If Quadrilateral Q is a scaled copy of Quadrilateral P created with a scale factor of 3, what is the perimeter of Q? **(Lesson 1-3)**

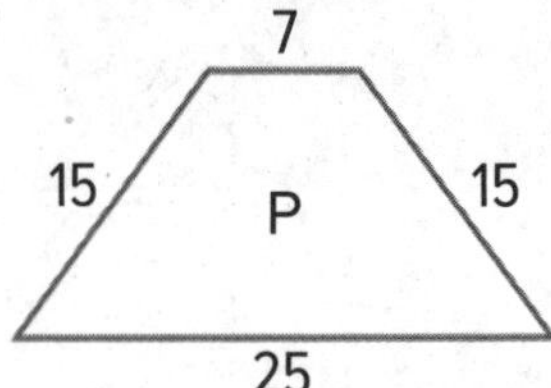

5. Triangle *DEF* is a scaled copy of triangle *ABC*. For each of the following parts of triangle *ABC*, identify the corresponding part of triangle *DEF*. **(Lesson 1-2)**

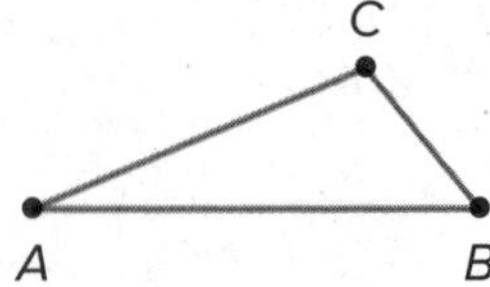

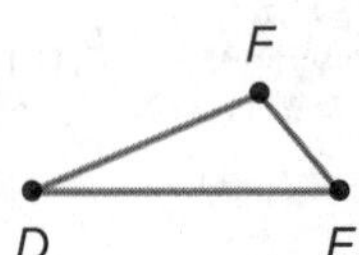

a. angle *ABC*

b. angle *BCA*

c. segment *AC*

d. segment *BA*

Lesson 1-11

Scales without Units

NAME ______________________ DATE ____________ PERIOD ____________

Learning Goal Let's explore a different way to express scales.

Warm Up

11.1 One to One Hundred

A map of a park says its scale is 1 to 100.

1. What do you think that means?

2. Give an example of how this scale could tell us about measurements in the park.

Activity

11.2 Apollo Lunar Module

Your teacher will give you a drawing of the Apollo Lunar Module. It is drawn at a scale of 1 to 50.

1. The "legs" of the spacecraft are its landing gear. Use the drawing to estimate the actual length of each leg on the sides. Write your answer to the nearest 10 centimeters. Explain or show your reasoning.

2. Use the drawing to estimate the actual height of the Apollo Lunar Module to the nearest 10 centimeters. Explain or show your reasoning.

3. Neil Armstrong was 71 inches tall when he went to the surface of the moon in the Apollo Lunar Module. How tall would he be in the drawing if he were drawn with his height to scale? Show your reasoning.

4. Sketch a stick figure to represent yourself standing next to the Apollo Lunar Module. Make sure the height of your stick figure is to scale. Show how you determined your height on the drawing.

Are you ready for more?

The table shows the distance between the sun and 8 planets in our solar system.

1. If you wanted to create a scale model of the solar system that could fit somewhere in your school, what scale would you use?

2. The diameter of the Earth is approximately 8,000 miles. What would the diameter of the Earth be in your scale model?

Planet	Average Distance (millions of miles)
Mercury	35
Venus	67
Earth	93
Mars	142
Jupiter	484
Saturn	887
Uranus	1784
Neptune	2795

NAME ______________________ DATE ____________ PERIOD ____________

Activity

11.3 Same Drawing, Different Scales

A rectangular parking lot is 120 feet long and 75 feet wide.

- Lin made a scale drawing of the parking lot at a scale of 1 inch to 15 feet. The drawing she produced is 8 inches by 5 inches.
- Diego made another scale drawing of the parking lot at a scale of 1 to 180. The drawing he produced is also 8 inches by 5 inches.

1. Explain or show how each scale would produce an 8-inch-by-5-inch drawing.

2. Make another scale drawing of the same parking lot at a scale of 1 inch to 20 feet. Be prepared to explain your reasoning.

3. Express the scale of 1 inch to 20 feet as a scale without units. Explain your reasoning.

Summary

Scales without Units

In some scale drawings, the scale specifies one unit for the distances on the drawing and a different unit for the actual distances represented. For example, a drawing could have a scale of 1 cm to 10 km.

In other scale drawings, the scale does not specify any units at all. For example, a map may simply say the scale is 1 to 1,000.

- In this case, the units for the scaled measurements and actual measurements can be any unit, so long as the same unit is being used for both.
- So if a map of a park has a scale 1 to 1,000, then 1 inch on the map represents 1,000 inches in the park, and 12 centimeters on the map represent 12,000 centimeters in the park.
- In other words, 1,000 is the scale factor that relates distances on the drawing to actual distances, and $\frac{1}{1{,}000}$ is the scale factor that relates an actual distance to its corresponding distance on the drawing.

A scale with units can be expressed as a scale without units by converting one measurement in the scale into the same unit as the other (usually the unit used in the drawing). For example, these scales are equivalent.

- 1 inch to 200 feet
- 1 inch to 2,400 inches
  (because there are 12 inches in 1 foot, and $200 \cdot 12 = 2{,}400$)
- 1 to 2,400

This scale tells us that all actual distances are 2,400 times their corresponding distances on the drawing, and distances on the drawing are $\frac{1}{2{,}400}$ times the actual distances they represent.

NAME ______________________ DATE ____________ PERIOD ____________

Practice

Scales without Units

1. A scale drawing of a car is presented in the following three scales. Order the scale drawings from smallest to largest. Explain your reasoning. (There are about 1.1 yards in a meter, and 2.54 cm in an inch.)

   A: 1 in. to 1 ft **B:** 1 in. to 1 m **C:** 1 in. to 1 yd

2. Which scales are equivalent to 1 inch to 1 foot? Select **all** that apply.

   (A.) 1 to 12

   (B.) $\frac{1}{12}$ to 1

   (C.) 100 to 0.12

   (D.) 5 to 60

   (E.) 36 to 3

   (F.) 9 to 108

3. A model airplane is built at a scale of 1 to 72. If the model plane is 8 inches long, how many feet long is the actual airplane?

4. Quadrilateral A has side lengths 3, 6, 6, and 9. Quadrilateral B is a scaled copy of A with a shortest side length equal to 2. Jada says, "Since the side lengths go down by 1 in this scaling, the perimeter goes down by 4 in total." Do you agree with Jada? Explain your reasoning. **(Lesson 1-3)**

5. Polygon B is a scaled copy of Polygon A using a scale factor of 5. Polygon A's area is what fraction of Polygon B's area? **(Lesson 1-6)**

6. Figures R, S, and T are all scaled copies of one another. Figure S is a scaled copy of R using a scale factor of 3. Figure T is a scaled copy of S using a scale factor of 2. Find the scale factors for each of the following: **(Lesson 1-5)**

   a. From T to S

   b. From S to R

   c. From R to T

   d. From T to R

Lesson 1-12

Units in Scale Drawings

NAME ______________________ DATE __________ PERIOD __________

Learning Goal Let's use different scales to describe the same drawing.

Warm Up

12.1 Centimeters in a Mile

There are 2.54 centimeters in an inch, 12 inches in a foot, and 5,280 feet in a mile. Which expression gives the number of centimeters in a mile? Explain your reasoning.

1. $\frac{2.54}{12 \cdot 5{,}280}$
2. $5{,}280 \cdot 12 \cdot (2.54)$
3. $\frac{1}{5{,}280 \cdot 12 \cdot (2.54)}$
4. $5{,}280 + 12 + 2.54$
5. $\frac{5{,}280 \cdot 12}{2.54}$

Activity

12.2 Card Sort: Scales

Your teacher will give you some cards with a scale on each card.

1. Sort the cards into sets of equivalent scales. Be prepared to explain how you know that the scales in each set are equivalent. Each set should have at least two cards.
2. Trade places with another group and check each other's work. If you disagree about how the scales should be sorted, work to reach an agreement.

Pause here so your teacher can review your work.

3. Next, record one of the sets with three equivalent scales and explain why they are equivalent.

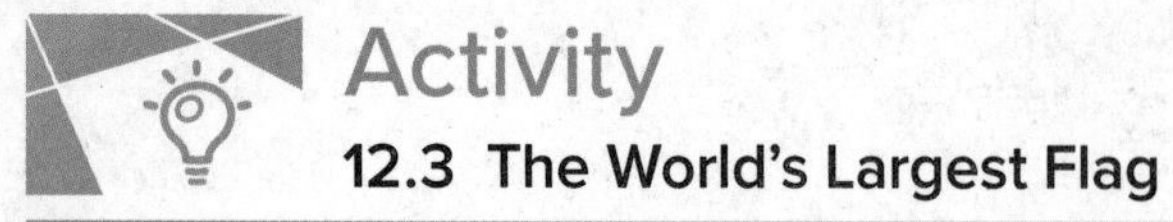

Activity

12.3 The World's Largest Flag

As of 2016, Tunisia holds the world record for the largest version of a national flag. It was almost as long as four soccer fields. The flag has a circle in the center, a crescent moon inside the circle, and a star inside the crescent moon.

1. Complete the table. Explain or show your reasoning.

	Flag Length	Flag Height	Height of Crescent Moon
Actual	396 m		99 m
At 1 to 2,000 Scale		13.2 cm	

2. Complete each scale with the value that makes it equivalent to the scale of 1 to 2,000. Explain or show your reasoning.

   a. 1 cm to ______________ cm

   b. 1 cm to ______________ m

   c. 1 cm to ______________ km

   d. 2 m to ______________ m

   e. 5 cm to ______________ m

   f. ______________ cm to 1,000 m

   g. ______________ mm to 20 m

3. a. What is the area of the large flag?

   b. What is the area of the smaller flag?

   c. The area of the large flag is how many times the area of the smaller flag?

NAME ______________________ DATE ______________ PERIOD ___________

Activity

12.4 Pondering Pools

Your teacher will give you a floor plan of a recreation center.

1. What is the scale of the floor plan if the actual side length of the square pool is 15 m? Express your answer both as a scale with units and without units.

2. Find the actual area of the large rectangular pool. Show your reasoning.

3. The kidney-shaped pool has an area of 3.2 cm^2 on the drawing. What is its actual area? Explain or show your reasoning.

Are you ready for more?

1. Square A is a scaled copy of Square B with scale factor 2. If the area of Square A is 10 units^2, what is the area of Square B?

2. Cube A is a scaled copy of Cube B with scale factor 2. If the volume of Cube A is 10 units^3, what is the volume of Cube B?

3. The four-dimensional Hypercube A is a scaled copy of Hypercube B with scale factor 2. If the "volume" of Hypercube A is 10 units^4, what do you think the "volume" of Hypercube B is?

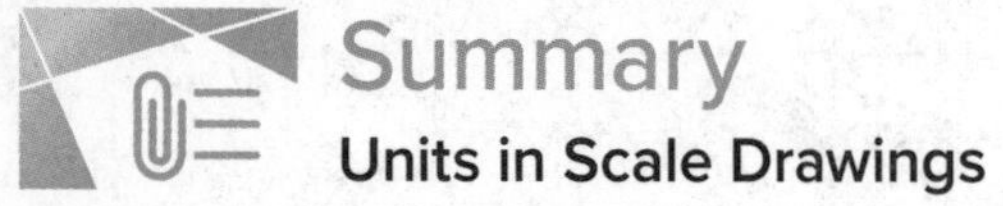

Summary

Units in Scale Drawings

Sometimes scales come with units, and sometimes they don't.

For example, a map of Nebraska may have a scale of 1 mm to 1 km.

- This means that each millimeter of distance on the map represents 1 kilometer of distance in Nebraska.
- Notice that there are 1,000 millimeters in 1 meter and 1,000 meters in 1 kilometer.
- This means there are 1,000 • 1,000 or 1,000,000 millimeters in 1 kilometer.

So, the same scale without units is 1 to 1,000,000, which means that each unit of distance on the map represents 1,000,000 units of distance in Nebraska. This is true for *any* choice of unit to express the scale of this map.

Sometimes when a scale comes with units, it is useful to rewrite it without units.

For example, let's say we have a different map of Rhode Island, and we want to use the two maps to compare the size of Nebraska and Rhode Island.

- It is important to know if the maps are at the same scale.
- The scale of the map of Rhode Island is 1 inch to 10 miles.
- There are 5,280 feet in 1 mile, and 12 inches in 1 foot, so there are 63,360 inches in 1 mile (because 5,280 • 12 = 63,360). Therefore, there are 633,600 inches in 10 miles.

The scale of the map of Rhode Island without units is 1 to 633,600. The two maps are not at the same scale, so we should not use these maps to compare the size of Nebraska to the size of Rhode Island.

Here is some information about equal lengths that you may find useful.

Customary Units

1 foot (ft) = 12 inches (in)

1 yard (yd) = 36 inches

1 yard = 3 feet

1 mile = 5,280 feet

Metric Units

1 meter (m) = 1,000 millimeters (mm)

1 meter = 100 centimeters

1 kilometer (km) = 1,000 meters

Equal Lengths in Different Systems

1 inch = 2.54 centimeters

1 foot ≈ 0.30 meter

1 mile ≈ 1.61 kilometers

1 centimeter ≈ 0.39 inch

1 meter ≈ 39.37 inches

1 kilometer ≈ 0.62 mile

NAME ______________________ DATE ____________ PERIOD ________

Practice

Units in Scale Drawings

1. The Empire State Building in New York City is about 1,450 feet high (including the antenna at the top) and 400 feet wide. Andre wants to make a scale drawing of the front view of the Empire State Building on an $8\frac{1}{2}$-inch-by-11-inch piece of paper. Select a scale that you think is the most appropriate for the scale drawing. Explain your reasoning.

   A: 1 inch to 1 foot

   B: 1 inch to 100 feet

   C: 1 inch to 1 mile

   D: 1 centimeter to 1 meter

   E: 1 centimeter to 50 meters

   F: 1 centimeter to 1 kilometer

2. Elena finds that the area of a house on a scale drawing is 25 square inches. The actual area of the house is 2,025 square feet. What is the scale of the drawing?

3. Which of these scales are equivalent to 3 cm to 4 km? Select **all** that apply. Recall that 1 inch is 2.54 centimeters.

   A. 0.75 cm to 1 km

   B. 1 cm to 12 km

   C. 6 mm to 2 km

   D. 0.3 mm to 40 m

   E. 1 inch to 7.62 km

4. These two triangles are scaled copies of one another. The area of the smaller triangle is 9 square units. What is the area of the larger triangle? Explain or show how you know. **(Lesson 1-6)**

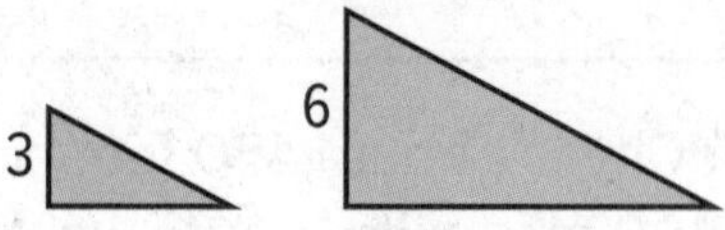

5. Water costs $1.25 per bottle. At this rate, what is the cost of:

   a. 10 bottles?

   b. 20 bottles?

   c. 50 bottles?

6. The first row of the table shows the amount of dish detergent and water needed to make a soap solution.

Number of Batches	Water (cups)	Detergent (cups)
1	6	1
2		
3		
4		

   a. Complete the table for 2, 3, and 4 batches.

   b. How much water and detergent is needed for 8 batches? Explain your reasoning.

Lesson 1-13

Draw It to Scale

NAME ______________________ DATE ______________ PERIOD ____________

Learning Goal Let's draw a floor plan.

Warm Up

13.1 Which Measurements Matter?

Which measurements would you need in order to draw a scale floor plan of your classroom? List which parts of the classroom you would measure and include in the drawing. Be as specific as possible.

Activity

13.2 Creating a Floor Plan (Part 1)

1. On a blank sheet of paper, make a *rough sketch* of a floor plan of the classroom. Include parts of the room that the class has decided to include or that you would like to include. Accuracy is not important for this rough sketch, but be careful not to omit important features like a door.
2. Trade sketches with a partner and check each other's work. Specifically, check if any parts are missing or incorrectly placed. Return their work and revise your sketch as needed.
3. Discuss with your group a plan for measuring. Work to reach an agreement on:
   - which classroom features must be measured and which are optional
   - the units to be used
   - how to record and organize the measurements (on the sketch, in a list, in a table, etc.)
   - how to share the measuring and recording work (or the role each group member will play)
4. Gather your tools, take your measurements, and record them as planned. Be sure to double-check your measurements.
5. Make your own copy of all the measurements that your group has gathered, if you haven't already done so. You will need them for the next activity.

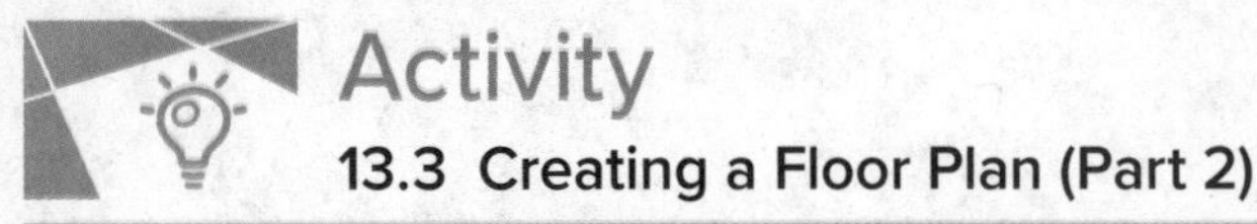

Activity

13.3 Creating a Floor Plan (Part 2)

Your teacher will give you several paper options for your scale floor plan.

1. Determine an appropriate scale for your drawing based on your measurements and your paper choice. Your floor plan should fit on the paper and not end up too small.
2. Use the scale and the measurements your group has taken to draw a scale floor plan of the classroom. Make sure to:
   - Show the scale of your drawing.
   - Label the key parts of your drawing (the walls, main openings, etc.) with their actual measurements.
   - Show your thinking and organize it so it can be followed by others.

Are you ready for more?

1. If the flooring material in your classroom is to be replaced with 10-inch-by-10-inch tiles, how many tiles would it take to cover the entire room? Use your scale drawing to approximate the number of tiles needed.
2. How would using 20-inch-by-20-inch tiles (instead of 10-inch-by-10-inch tiles) change the number of tiles needed? Explain your reasoning.

Activity

13.4 Creating a Floor Plan (Part 3)

1. Trade floor plans with another student who used the same paper size as you. Discuss your observations and thinking.
2. Trade floor plans with another student who used a different paper size than you. Discuss your observations and thinking.
3. Based on your discussions, record ideas for how your floor plan could be improved.

Learning Targets

Lesson	Learning Target(s)
1-1 What Are Scaled Copies?	• I can describe some characteristics of a scaled copy. • I can tell whether or not a figure is a scaled copy of another figure.
1-2 Corresponding Parts and Scale Factors	• I can describe what the scale factor has to do with a figure and its scaled copy. • In a pair of figures, I can identify corresponding points, corresponding segments, and corresponding angles.
1-3 Making Scaled Copies	• I can draw a scaled copy of a figure using a given scale factor. • I know what operation to use on the side lengths of a figure to produce a scaled copy.

(continued on the next page)

(continued from the previous page)

Lesson	Learning Target(s)
1-4 Scaled Relationships	• I can use corresponding distances and corresponding angles to tell whether one figure is a scaled copy of another. • When I see a figure and its scaled copy, I can explain what is true about corresponding angles. • When I see a figure and its scaled copy, I can explain what is true about corresponding distances.
1-5 The Size of the Scale Factor	• I can describe the effect on a scaled copy when I use a scale factor that is greater than 1, less than 1, or equal to 1. • I can explain how the scale factor that takes Figure A to its copy Figure B is related to the scale factor that takes Figure B to Figure A.
1-6 Scaling and Area	• I can describe how the area of a scaled copy is related to the area of the original figure and the scale factor that was used.
1-7 Scale Drawings	• I can explain what a scale drawing is, and I can explain what its scale means. • I can use actual distances and a scale to find scaled distances. • I can use a scale drawing and its scale to find actual distances.

Lesson	Learning Target(s)
1-8 Scale Drawings and Maps	• I can use a map and its scale to solve problems about traveling.
1-9 Creating Scale Drawings	• I can determine the scale of a scale drawing when I know lengths on the drawing and corresponding actual lengths. • I know how different scales affect the lengths in the scale drawing. • When I know the actual measurements, I can create a scale drawing at a given scale.
1-10 Changing Scales in Scale Drawings	• Given a scale drawing, I can create another scale drawing that shows the same thing at a different scale. • I can use a scale drawing to find actual areas.
1-11 Scales without Units	• I can explain the meaning of scales expressed without units. • I can use scales without units to find scaled distances or actual distances.

(continued on the next page)

(continued from the previous page)

Lesson	Learning Target(s)
1-12 Units in Scale Drawings	• I can tell whether two scales are equivalent. • I can write scales with units as scales without units.
1-13 Let's Draw It to Scale	• I can create a scale drawing of my classroom. • When given requirements on drawing size, I can choose an appropriate scale to represent an actual object.

Notes:

Unit 2

Introducing Proportional Relationships

Collins93/Shutterstock

Peregrine falcons typically fly 60 miles per hour. How does their flying time relate to the distance flown? You'll learn more about proportional relationships in this unit.

Topics

- Representing Proportional Relationships with Tables
- Representing Proportional Relationships with Equations
- Comparing Proportional and Nonproportional Relationships
- Representing Proportional Relationships with Graphs
- Let's Put It to Work

Unit 2

Introducing Proportional Relationships

Representing Proportional Relationships with Tables

Representing Proportional Relationships with Equations

Comparing Proportional and Nonproportional Relationships

Representing Proportional Relationships with Graphs

Let's Put It to Work

Lesson 2-1

One of These Things is Not Like the Others

NAME ______________________ DATE ____________ PERIOD ____________

Learning Goal Let's remember what equivalent ratios are.

Warm Up

1.1 Remembering Double Number Lines

1. Complete the double number line diagram with the missing numbers.

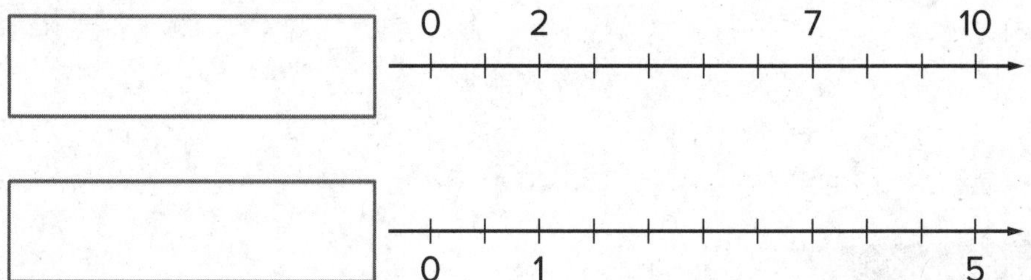

2. What could each of the number lines represent? Invent a situation and label the diagram.

3. Make sure your labels include appropriate units of measure.

Activity

1.2 Mystery Mixtures

Your teacher will show you three mixtures. Two taste the same, and one is different.

1. Which mixture tastes different? Describe how it is different.

2. Here are the recipes that were used to make the three mixtures:

- 1 cup of water with $1\frac{1}{2}$ teaspoons of powdered drink mix
- 2 cups of water with $\frac{1}{2}$ teaspoon of powdered drink mix
- 1 cup of water with $\frac{1}{4}$ teaspoon of powdered drink mix

Which of these recipes is for the stronger tasting mixture?
Explain how you know.

Are you ready for more?

Salt and sugar give two distinctly different tastes, one salty and the other sweet. In a mixture of salt and sugar, it is possible for the mixture to be salty, sweet or both? Will any of these mixtures taste exactly the same?

- **Mixture A:** 2 cups water, 4 teaspoons salt, 0.25 cup sugar
- **Mixture B:** 1.5 cups water, 3 teaspoons salt, 0.2 cup sugar
- **Mixture C:** 1 cup water, 2 teaspoons salt, 0.125 cup sugar

NAME ______________________ DATE ____________ PERIOD ____________

Activity
1.3 Crescent Moons

Here are four different crescent moon shapes.

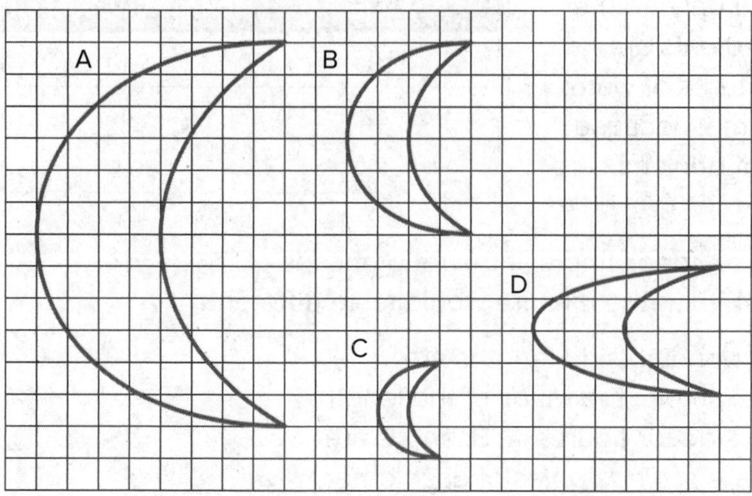

1. What do Moons A, B, and C all have in common that Moon D doesn't?

2. Use numbers to describe how Moons A, B, and C are different from Moon D.

3. Use a table or a double number line to show how Moons A, B, and C are different from Moon D.

Summary

One of These Things is Not Like the Others

When two different situations can be described by **equivalent ratios**, that means they are alike in some important way.

An example is a recipe. If two people make something to eat or drink, the taste will only be the same as long as the ratios of the ingredients are equivalent. For example, all of the mixtures of water and drink mix in this table taste the same, because the ratios of cups of water to scoops of drink mix are all equivalent ratios.

Water (cups)	Drink Mix (scoops)
3	1
12	4
1.5	0.5

If a mixture were not equivalent to these, for example, if the ratio of cups of water to scoops of drink mix were 6 : 4, then the mixture would taste different.

Notice that the ratios of pairs of corresponding side lengths are equivalent in figures A, B, and C. For example, the ratios of the length of the top side to the length of the left side for figures A, B, and C are equivalent ratios. Figures A, B, and C are *scaled copies* of each other; this is the important way in which they are alike.

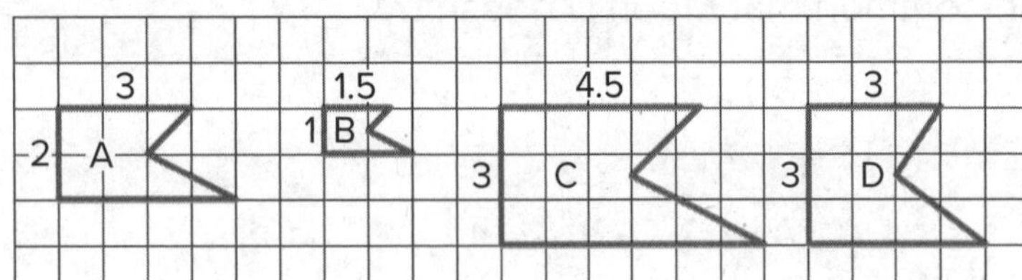

If a figure has corresponding sides that are not in a ratio equivalent to these, like figure D, then it's not a scaled copy. In this unit, you will study relationships like these that can be described by a set of equivalent ratios.

Glossary

equivalent ratios

NAME ______________________ DATE ____________ PERIOD ________

Practice

One of These Things is Not Like the Others

1. Which one of these shapes is not like the others? Explain what makes it different by representing each width and height pair with a ratio.

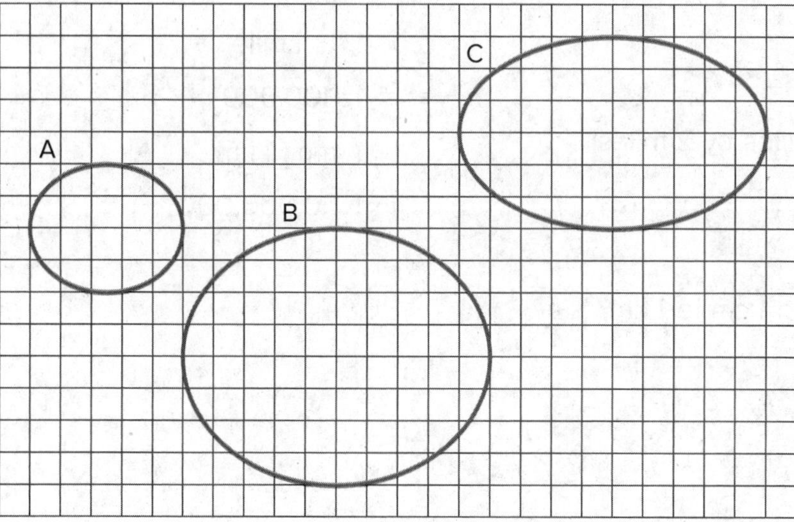

2. In one version of a trail mix, there are 3 cups of peanuts mixed with 2 cups of raisins. In another version of trail mix, there are 4.5 cups of peanuts mixed with 3 cups of raisins. Are the ratios equivalent for the two mixes? Explain your reasoning.

3. For each object, choose an appropriate scale for a drawing that fits on a regular sheet of paper. Not all of the scales on the list will be used. **(Lesson 1-12)**

Objects	Scales
a. A person	1 in : 1 ft
b. A football field (120 yards by $53\frac{1}{3}$ yards)	1 cm : 1 m
c. The state of Washington (about 240 miles by 360 miles)	1 : 1,000
d. The floor plan of a house	1 ft : 1 mile
e. A rectangular farm (6 miles by 2 miles)	1 : 100,000
	1 mm : 1 km
	1 : 10,000,000

4. Which scale is equivalent to 1 cm to 1 km? **(Lesson 1-11)**

(A.) 1 to 1,000

(B.) 10,000 to 1

(C.) 1 to 100,000

(D.) 100,000 to 1

(E.) 1 to 1,000,000

5. Respond to each of the following.

   a. Find 3 different ratios that are equivalent to 7 : 3.

   b. Explain why these ratios are equivalent.

Lesson 2-2

Introducing Proportional Relationships with Tables

NAME ____________________ DATE ____________ PERIOD ____________

Learning Goal Let's solve problems involving proportional relationships using tables.

Warm Up

2.1 Notice and Wonder: Paper Towels by the Case

Here is a table that shows how many rolls of paper towels a store receives when they order different numbers of cases.

What do you notice about the table? What do you wonder?

Number of Cases They Order	Number of Rolls of Paper Towels
1	12
3	36
5	60
10	120

$\cdot 2$ (from 5 to 10) $\cdot 2$ (from 60 to 120)

Activity

2.2 Feeding a Crowd

1. A recipe says that 2 cups of dry rice will serve 6 people. Complete the table as you answer the questions. Be prepared to explain your reasoning.

   a. How many people will 10 cups of rice serve?

   b. How many cups of rice are needed to serve 45 people?

Cups of Rice	Number of People
2	6
3	9
10	
	45

2. A recipe says that 6 spring rolls will serve 3 people. Complete the table.

Number of Spring Rolls	Number of People
6	3
30	
40	
	28

Activity
2.3 Making Bread Dough

A bakery uses 8 tablespoons of honey for every 10 cups of flour to make bread dough. Some days they bake bigger batches and some days they bake smaller batches, but they always use the same ratio of honey to flour.

Honey (tablespoons)	Flour (cups)
8	10
20	
13	
	20

Complete the table as you answer the questions. Be prepared to explain your reasoning.

1. How many cups of flour do they use with 20 tablespoons of honey?

2. How many cups of flour do they use with 13 tablespoons of honey?

3. How many tablespoons of honey do they use with 20 cups of flour?

4. What is the **proportional relationship** represented by this table?

NAME ______________________ DATE ____________ PERIOD ____________

Activity
2.4 Quarters and Dimes

4 quarters are equal in value to 10 dimes.

Number of Quarters	Number of Dimes
1	
4	10
6	
14	

1. How many dimes equal the value of 6 quarters?

2. How many dimes equal the value of 14 quarters?

3. What value belongs next to the 1 in the table? What does it mean in this context?

Are you ready for more?

Pennies made before 1982 are 95% copper and weigh about 3.11 grams each. (Pennies made after that date are primarily made of zinc). Some people claim that the value of the copper in one of these pennies is greater than the face value of the penny. Find out how much copper is worth right now and decide if this claim is true.

Summary

Introducing Proportional Relationships with Tables

If the ratios between two corresponding quantities are always equivalent, the relationship between the quantities is called a **proportional relationship**.

This table shows different amounts of milk and chocolate syrup. The ingredients in each row, when mixed together, would make a different total amount of chocolate milk, but these mixtures would all taste the same.

Tablespoons of Chocolate Syrup	Cups of Milk
4	1
6	$1\frac{1}{2}$
8	2
$\frac{1}{2}$	$\frac{1}{8}$
12	3
1	$\frac{1}{4}$

Notice that each row in the table shows a ratio of tablespoons of chocolate syrup to cups of milk that is equivalent to 4 : 1.

About the relationship between these quantities, we could say:

- The relationship between amount of chocolate syrup and amount of milk is proportional.
- The relationship between the amount of chocolate syrup and the amount of milk is a proportional relationship.
- The table represents a proportional relationship between the amount of chocolate syrup and amount of milk.
- The amount of milk is proportional to the amount of chocolate syrup.

We could multiply any value in the chocolate syrup column by $\frac{1}{4}$ to get the value in the milk column. We might call $\frac{1}{4}$ a *unit rate*, because $\frac{1}{4}$ cup of milk are needed for 1 tablespoon of chocolate syrup. We also say that $\frac{1}{4}$ is the **constant of proportionality** for this relationship. It tells us how many cups of milk we would need to mix with 1 tablespoon of chocolate syrup.

Glossary

constant of proportionality
equivalent ratios

NAME ____________ DATE ____________ PERIOD ____________

Practice

Introducing Proportional Relationships with Tables

1. When Han makes chocolate milk, he mixes 2 cups of milk with 3 tablespoons of chocolate syrup. Here is a table that shows how to make batches of different sizes.

Cups of Milk	Tablespoons of Chocolate Syrup
2	3
8	12
1	$\frac{3}{2}$
10	15

($\cdot 4$ from 2 to 8; $\cdot 4$ from 3 to 12)

Use the information in the table to complete the statements. Some terms are used more than once.

a. The table shows a proportional relationship between

____________ and ____________.

b. The scale factor shown is ______.

c. The constant of proportionality for this relationship is ______.

d. The units for the constant of proportionality are ____________

per ____________.

Bank of Terms: tablespoons of chocolate syrup, 4, cup(s) of milk, $\frac{3}{2}$

2. A certain shade of pink is created by adding 3 cups of red paint to 7 cups of white paint.

White Paint (cups)	Red Paint (cups)
1	
7	3

a. How many cups of red paint should be added to 1 cup of white paint?

b. What is the constant of proportionality?

3. A map of a rectangular park has a length of 4 inches and a width of 6 inches. It uses a scale of 1 inch for every 30 miles.

   a. What is the actual area of the park? Show how you know.

   b. The map needs to be reproduced at a different scale so that it has an area of 6 square inches and can fit in a brochure. At what scale should the map be reproduced so that it fits on the brochure? Show your reasoning. **(Lesson 1-12)**

4. Noah drew a scaled copy of Polygon P and labeled it Polygon Q. **(Lesson 1-6)**

   If the area of Polygon P is 5 square units, what scale factor did Noah apply to Polygon P to create Polygon Q? Explain or show how you know.

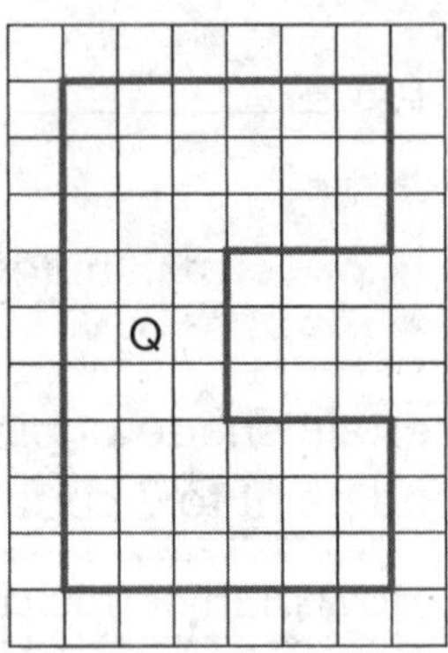

5. Select **all** the ratios that are equivalent to each other.

   A. 4 : 7

   B. 8 : 15

   C. 16 : 28

   D. 2 : 3

   E. 20 : 35

Lesson 2-3

More about Constant of Proportionality

NAME ______________________ DATE __________ PERIOD __________

Learning Goal Let's solve more problems involving proportional relationships using tables.

Warm Up

3.1 Equal Measures

Use the numbers and units from the list to find as many equivalent measurements as you can. For example, you might write "30 minutes is $\frac{1}{2}$ hour." You can use the numbers and units more than once.

1	$\frac{1}{2}$	0.3	12
40	24	0.4	0.01
$\frac{1}{10}$	60	$3\frac{1}{3}$	6
50	30	2	$\frac{2}{5}$

centimeter	meter
hour	feet
minute	inch

Activity

3.2 Centimeters and Millimeters

There is a proportional relationship between any length measured in centimeters and the same length measured in millimeters.

There are two ways of thinking about this proportional relationship.

centimeters

0 1 2 3 4 5

0 10 20 30 40 50

millimeters

1. If you know the length of something in centimeters, you can calculate its length in millimeters.
   a. Complete the table.
   b. What is the constant of proportionality?

Length (cm)	Length (mm)
9	
12.5	
50	
88.49	

2. If you know the length of something in millimeters, you can calculate its length in centimeters.
   a. Complete the table.
   b. What is the constant of proportionality?

Length (mm)	Length (cm)
70	
245	
4	
699.1	

3. How are these two constants of proportionality related to each other?

NAME ______________ DATE ______________ PERIOD ______________

4. Complete each sentence:

   a. To convert from centimeters to millimeters, you can multiply by __________.

   b. To convert from millimeters to centimeters, you can divide by __________ *or* multiply by __________.

Are you ready for more?

1. How many square millimeters are there in a square centimeter?

2. How do you convert square centimeters to square millimeters? How do you convert the other way?

Activity

3.3 Pittsburgh to Phoenix

On its way from New York to San Diego, a plane flew over Pittsburgh, St. Louis, Albuquerque, and Phoenix traveling at a constant speed.

Complete the table as you answer the questions. Be prepared to explain your reasoning.

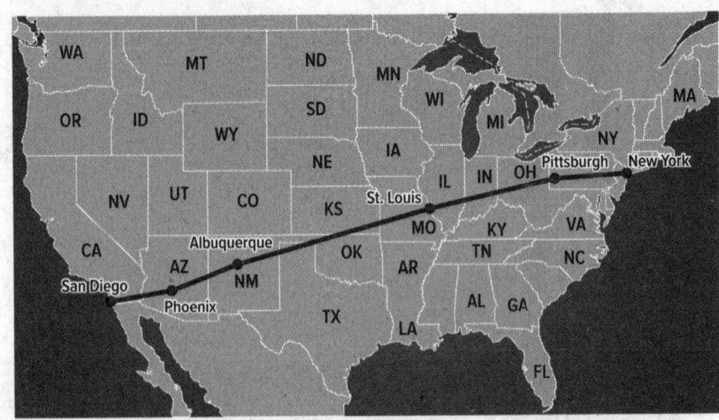

Segment	Time	Distance	Speed
Pittsburgh to St. Louis	1 hour	550 miles	
St. Louis to Albuquerque	1 hour 42 minutes		
Albuquerque to Phoenix		330 miles	

1. What is the distance between St. Louis and Albuquerque?

2. How many minutes did it take to fly between Albuquerque and Phoenix?

3. What is the proportional relationship represented by this table?

4. Diego says the constant of proportionality is 550. Andre says the constant of proportionality is $9\frac{1}{6}$. Do you agree with either of them? Explain your reasoning.

NAME ______________________ DATE ______________ PERIOD ______________

Summary

More about Constant of Proportionality

When something is traveling at a constant speed, there is a proportional relationship between the time it takes and the distance traveled. The table shows the distance traveled and elapsed time for a bug crawling on a sidewalk.

Distance Traveled (cm)	Elapsed Time (sec)
$\frac{3}{2}$	1
1	$\frac{2}{3}$
3	2
10	$\frac{20}{3}$

$\cdot \frac{2}{3}$

We can multiply any number in the first column by $\frac{2}{3}$ to get the corresponding number in the second column. We can say that the elapsed time is proportional to the distance traveled, and the constant of proportionality is $\frac{2}{3}$.

This means that the bug's *pace* is $\frac{2}{3}$ second per centimeter.

This table represents the same situation, except the columns are switched.

Elapsed Time (sec)	Distance Traveled (cm)
1	$\frac{3}{2}$
$\frac{2}{3}$	1
2	3
$\frac{20}{3}$	10

$\cdot \frac{3}{2}$

We can multiply any number in the first column by $\frac{3}{2}$ to get the corresponding number in the second column. We can say that the distance traveled is proportional to the elapsed time, and the constant of proportionality is $\frac{3}{2}$.

This means that the bug's *speed* is $\frac{3}{2}$ centimeters per second.

Notice that $\frac{3}{2}$ is the reciprocal of $\frac{2}{3}$. When two quantities are in a proportional relationship, there are two constants of proportionality, and they are always reciprocals of each other.

When we represent a proportional relationship with a table, we say the quantity in the second column is proportional to the quantity in the first column, and the corresponding constant of proportionality is the number we multiply by values in the first column to get the values in the second.

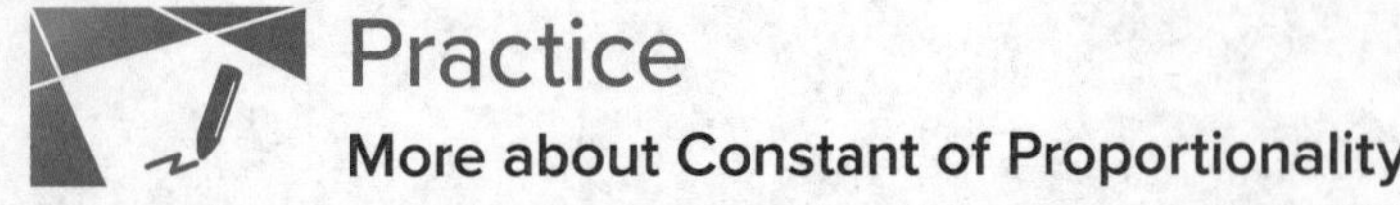

Practice

More about Constant of Proportionality

1. Noah is running a portion of a marathon at a constant speed of 6 miles per hour.

   Complete the table to predict how long it would take him to run different distances at that speed, and how far he would run in different time intervals.

Time (hours)	Miles Traveled at 6 Miles per Hour
1	
$\frac{1}{2}$	
$1\frac{1}{3}$	
	$1\frac{1}{2}$
	9
	$4\frac{1}{2}$

2. One kilometer is 1,000 meters.

   a. Complete the tables. What is the interpretation of the constant of proportionality in each case?

Meters	Kilometers
1,000	1
250	
12	
1	

The constant of proportionality tells us that:

Kilometers	Meters
1	1,000
5	
20	
0.3	

The constant of proportionality tells us that:

   b. What is the relationship between the two constants of proportionality?

NAME ______________________ DATE ____________ PERIOD ________

3. Jada and Lin are comparing inches and feet. Jada says that the constant of proportionality is 12. Lin says it is $\frac{1}{12}$. Do you agree with either of them? Explain your reasoning.

4. The area of the Mojave Desert is 25,000 square miles. A scale drawing of the Mojave Desert has an area of 10 square inches. What is the scale of the map? **(Lesson 1-12)**

5. Which of these scales is equivalent to the scale 1 cm to 5 km? Select **all** that apply. **(Lesson 1-11)**

A. 3 cm to 15 km

B. 1 mm to 150 km

C. 5 cm to 1 km

D. 5 mm to 2.5 km

E. 1 mm to 500 m

6. Which one of these pictures is not like the others? Explain what makes it different using ratios. **(Lesson 2-1)**

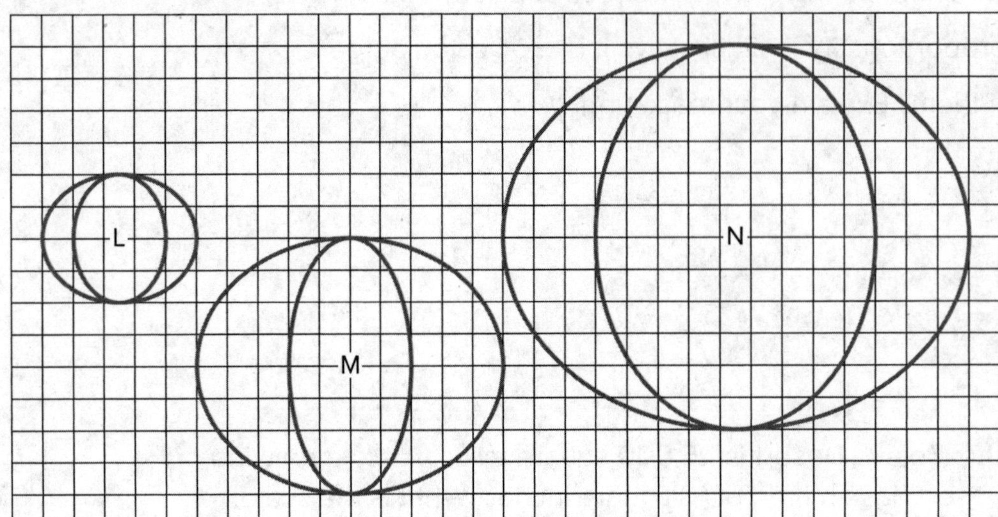

Lesson 2-4

Proportional Relationships and Equations

NAME ______________________ DATE ____________ PERIOD ____________

Learning Goal Let's write equations describing proportional relationships.

Warm Up

4.1 Number Talk: Division

Find each quotient mentally.

1. $645 \div 100$

2. $645 \div 50$

3. $48.6 \div 30$

4. $48.6 \div x$

Activity

4.2 Feeding a Crowd, Revisited

1. A recipe says that 2 cups of dry rice will serve 6 people. Complete the table as you answer the questions. Be prepared to explain your reasoning.

   a. How many people will 1 cup of rice serve?

   b. How many people will 3 cups of rice serve? 12 cups? 43 cups?

   c. How many people will x cups of rice serve?

Cups of Dry Rice	Number of People
1	
2	6
3	
12	
43	
x	

2. A recipe says that 6 spring rolls will serve 3 people. Complete the table as you answer the questions. Be prepared to explain your reasoning.

   a. How many people will 1 spring roll serve?

   b. How many people will 10 spring rolls serve? 16 spring rolls? 25 spring rolls?

   c. How many people will n spring rolls serve?

Number of Spring Rolls	Number of People
1	
6	3
10	
16	
25	
n	

3. How was completing this table different from the previous table? How was it the same?

Activity

4.3 Denver to Chicago

A plane flew at a constant speed between Denver and Chicago. It took the plane 1.5 hours to fly 915 miles.

1. Complete the table.

Time (hours)	Distance (miles)	Speed (miles per hour)
1		
1.5	915	
2		
2.5		
t		

2. How far does the plane fly in one hour?

3. How far would the plane fly in t hours at this speed?

4. If d represents the distance that the plane flies at this speed for t hours, write an equation that relates t and d.

5. How far would the plane fly in 3 hours at this speed? in 3.5 hours? Explain or show your reasoning.

NAME ______________________ DATE __________ PERIOD __________

Are you ready for more?

A rocky planet orbits Proxima Centauri, a star that is about 1.3 parsecs from Earth. This planet is the closest planet outside of our solar system.

1. How long does it take light from Proxima Centauri to reach Earth? (A parsec is about 3.26 light years. A light year is the distance light travels in one year.)

2. There are two twins. One twin leaves on a spaceship to explore the planet near Proxima Centauri traveling at 90% of the speed of light, while the other twin stays home on Earth. How much does the twin on Earth age while the other twin travels to Proxima Centauri? (Do you think the answer would be the same for the other twin? Consider researching "The Twin Paradox" to learn more.)

A bakery uses 8 tablespoons of honey for every 10 cups of flour to make bread dough. Some days they bake bigger batches, and some days they bake smaller batches, but they always use the same ratio of honey to flour.

1. Complete the table.

2. If f is the cups of flour needed for h tablespoons of honey, write an equation that relates f and h.

3. How much flour is needed for 15 tablespoons of honey? 17 tablespoons? Explain or show your reasoning.

Honey (tbsp)	Flour (c)
1	
8	10
16	
20	
h	

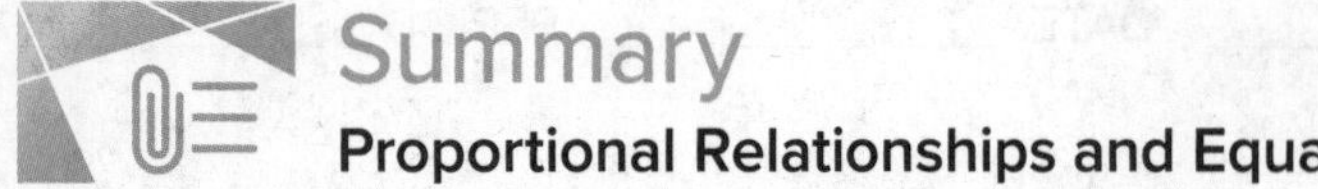

Summary

Proportional Relationships and Equations

The table shows the amount of red paint and blue paint needed to make a certain shade of purple paint, called Venusian Sunset.

Red Paint (parts)	Blue Paint (parts)
3	12
1	4
7	28
$\frac{1}{4}$	1
r	$4r$

Note that "parts" can be *any* unit for volume. If we mix 3 cups of red with 12 cups of blue, you will get the same shade as if we mix 3 teaspoons of red with 12 teaspoons of blue.

The last row in the table says that if we know the amount of red paint needed, r, we can always multiply it by 4 to find the amount of blue paint needed, b, to mix with it to make Venusian Sunset.

We can say this more succinctly with the equation $b = 4r$. So the amount of blue paint is proportional to the amount of red paint and the constant of proportionality is 4.

We can also look at this relationship the other way around.

Blue Paint (parts)	Red Paint (parts)
12	3
4	1
28	7
1	$\frac{1}{4}$
b	$\frac{1}{4}b$

If we know the amount of blue paint needed, b, we can always multiply it by $\frac{1}{4}$ to find the amount of red paint needed, r, to mix with it to make Venusian Sunset.

So $r = \frac{1}{4}b$. The amount of blue paint is proportional to the amount of red paint and the constant of proportionality $\frac{1}{4}$.

In general, when y is proportional to x, we can always multiply x by the same number k—the constant of proportionality—to get y.

We can write this much more succinctly with the equation $y = kx$.

NAME ______________________ DATE ____________ PERIOD ____________

Practice

Proportional Relationships and Equations

1. A certain ceiling is made up of tiles. Every square meter of ceiling requires 10.75 tiles. Fill in the table with the missing values.

Square Meters of Ceiling	Number of Tiles
1	
10	
	100
a	

2. On a flight from New York to London, an airplane travels at a constant speed. An equation relating the distance traveled in miles, d, to the number of hours flying, t, is $t = \frac{1}{500}d$. How long will it take the airplane to travel 800 miles?

3. Each table represents a proportional relationship. For each, find the constant of proportionality, and write an equation that represents the relationship.

s	P
2	8
3	12
5	20
10	40

Constant of proportionality: __________

Equation: $P =$ __________

d	C
2	6.28
3	9.42
5	15.7
10	31.4

Constant of proportionality: __________

Equation: $C =$ __________

4. A map of Colorado says that the scale is 1 inch to 20 miles or 1 to 1,267,200. Are these two ways of reporting the scale the same? Explain your reasoning. **(Lesson 1-11)**

5. Here is a polygon on a grid. **(Lesson 1-3)**

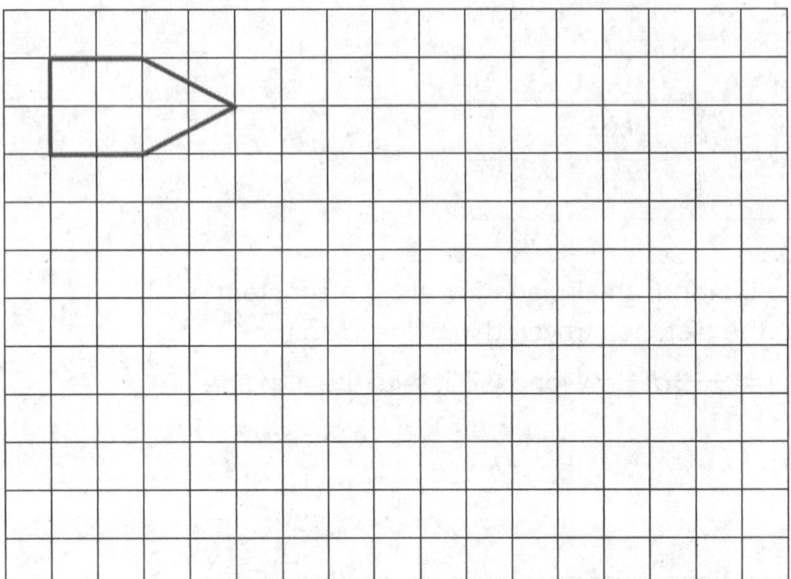

a. Draw a scaled copy of the polygon using a scale factor 3. Label the copy A.

b. Draw a scaled copy of the polygon with a scale factor $\frac{1}{2}$. Label it B.

c. Is Polygon A a scaled copy of Polygon B? If so, what is the scale factor that takes B to A?

Lesson 2-5

Two Equations for Each Relationship

NAME ______________________ DATE ______________ PERIOD ____________

Learning Goal Let's investigate the equations that represent proportional relationships.

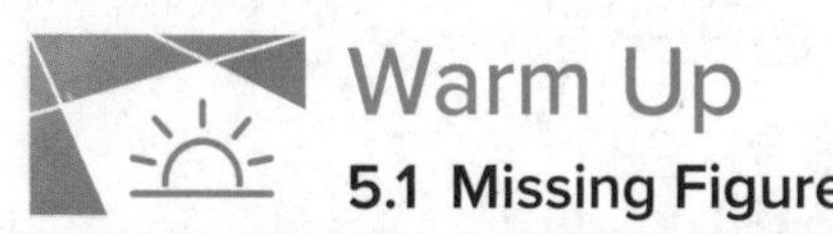

Warm Up

5.1 Missing Figures

Here are the second and fourth figures in a pattern.

? 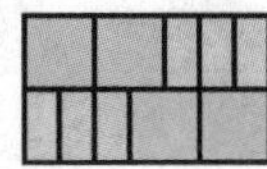 ? 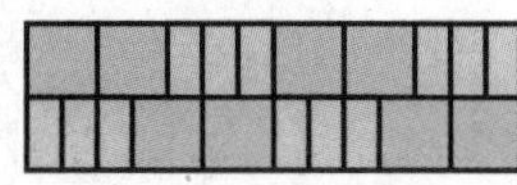

Figure 1 Figure 2 Figure 3 Figure 4

1. What do you think the first and third figures in the pattern look like?

2. Describe the 10th figure in the pattern.

Activity

5.2 Meters and Centimeters

There are 100 centimeters (cm) in every meter (m).

1. Complete each of the tables.

Length (m)	Length (cm)
1	100
0.94	
1.67	
57.24	
x	

Length (cm)	Length (m)
100	1
250	
78.2	
123.9	
y	

2. For each table, find the constant of proportionality.

3. What is the relationship between these constants of proportionality?

4. For each table, write an equation for the proportional relationship. Let x represent a length measured in meters and y represent the same length measured in centimeters.

NAME ____________________ DATE ____________ PERIOD ____________

Are you ready for more?

1. How many cubic centimeters are there in a cubic meter?

2. How do you convert cubic centimeters to cubic meters?

3. How do you convert the other way?

Activity

5.3 Filling a Water Cooler

It took Priya 5 minutes to fill a cooler with 8 gallons of water from a faucet that was flowing at a steady rate. Let w be the number of gallons of water in the cooler after t minutes.

1. Which of the following equations represent the relationship between w and t? Select **all** that apply.

   A. $w = 1.6t$

   B. $w = 0.625t$

   C. $t = 1.6w$

   D. $t = 0.625w$

2. What does 1.6 tell you about the situation?

3. What does 0.625 tell you about the situation?

4. Priya changed the rate at which water flowed through the faucet. Write an equation that represents the relationship of w and t when it takes 3 minutes to fill the cooler with 1 gallon of water.

5. Was the cooler filling faster before or after Priya changed the rate of water flow? Explain how you know.

Activity

5.4 Feeding Shrimp

At an aquarium, a shrimp is fed $\frac{1}{5}$ gram of food each feeding and is fed 3 times each day.

Ales Veluscek/E+/Getty Images

1. How much food does a shrimp get fed in one day?

2. Complete the table to show how many grams of food the shrimp is fed over different numbers of days.

Number of Days	Grams of Food
1	
7	
30	

3. What is the constant of proportionality? What does it tell us about the situation?

4. If we switched the columns in the table, what would be the constant of proportionality? Explain your reasoning.

5. Use d for number of days and f for amount of food in grams that a shrimp eats to write *two* equations that represent the relationship between d and f.

6. If a tank has 10 shrimp in it, how much food is added to the tank each day?

7. If the aquarium manager has 300 grams of shrimp food for this tank of 10 shrimp, how many days will it last? Explain or show your reasoning.

NAME ______________________ DATE ______________ PERIOD ____________

Summary

Two Equations for Each Relationship

If Kiran rode his bike at a constant 10 miles per hour, his distance in miles, *d*, is proportional to the number of hours, *t*, that he rode.

We can write the equation:

$$d = 10t$$

With this equation, it is easy to find the distance Kiran rode when we know how long it took because we can just multiply the time by 10.

We can rewrite the equation:

$$d = 10t$$

$$\left(\frac{1}{10}\right)d = t$$

$$t = \left(\frac{1}{10}\right)d$$

This version of the equation tells us that the amount of time he rode is proportional to the distance he traveled, and the constant of proportionality is $\frac{1}{10}$. That form is easier to use when we know his distance and want to find how long it took because we can just multiply the distance by $\frac{1}{10}$.

When two quantities x and y are in a proportional relationship, we can write the equation:

$$y = kx$$

and say, "y is proportional to x." In this case, the number k is the corresponding constant of proportionality.

We can also write the equation:

$$x = \frac{1}{k}y$$

and say, "x is proportional to y." In this case, the number $\frac{1}{k}$ is the corresponding constant of proportionality.

Each one can be useful depending on the information we have and the quantity we are trying to figure out.

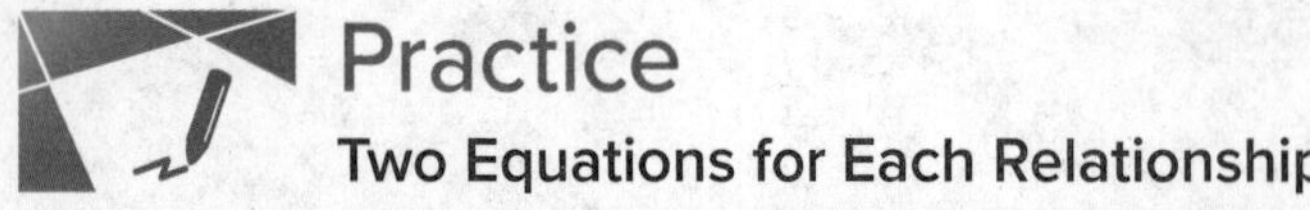

Practice

Two Equations for Each Relationship

1. The table represents the relationship between a length measured in meters and the same length measured in kilometers.

Meters	Kilometers
1,000	1
3,500	
500	
75	
1	
x	

a. Complete the table.

b. Write an equation for converting the number of meters to kilometers. Use x for number of meters and y for number of kilometers.

2. Concrete building blocks weigh 28 pounds each. Using b for the number of concrete blocks and w for the weight, write two equations that relate the two variables. One equation should begin with $w =$ and the other should begin with $b =$.

NAME ______________________ DATE ____________ PERIOD ________

3. A store sells rope by the meter. The equation $p = 0.8L$ represents the price p (in dollars) of a piece of nylon rope that is L meters long.

   a. How much does the nylon rope cost per meter?

   b. How long is a piece of nylon rope that costs $1.00?

4. The table represents a proportional relationship. Find the constant of proportionality and write an equation to represent the relationship. **(Lesson 2-4)**

a	y
2	$\frac{2}{3}$
3	1
10	$\frac{10}{3}$
12	4

Constant of proportionality: __________

Equation: $y =$ __________

5. On a map of Chicago, 1 cm represents 100 m. Select **all** statements that express the same scale. **(Lesson 1-8)**

A. 5 cm on the map represents 50 m in Chicago.

B. 1 mm on the map represents 10 m in Chicago.

C. 1 km in Chicago is represented by 10 cm on the map.

D. 100 cm in Chicago is represented by 1 m on the map.

Lesson 2-6

Using Equations to Solve Problems

NAME ______________________ DATE ____________ PERIOD ____________

Learning Goal Let's use equations to solve problems involving proportional relationships.

Warm Up

6.1 Number Talk: Quotients with Decimal Points

1. Without calculating, order the quotients of these expressions from least to greatest.

   $42.6 \div 0.07$ $\quad$ $42.6 \div 70$ $\quad$ $42.6 \div 0.7$ $\quad$ $426 \div 70$

2. Place the decimal point in the appropriate location in the quotient:
   $42.6 \div 7 = 608571$

3. Use this answer to find the quotient of *one* of the previous expressions.

Activity

6.2 Concert Ticket Sales

A performer expects to sell 5,000 tickets for an upcoming concert. They want to make a total of $311,000 in sales from these tickets.

1. Assuming that all tickets have the same price, what is the price for one ticket?

2. How much will they make if they sell 7,000 tickets?

3. How much will they make if they sell 10,000 tickets? 50,000? 120,000? a million? x tickets?

4. If they make $404,300, how many tickets have they sold?

5. How many tickets will they have to sell to make $5,000,000?

Activity
6.3 Recycling

Aluminum cans can be recycled instead of being thrown in the garbage. The weight of 10 aluminum cans is 0.16 kilogram. The aluminum in 10 cans that are recycled has a value of $0.14.

1. If a family threw away 2.4 kg of aluminum in a month, how many cans did they throw away? Explain or show your reasoning.

2. What would be the recycled value of those same cans? Explain or show your reasoning.

3. Write an equation to represent the number of cans c given their weight w.

4. Write an equation to represent the recycled value r of c cans.

5. Write an equation to represent the recycled value r of w kilograms of aluminum.

NAME ______________________ DATE ____________ PERIOD ____________

Are you ready for more?

The EPA estimated that in 2013, the average amount of garbage produced in the United States was 4.4 pounds per person per day. At that rate, how long would it take your family to produce a ton of garbage? (A ton is 2,000 pounds.)

Summary

Using Equations to Solve Problems

Remember that if there is a proportional relationship between two quantities, their relationship can be represented by an equation of the form $y = kx$.

Sometimes writing an equation is the easiest way to solve a problem.

For example, we know that Denali, the highest mountain peak in North America, is 20,310 feet above sea level. How many miles is that?

There are 5,280 feet in 1 mile. This relationship can be represented by the equation $f = 5{,}280m$, where f represents a distance measured in feet and m represents the same distance measured in miles.

Since we know Denali is 20,310 feet above sea level, we can write $20{,}310 = 5{,}280m$.

So $m = \frac{20{,}310}{5{,}280}$, which is approximately 3.85 miles.

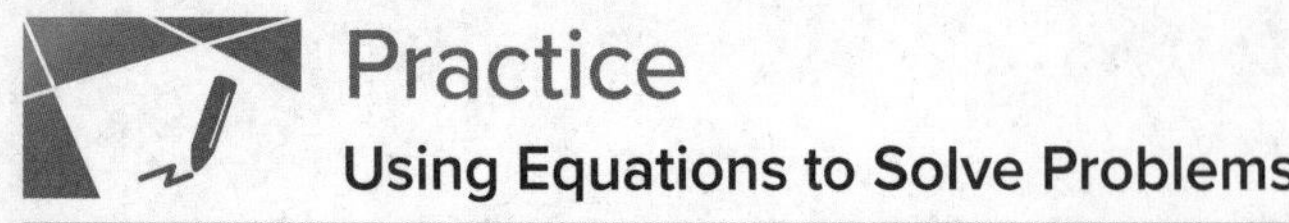

Practice

Using Equations to Solve Problems

1. A car is traveling down a highway at a constant speed, described by the equation $d = 65t$, where d represents the distance, in miles, that the car travels at this speed in t hours.

   a. What does the 65 tell us in this situation?

   b. How many miles does the car travel in 1.5 hours?

   c. How long does it take the car to travel 26 miles at this speed?

2. Elena has some bottles of water that hold 17 fluid ounces each.

   a. Write an equation that relates the number of bottles of water (b) to the total volume of water (w) in fluid ounces.

   b. How much water is in 51 bottles?

   c. How many bottles does it take to hold 51 fluid ounces of water?

NAME ______________________ DATE ____________ PERIOD ____________

3. There are about 1.61 kilometers in 1 mile. Let x represent a distance measured in kilometers and y represent the same distance measured in miles. Write two equations that relate a distance measured in kilometers and the same distance measured in miles. **(Lesson 2-5)**

4. In Canadian coins, 16 quarters is equal in value to 2 toonies. **(Lesson 2-2)**

Number of Quarters	Number of Toonies
1	
16	2
20	
24	

a. Complete the table.

b. What does the value next to 1 mean in this situation?

5. Each table represents a proportional relationship. For each table:

   a. Fill in the missing parts of the table.

   b. Draw a circle around the constant of proportionality. **(Lesson 2-2)**

x	y
2	10
	15
7	
1	

a	b
12	3
20	
	10
1	

m	n
5	3
10	
	18
1	

6. Describe some things you could notice in two polygons that would help you decide that they were not scaled copies. **(Lesson 1-4)**

Lesson 2-7

Comparing Relationships with Tables

NAME ______________________ DATE ____________ PERIOD ____________

Learning Goal Let's explore how proportional relationships are different from other relationships.

Warm Up

7.1 Adjusting a Recipe

A lemonade recipe calls for the juice of 5 lemons, 2 cups of water, and 2 tablespoons of honey.

Invent four new versions of this lemonade recipe:

1. One that would make more lemonade but taste the same as the original recipe.

2. One that would make less lemonade but taste the same as the original recipe.

3. One that would have a stronger lemon taste than the original recipe.

4. One that would have a weaker lemon taste than the original recipe.

Activity

7.2 Visiting the State Park

Entrance to a state park costs $6 per vehicle, plus $2 per person in the vehicle.

1. How much would it cost for a car with 2 people to enter the park? 4 people? 10 people? Record your answers in the table.

Number of People in Vehicle	Total Entrance Cost in Dollars
2	
4	
10	

2. For each row in the table, if each person in the vehicle splits the entrance cost equally, how much will each person pay?

3. How might you determine the entrance cost for a bus with 50 people?

4. Is the relationship between the number of people and the total entrance cost a proportional relationship? Explain how you know.

Are you ready for more?

What equation could you use to find the total entrance cost for a vehicle with any number of people?

NAME ______________________________ DATE ______________ PERIOD ____________

Activity
7.3 Running Laps

Han and Clare were running laps around the track. The coach recorded their times at the end of laps 2, 4, 6, and 8.

Han's run:

Distance (laps)	Time (minutes)	Minutes per Lap
2	4	
4	9	
6	15	
8	23	

Clare's run:

Distance (laps)	Time (minutes)	Minutes per Lap
2	5	
4	10	
6	15	
8	20	

1. Is Han running at a constant pace? Is Clare? How do you know?

2. Write an equation for the relationship between distance and time for anyone who is running at a constant pace.

Summary

Comparing Relationships with Tables

Here are the prices for some smoothies at two different smoothie shops:

Smoothie Shop A:

Smoothie Size (oz)	Price ($)	Dollars per Ounce
8	6	0.75
12	9	0.75
16	12	0.75
s	$0.75s$	0.75

Smoothie Shop B:

Smoothie Size (oz)	Price ($)	Dollars per Ounce
8	6	0.75
12	8	0.67
16	10	0.625
s	???	???

For Smoothie Shop A, smoothies cost $0.75 per ounce no matter which size we buy. There could be a proportional relationship between smoothie size and the price of the smoothie. An equation representing this relationship is $p = 0.75s$ where s represents size in ounces and p represents price in dollars. (The relationship could still not be proportional, if there were a different size on the menu that did not have the same price per ounce.)

For Smoothie Shop B, the cost per ounce is different for each size. Here the relationship between smoothie size and price is definitely *not* proportional.

In general, two quantities in a proportional relationship will always have the same quotient. When we see some values for two related quantities in a table and we get the same quotient when we divide them, that means they might be in a proportional relationship—but if we can't see all of the possible pairs, we can't be completely sure.

However, if we know the relationship can be represented by an equation is of the form $y = kx$, then we are sure it is proportional.

NAME ______________________ DATE ____________ PERIOD ________

Practice

Comparing Relationships with Tables

1. Decide whether each table could represent a proportional relationship. If the relationship could be proportional, what would the constant of proportionality be?

a. How loud a sound is depending on how far away you are.

Distance to Listener (ft)	Sound Level (dB)
5	85
10	79
20	73
40	67

b. The cost of fountain drinks at Hot Dog Hut.

Volume (fluid ounces)	Cost ($)
16	$1.49
20	$1.59
30	$1.89

2. A taxi service charges $1.00 for the first $\frac{1}{10}$ mile then $0.10 for each additional $\frac{1}{10}$ mile after that. Fill in the table with the missing information. Then determine if this relationship between distance traveled and price of the trip is a proportional relationship.

Distance Traveled (mi)	Price (dollars)
$\frac{9}{10}$	
2	
$3\frac{1}{10}$	
10	

3. A rabbit and turtle are in a race. Is the relationship between distance traveled and time proportional for either one? If so, write an equation that represents the relationship.

Turtle's run:

Distance (meters)	Time (minutes)
108	2
405	7.5
540	10
1,768.5	32.75

Rabbit's run:

Distance (meters)	Time (minutes)
800	1
900	5
1,107.5	20
1,524	32.5

4. For each table, answer: What is the constant of proportionality? **(Lesson 2-2)**

a	b
2	14
5	35
9	63
$\frac{1}{3}$	$\frac{7}{3}$

a	b
3	360
5	600
8	960
12	1,440

a	b
75	3
200	8
1,525	61
10	0.4

a	b
4	10
6	15
22	55
3	$7\frac{1}{2}$

5. Kiran and Mai are standing at one corner of a rectangular field of grass looking at the diagonally opposite corner. Kiran says that if the field were twice as long and twice as wide, then it would be twice the distance to the far corner. Mai says that it would be more than twice as far, since the diagonal is even longer than the side lengths. Do you agree with either of them? **(Lesson 1-4)**

Lesson 2-8

Comparing Relationships with Equations

NAME ______________________ DATE ______________ PERIOD ______________

Learning Goal Let's develop methods for deciding if a relationship is proportional.

Warm Up

8.1 Notice and Wonder: Patterns with Rectangles

Do you see a pattern? What predictions can you make about future rectangles in the set if your pattern continues?

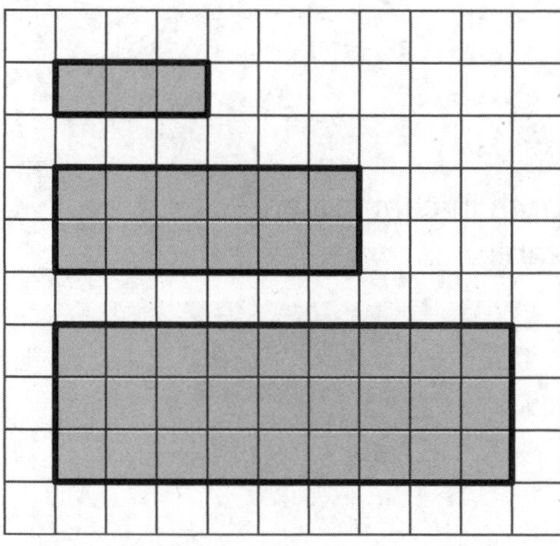

Activity

8.2 More Conversions

The other day you worked with converting meters, centimeters, and millimeters. Here are some more unit conversions.

1. Use the equation $F = \frac{9}{5}C + 32$, where F represents degrees Fahrenheit and C represents degrees Celsius, to complete the table.

Temperature (°C)	Temperature (°F)
20	
4	
175	

2. Use the equation $c = 2.54n$, where c represents the length in centimeters and n represents the length in inches, to complete the table.

Length (in)	Length (cm)
10	
8	
$3\frac{1}{2}$	

3. Are these proportional relationships? Explain why or why not.

NAME ______________________ DATE ____________ PERIOD ________

Activity

8.3 Total Edge Length, Surface Area, and Volume

Here are some cubes with different side lengths. Complete each table. Be prepared to explain your reasoning.

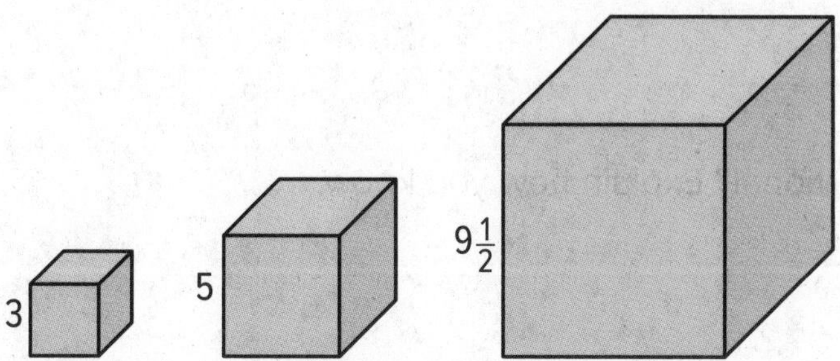

1. How long is the total edge length of each cube?

Side Length	Total Edge Length
3	
5	
$9\frac{1}{2}$	
s	

2. What is the surface area of each cube?

Side Length	Surface Area
3	
5	
$9\frac{1}{2}$	
s	

3. What is the volume of each cube?

Side Length	Volume
3	
5	
$9\frac{1}{2}$	
s	

4. Which of these relationships is proportional? Explain how you know.

5. Write equations for the total edge length E, total surface area A, and volume V of a cube with side length s.

Are you ready for more?

1. A rectangular solid has a square base with side length ℓ, height 8, and volume V. Is the relationship between ℓ and V a proportional relationship?

2. A different rectangular solid has length ℓ, width 10, height 5, and volume V. Is the relationship between ℓ and V a proportional relationship?

3. Why is the relationship between the side length and the volume proportional in one situation and not the other?

NAME ______________________ DATE ____________ PERIOD ________

Activity

8.4 All Kinds of Equations

Here are six different equations.

$y = 4 + x$ $\quad$ $y = 4x$ $\quad$ $y = \frac{4}{x}$ $\quad$ $y = \frac{x}{4}$ $\quad$ $y = 4^x$ $\quad$ $y = x^4$

1. Predict which of these equations represent a proportional relationship.

2. Complete each table using the equation that represents the relationship.

$y = 4 + x$

x	y	$\frac{y}{x}$
2		
3		
4		
5		

$y = 4x$

x	y	$\frac{y}{x}$
2		
3		
4		
5		

$y = \frac{4}{x}$

x	y	$\frac{y}{x}$
2		
3		
4		
5		

$y = \frac{x}{4}$

x	y	$\frac{y}{x}$
2		
3		
4		
5		

$y = 4^x$

x	y	$\frac{y}{x}$
2		
3		
4		
5		

$y = x^4$

x	y	$\frac{y}{x}$
2		
3		
4		
5		

3. Do these results change your answer to the first question? Explain your reasoning.

4. What do the equations of the proportional relationships have in common?

Summary

Comparing Relationships with Equations

If two quantities are in a proportional relationship, then their quotient is always the same.

This table represents different values of a and b, two quantities that are in a proportional relationship.

a	b	$\frac{b}{a}$
20	100	5
3	15	5
11	55	5
1	5	5

Notice that the quotient of b and a is always 5. To write this as an equation, we could say $\frac{b}{a} = 5$. If this is true, then $b = 5a$. (This doesn't work if $a = 0$, but it works otherwise.)

If quantity y is proportional to quantity x, we will always see this pattern: $\frac{y}{x}$ will always have the same value.

- This value is the constant of proportionality, which we often refer to as k.
- We can represent this relationship with the equation $\frac{y}{x} = k$ (as long as x is not 0) or $y = kx$.

Note that if an equation cannot be written in this form, then it does not represent a proportional relationship.

NAME ______________________ DATE ____________ PERIOD ________

Practice

Comparing Relationships with Equations

1. The relationship between a distance in yards (y) and the same distance in miles (m) is described by the equation $y = 1{,}760m$.

   a. Find measurements in yards and miles for distances by completing the table.

Distance Measured in Miles	Distance Measured in Yards
1	
5	
	3,520
	17,600

   b. Is there a proportional relationship between a measurement in yards and a measurement in miles for the same distance? Explain why or why not.

2. Decide whether or not each equation represents a proportional relationship.

   a. The remaining length (L) of 120-inch rope after x inches have been cut off: $120 - x = L$

   b. The total cost (t) after 8% sales tax is added to an item's price (p): $1.08p = t$

   c. The number of marbles each sister gets (x) when m marbles are shared equally among four sisters: $x = \frac{m}{4}$

   d. The volume (V) of a rectangular prism whose height is 12 cm and base is a square with side lengths s cm: $V = 12s^2$

3. Here are two relationships.

a. Use the equation $y = \frac{5}{2}x$ to complete the table. Is y proportional to x and y? Explain why or why not.

x	y
2	
3	
6	

b. Use the equation $y = 3.2x + 5$ to complete the table. Is y proportional to x and y? Explain why or why not.

x	y
1	
2	
4	

4. To transmit information on the internet, large files are broken into packets of smaller sizes. Each packet has 1,500 bytes of information. An equation relating packets to bytes of information is given by $b = 1{,}500p$, where p represents the number of packets and b represents the number of bytes of information. **(Lesson 2-6)**

a. How many packets would be needed to transmit 30,000 bytes of information?

b. How much information could be transmitted in 30,000 packets?

c. Each byte contains 8 bits of information. Write an equation to represent the relationship between the number of packets and the number of bits.

Lesson 2-9

Solving Problems about Proportional Relationships

NAME ______________________ DATE ____________ PERIOD ____________

Learning Goal Let's solve problems about proportional relationships.

Warm Up

9.1 What Do You Want to Know?

Consider the problem: A person is running a distance race at a constant rate. What time will they finish the race?

What information would you need to be able to solve the problem?

Activity

9.2 Info Gap: Biking and Rain

Your teacher will give you either a *problem card* or a *data card*. Do not show or read your card to your partner.

If your teacher gives you the *problem card*:	If your teacher gives you the *data card*:
1. Silently read your card and think about what information you need to be able to answer the question. 2. Ask your partner for the specific information that you need. 3. Explain how you are using the information to solve the problem. Continue to ask questions until you have enough information to solve the problem. 4. Share the *problem card* and solve the problem independently. 5. Read the *data card* and discuss your reasoning.	1. Silently read your card. 2. Ask your partner *"What specific information do you need?"* and wait for them to *ask* for information. If your partner asks for information that is not on the card, do not do the calculations for them. Tell them you don't have that information. 3. Before sharing the information, ask "*Why do you need that information?*" Listen to your partner's reasoning and ask clarifying questions. 4. Read the *problem card* and solve the problem independently. 5. Share the *data card* and discuss your reasoning.

Pause here so your teacher can review your work. Ask your teacher for a new set of cards and repeat the activity, trading roles with your partner.

NAME ______________________ DATE ____________ PERIOD ____________

Activity

9.3 Moderating Comments

A company is hiring people to read through all the comments posted on their website to make sure they are appropriate. Four people applied for the job and were given one day to show how quickly they could check comments.

- Person 1 worked for 210 minutes and checked a total of 50,000 comments.
- Person 2 worked for 200 minutes and checked 1,325 comments every 5 minutes.
- Person 3 worked for 120 minutes, at a rate represented by $c = 331t$, where c is the number of comments checked and t is the time in minutes.
- Person 4 worked for 150 minutes, at a rate represented by $t = \left(\frac{3}{800}\right)c$.

1. Order the people from greatest to least in terms of total number of comments checked.

2. Order the people from greatest to least in terms of how fast they checked the comments.

Are you ready for more?

1. Write equations for each job applicant that allow you to easily decide who is working the fastest.

2. Make a table that allows you to easily compare how many comments the four job applicants can check.

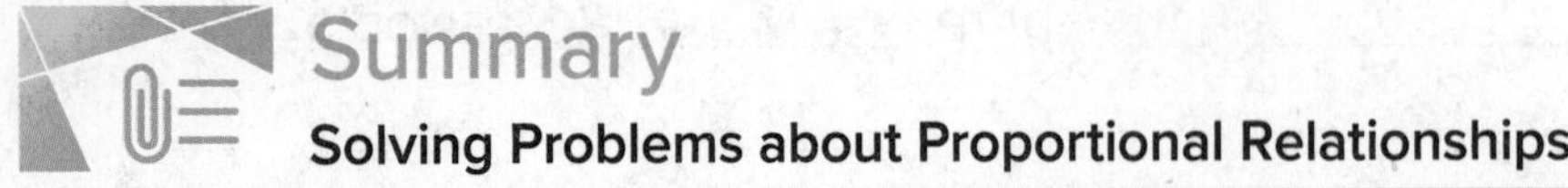

Summary

Solving Problems about Proportional Relationships

Whenever we have a situation involving constant rates, we are likely to have a proportional relationship between quantities of interest.

- When a bird is flying at a constant speed, then there is a proportional relationship between the flying time and distance flown.
- If water is filling a tub at a constant rate, then there is a proportional relationship between the amount of water in the tub and the time the tub has been filling up.
- If an aardvark is eating termites at a constant rate, then there is a proportional relationship between the number of termites the aardvark has eaten and the time since it started eating.

Sometimes we are presented with a situation, and it is not so clear whether a proportional relationship is a good model. How can we decide if a proportional relationship is a good representation of a particular situation?

- If you aren't sure where to start, look at the quotients of corresponding values. If they are not always the same, then the relationship is definitely not a proportional relationship.
- If you can see that there is a single value that we always multiply one quantity by to get the other quantity, it is definitely a proportional relationship.

After establishing that it is a proportional relationship, setting up an equation is often the most efficient way to solve problems related to the situation.

NAME ______________________ DATE ______________ PERIOD __________

Practice

Solving Problems about Proportional Relationships

1. For each situation, explain whether you think the relationship is proportional or not. Explain your reasoning.

   a. The weight of a stack of standard 8.5×11 copier paper vs. number of sheets of paper.

   b. The weight of a stack of different-sized books vs. the number of books in the stack.

2. Every package of a certain toy also includes 2 batteries.

   a. Are the number of toys and number of batteries in a proportional relationship? If so, what are the two constants of proportionality? If not, explain your reasoning.

   b. Use t for the number of toys and b for the number of batteries to write two equations relating the two variables.

   $b =$ __________

   $t =$ __________

3. Lin and her brother were born on the same date in different years. Lin was 5 years old when her brother was 2.

   a. Find their ages in different years by filling in the table.

   b. Is there a proportional relationship between Lin's age and her brother's age? Explain your reasoning.

Lin's Age	Her Brother's Age
5	2
6	
15	
	25

4. A student argues that $y = \frac{x}{9}$ does not represent a proportional relationship between x and y because we need to multiply one variable by the same constant to get the other one and not divide it by a constant. Do you agree or disagree with this student? **(Lesson 2-8)**

5. Quadrilateral A has side lengths 3, 4, 5, and 6. Quadrilateral B is a scaled copy of Quadrilateral A with a scale factor of 2. Select **all** of the following that are side lengths of Quadrilateral B. **(Lesson 1-3)**

   A. 5

   B. 6

   C. 7

   D. 8

   E. 9

Lesson 2-10

Introducing Graphs of Proportional Relationships

NAME _______________ DATE _______ PERIOD _______

Learning Goal Let's see how graphs of proportional relationships differ from graphs of other relationships.

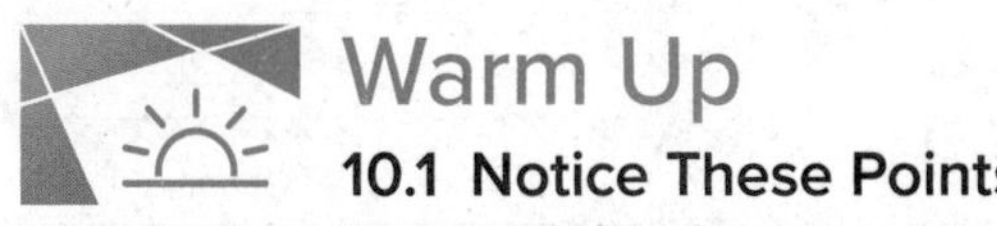

Warm Up

10.1 Notice These Points

1. Plot the points (0, 10), (1, 8), (2, 6), (3, 4), (4, 2).

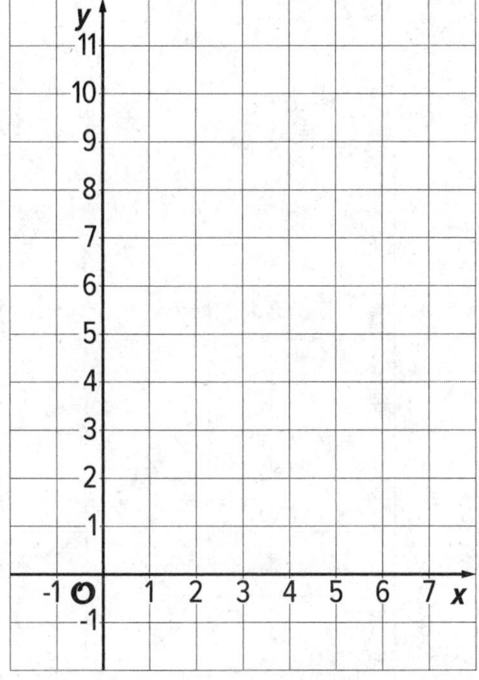

2. What do you notice about the graph?

Activity

10.2 T-shirts for Sale

Some T-shirts cost $8 each.

1. Use the table to answer these questions.

x	y
1	8
2	16
3	24
4	32
5	40
6	48

   a. What does x represent?

   b. What does y represent?

   c. Is there a proportional relationship between x and y?

2. Plot the pairs in the table on the **coordinate plane.**

3. What do you notice about the graph?

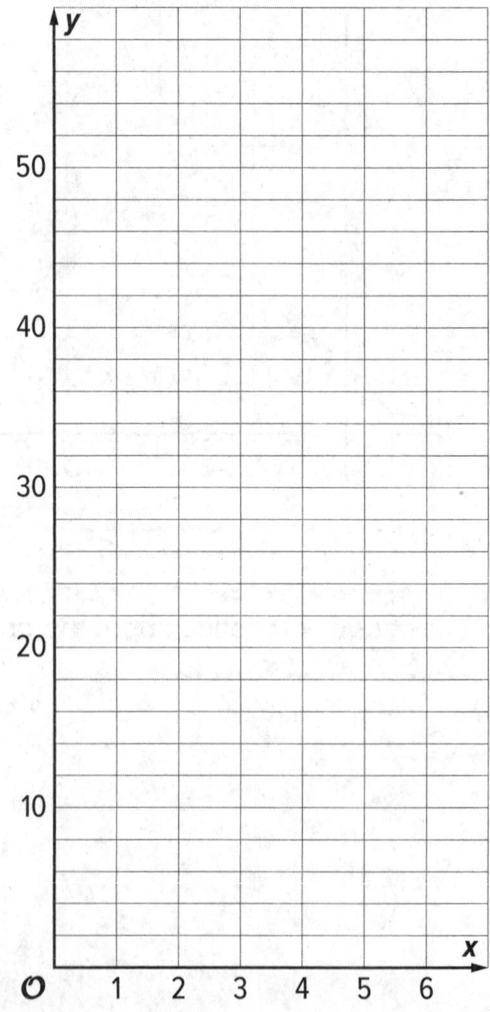

NAME ______________________ DATE ______________ PERIOD __________

Activity

10.3 Matching Tables and Graphs

Your teacher will give you papers showing tables and graphs.

1. Examine the graphs closely. What is the same and what is different about the graphs?

2. Sort the graphs into categories of your choosing. Label each category. Be prepared to explain why you sorted the graphs the way you did.

3. Take turns with a partner to match a table with a graph.

    a. For each match you find, explain to your partner how you know it is a match.

    b. For each match your partner finds, listen carefully to their explanation. If you disagree, work to reach an agreement.

    Pause here so your teacher can review your work.

4. Trade places with another group. How are their categories the same as your group's categories? How are they different?

5. Return to your original place. Discuss any changes you may wish to make to your categories based on what the other group did.

6. Which of the relationships are proportional?

7. What have you noticed about the graphs of proportional relationships? Do you think this will hold true for *all* graphs of proportional relationships?

Are you ready for more?

1. All the graphs in this activity show points where both coordinates are positive. Would it make sense for any of them to have one or more coordinates that are negative?

2. The equation of a proportional relationship is of the form $y = kx$, where k is a positive number, and the graph is a line through (0, 0). What would the graph look like if k were a negative number?

Summary

Introducing Graphs of Proportional Relationships

One way to represent a proportional relationship is with a graph.

Here is a graph that represents different amounts that fit the situation, "Blueberries cost $6 per pound."

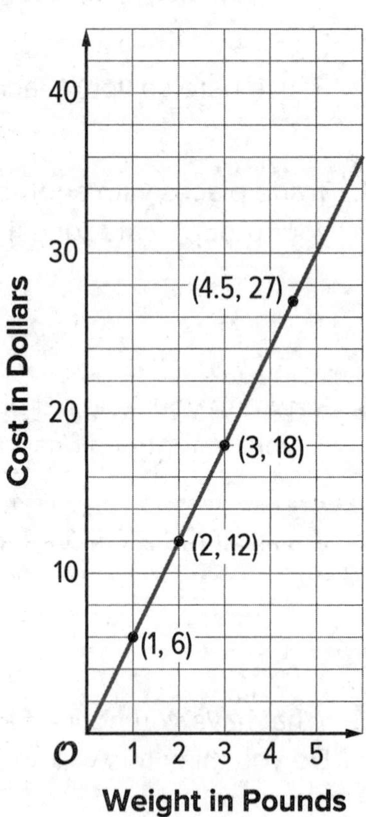

Different points on the graph tell us, for example, that 2 pounds of blueberries cost $12, and 4.5 pounds of blueberries cost $27.

Sometimes it makes sense to connect the points with a line, and sometimes it doesn't.

- We could buy, for example, 4.5 pounds of blueberries or 1.875 pounds of blueberries. All the points in between the whole numbers make sense in the situation, so any point on the line is meaningful.
- If the graph represented the cost for different *numbers of sandwiches* (instead of pounds of blueberries), it might not make sense to connect the points with a line, because it is often not possible to buy 4.5 sandwiches or 1.875 sandwiches.

Even if only points make sense in the situation, though, sometimes we connect them with a line anyway to make the relationship easier to see.

NAME _______________ DATE _______________ PERIOD _______________

Graphs that represent proportional relationships all have a few things in common:

- Points that satisfy the relationship lie on a straight line.
- The line that they lie on passes through the **origin**, (0, 0).

Here are some graphs that do *not* represent proportional relationships:

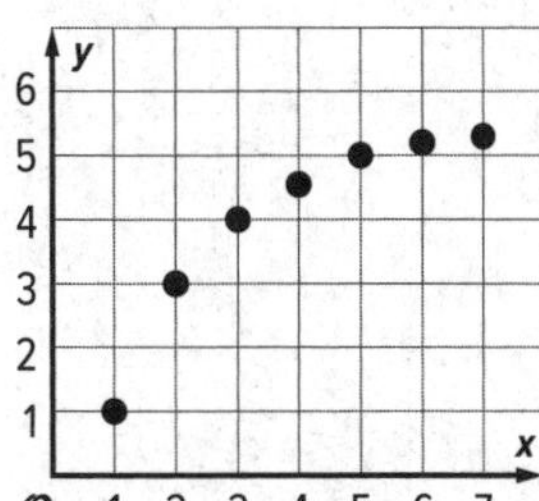

These points do not lie on a line.

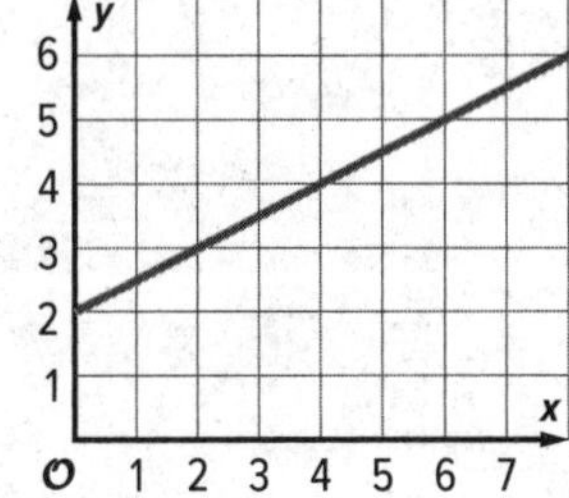

This is a line, but it doesn't go through the origin.

Glossary

coordinate plane
origin

Practice

Introducing Graphs of Proportional Relationships

1. Which graphs could represent a proportional relationship? Explain how you decided.

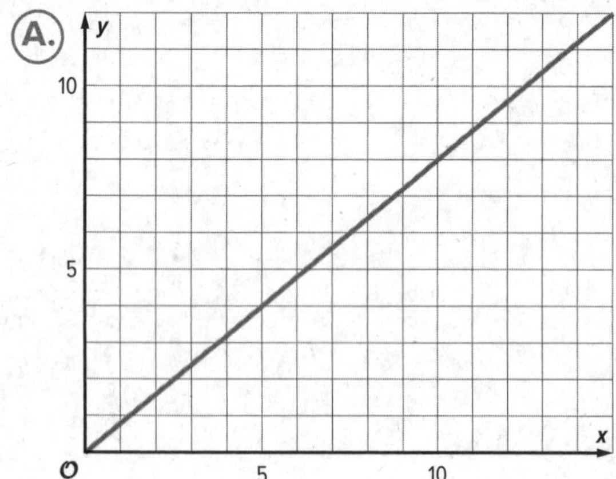

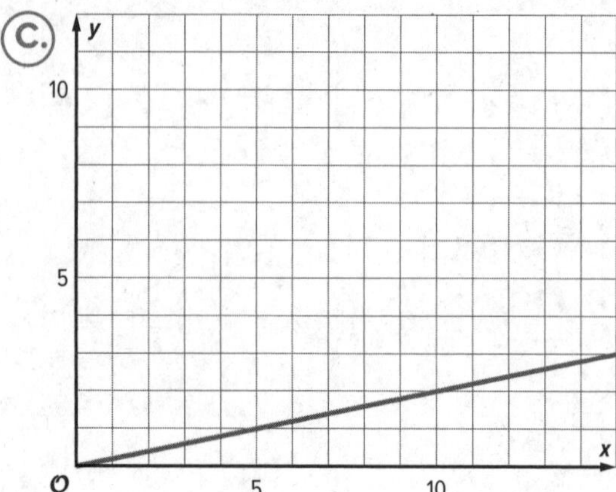

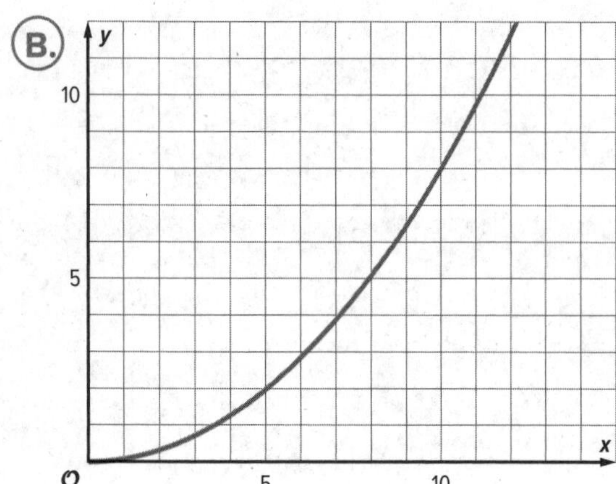

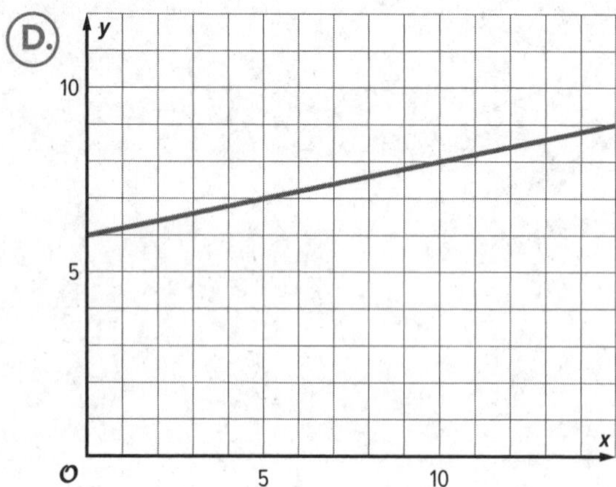

NAME ______________________ DATE ______________ PERIOD ____________

2. A lemonade recipe calls for $\frac{1}{4}$ cup of lemon juice for every cup of water.

a. Use the table to answer these questions.

x	y
1	$\frac{1}{4}$
2	$\frac{1}{2}$
3	$\frac{3}{4}$
4	1
5	$1\frac{1}{4}$
6	$1\frac{1}{2}$

i. What does x represent?

ii. What does y represent?

iii. Is there a proportional relationship between x and y?

b. Plot the pairs in the table in a coordinate plane.

3. Select **all** of the pieces of information that would tell you x and y have a proportional relationship. Let y represent the distance between a rock and a turtle's current position in meters and x represent the number of minutes the turtle has been moving. **(Lesson 2-9)**

A. $y = 3x$

B. After 4 minutes, the turtle has walked 12 feet away from the rock.

C. The turtle walks for a bit, then stops for a minute before walking again.

D. The turtle walks away from the rock at a constant rate.

4. Decide whether each table could represent a proportional relationship. If the relationship could be proportional, what would be the constant of proportionality? **(Lesson 2-7)**

a. The sizes you can print a photo

Width of Photo (inches)	Height of Photo (inches)
2	3
4	6
5	7
8	10

b. The distance from which a lighthouse is visible.

Height of a Lighthouse (feet)	Distance It Can Be Seen (miles)
20	6
45	9
70	11
95	13
150	16

Lesson 2-11

Interpreting Graphs of Proportional Relationships

NAME ______________________ DATE ______________ PERIOD ______________

Learning Goal Let's read stories from the graphs of proportional relationships.

Warm Up

11.1 What Could the Graph Represent?

Here is a graph that represents a proportional relationship.

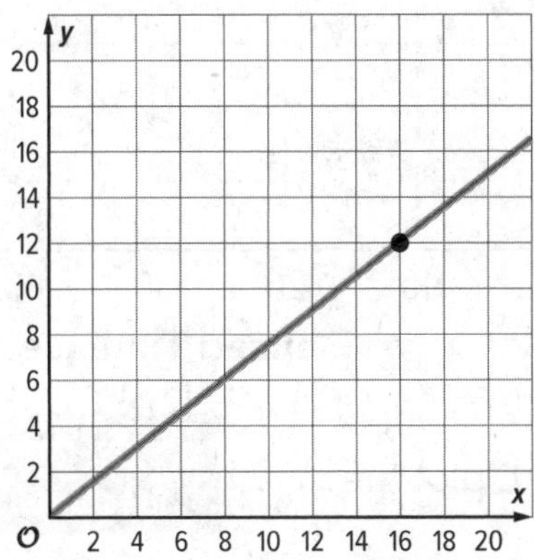

1. Invent a situation that could be represented by this graph.

2. Label the axes with the quantities in your situation.

3. Give the graph a title.

4. There is a point on the graph. What are its coordinates? What does it represent in your situation?

Activity
11.2 Tyler's Walk

Tyler was at the amusement park. He walked at a steady pace from the ticket booth to the bumper cars.

1. The point on the graph shows his arrival at the bumper cars. What do the coordinates of the point tell us about the situation?

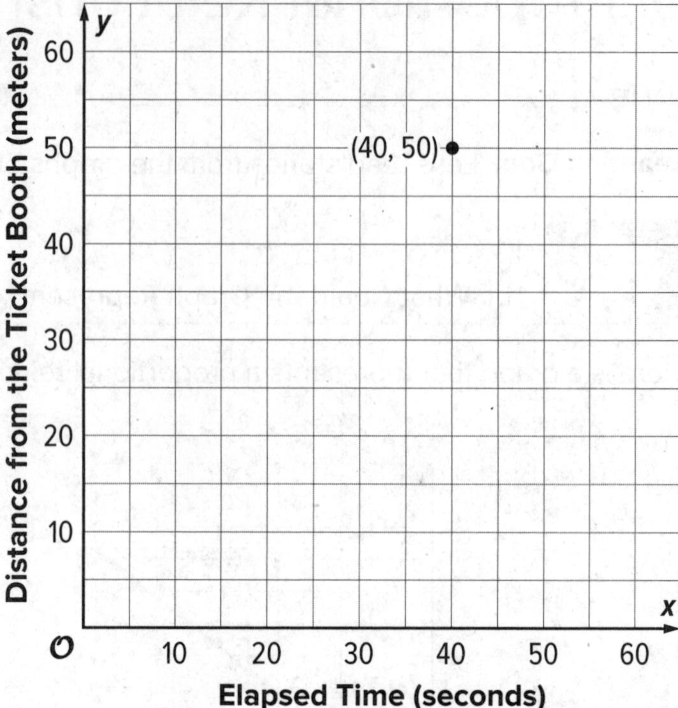

2. The table representing Tyler's walk shows other values of time and distance. Complete the table. Next, plot the pairs of values on the grid.

Time (seconds)	Distance (meters)
0	0
20	25
30	37.5
40	50
1	

3. What does the point (0, 0) mean in this situation?

4. How far away from the ticket booth was Tyler after 1 second? Label the point on the graph that shows this information with its coordinates.

5. What is the constant of proportionality for the relationship between time and distance? What does it tell you about Tyler's walk? Where do you see it in the graph?

NAME ______________________ DATE ____________ PERIOD ____________

If Tyler wanted to get to the bumper cars in half the time, how would the graph representing his walk change? How would the table change? What about the constant of proportionality?

Activity

11.3 Seagulls Eat What?

4 seagulls ate 10 pounds of garbage. Assume this information describes a proportional relationship.

1. Plot a point that shows the number of seagulls and the amount of garbage they ate.

2. Use a straight edge to draw a line through this point and (0, 0).

3. Plot the point (1, k) on the line. What is the value of k? What does the value of k tell you about this context?

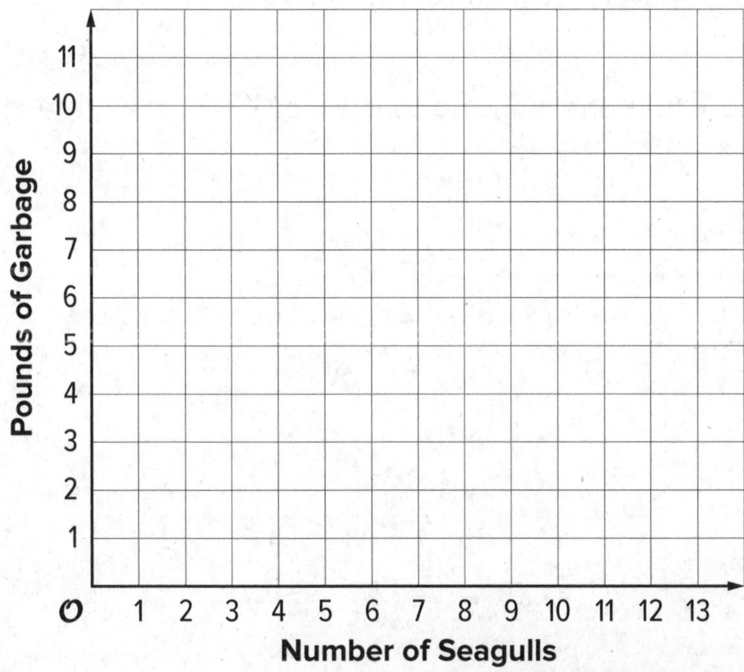

Summary

Interpreting Graphs of Proportional Relationships

For the relationship represented in this table, y is proportional to x. We can see in the table that $\frac{5}{4}$ is the constant of proportionality because it's the y value when x is 1.

x	y
4	5
5	$\frac{25}{4}$
8	10
1	$\frac{5}{4}$

The equation $y = \frac{5}{4}x$ also represents this relationship.

Here is the graph of this relationship.

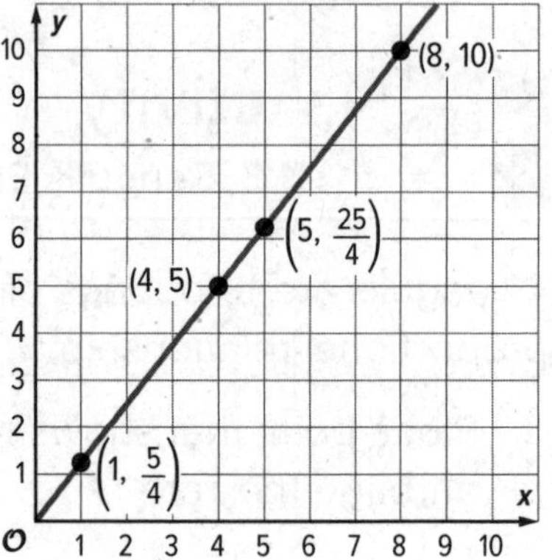

- If y represents the distance in feet that a snail crawls in x minutes, then the point (4, 5) tells us that the snail can crawl 5 feet in 4 minutes.
- If y represents the cups of yogurt and x represents the teaspoons of cinnamon in a recipe for fruit dip, then the point (4, 5) tells us that you can mix 4 teaspoons of cinnamon with 5 cups of yogurt to make this fruit dip.

We can find the constant of proportionality by looking at the graph, because $\frac{5}{4}$ is the y-coordinate of the point on the graph where the x-coordinate is 1. This could mean the snail is traveling $\frac{5}{4}$ feet per minute or that the recipe calls for $1\frac{1}{4}$ cups of yogurt for every teaspoon of cinnamon.

In general, when y is proportional to x, the corresponding constant of proportionality is the y-value when $x = 1$.

NAME ______________________ DATE ____________ PERIOD ________

Practice

Interpreting Graphs of Proportional Relationships

1. There is a proportional relationship between the number of months a person has had a streaming movie subscription and the total amount of money they have paid for the subscription. The cost for 6 months is \$47.94. The point (6, 47.94) is shown on the graph below.

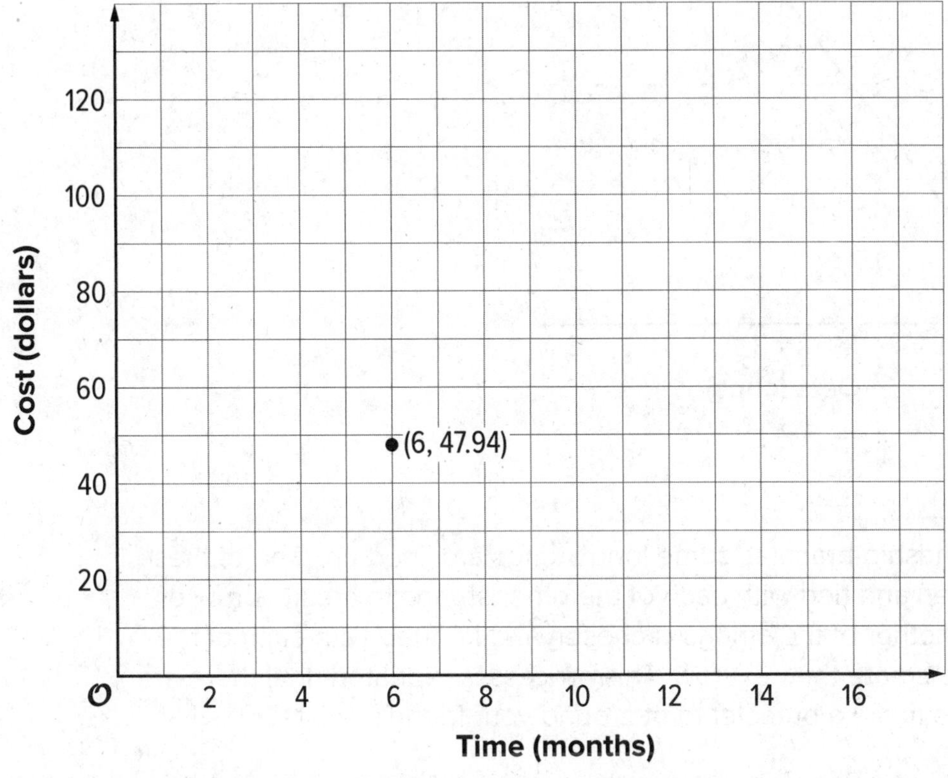

a. What is the constant of proportionality in this relationship?

b. What does the constant of proportionality tell us about the situation?

c. Add at least three more points to the graph and label them with their coordinates.

d. Write an equation that represents the relationship between C, the total cost of the subscription, and m, the number of months.

2. The graph shows the amounts of almonds, in grams, for different amounts of oats, in cups, in a granola mix. Label the point $(1, k)$ on the graph, find the value of k, and explain its meaning.

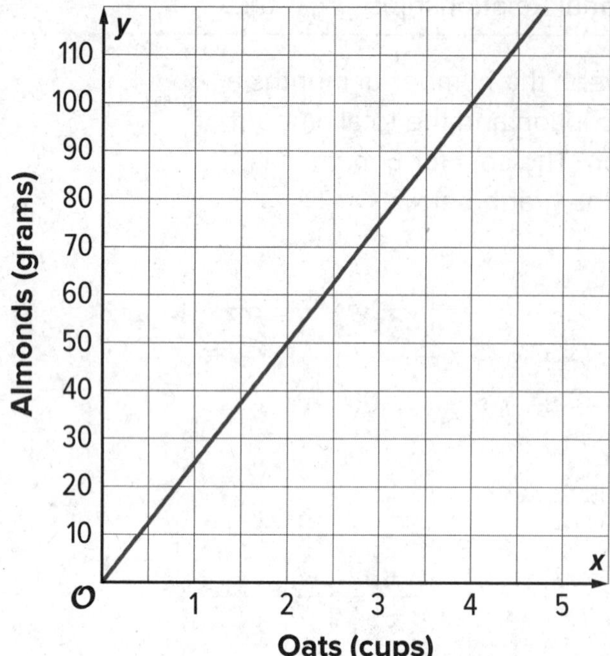

3. To make a friendship bracelet, some long strings are lined up. One of these strings is chosen and tied with each of the other strings to create a row of knots. Then, another of the strings is chosen and knotted with each of the other strings to create the next row. This process is repeated until there are enough rows to make a bracelet to fit around your friend's wrist.

   Are the number of knots proportional to the number of rows? Explain your reasoning. **(Lesson 2-9)**

4. What information do you need to know to write an equation relating two quantities that have a proportional relationship? **(Lesson 2-9)**

Lesson 2-12

Using Graphs to Compare Relationships

NAME ______________________ DATE ____________ PERIOD ____________

Learning Goal Let's graph more than one relationship on the same grid.

Warm Up

12.1 Number Talk: Fraction Multiplication and Division

Find each product or quotient mentally.

1. $\frac{2}{3} \cdot \frac{1}{2}$
2. $\frac{4}{3} \cdot \frac{1}{4}$
3. $4 \div \frac{1}{5}$
4. $\frac{9}{6} \div \frac{1}{2}$

Activity

12.2 Race to the Bumper Cars

Diego, Lin, and Mai went from the ticket booth to the bumper cars.

1. Use each description to complete the table representing that person's journey.
   a. Diego left the ticket booth at the same time as Tyler. Diego jogged ahead at a steady pace and reached the bumper cars in 30 seconds.
   b. Lin left the ticket booth at the same time as Tyler. She ran at a steady pace and arrived at the bumper cars in 20 seconds.
   c. Mai left the booth 10 seconds later than Tyler. Her steady jog enabled her to catch up with Tyler just as he arrived at the bumper cars.

Diego's journey:

Time (seconds)	Distance (meters)
0	
15	
30	50
1	

Lin's journey:

Time (seconds)	Distance (meters)
	0
	25
20	50
1	

Mai's journey:

Time (seconds)	Distance (meters)
	0
	25
40	50
1	

2. Using a different color for each person, draw a graph of all four people's journeys (including Tyler's from the other day).

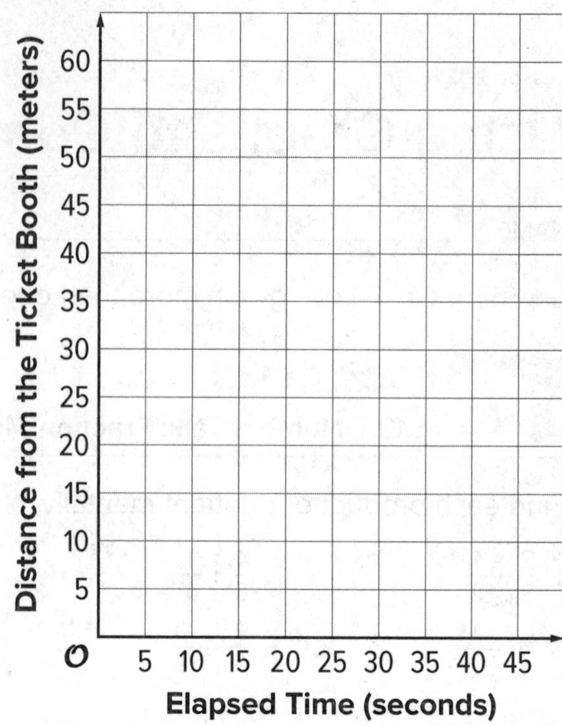

3. Which person is moving the most quickly? How is that reflected in the graph?

Are you ready for more?

Write equations to represent each person's relationship between time and distance.

Activity

12.3 Space Rocks and the Price of Rope

1. Meteoroid Perseid 245 and Asteroid X travel through the solar system. The graph shows the distance each traveled after a given point in time.

   Is Asteroid X traveling faster or slower than Perseid 245? Explain how you know.

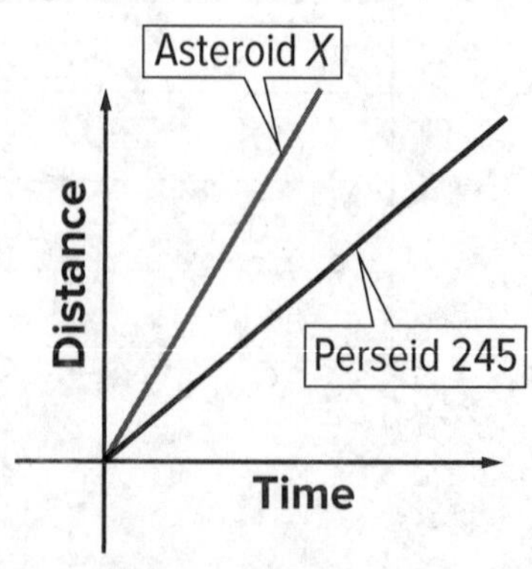

NAME ______________________ DATE ____________ PERIOD ____________

2. The graph shows the price of different lengths of two types of rope.

   If you buy $1.00 of each kind of rope, which one will be longer? Explain how you know.

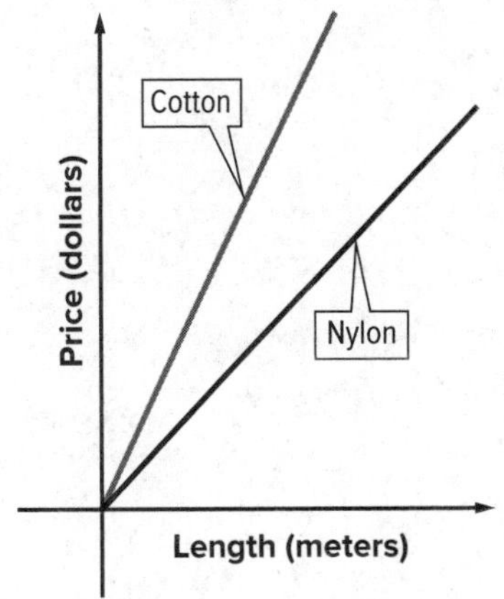

Summary
Using Graphs to Compare Relationships

Here is a graph that shows the price of blueberries at two different stores. Which store has a better price?

We can compare points that have the same *x* value or the same *y* value.

For example, the points (2, 12) and (3, 12) tell us that at store B you can get more pounds of blueberries for the same price.

The points (3, 12) and (3, 18) tell us that at store A you have to pay more for the same quantity of blueberries. This means store B has the better price.

We can also use the graphs to compare the constants of proportionality. The line representing store B goes through the point (1, 4), so the constant of proportionality is 4. This tells us that at store B the blueberries cost $4 per pound. This is cheaper than the $6 per pound unit price at store A.

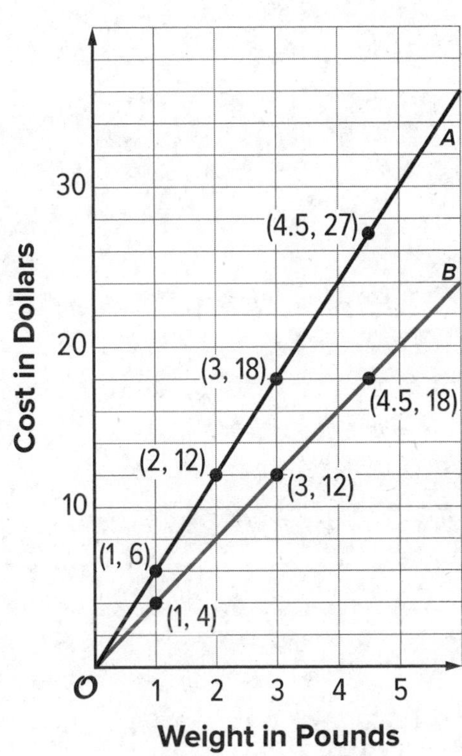

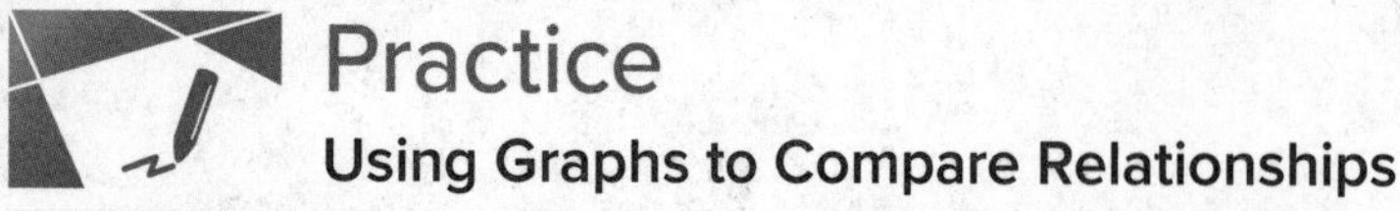

Practice
Using Graphs to Compare Relationships

1. The graphs below show some data from a coffee shop menu. One of the graphs shows cost (in dollars) vs. drink volume (in ounces), and one of the graphs shows calories vs. drink volume (in ounces).

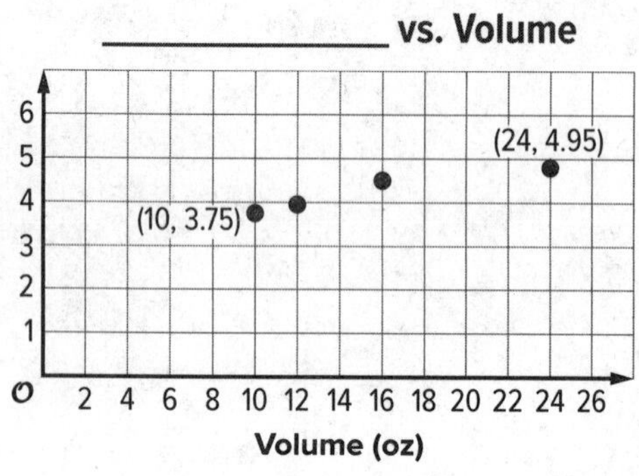

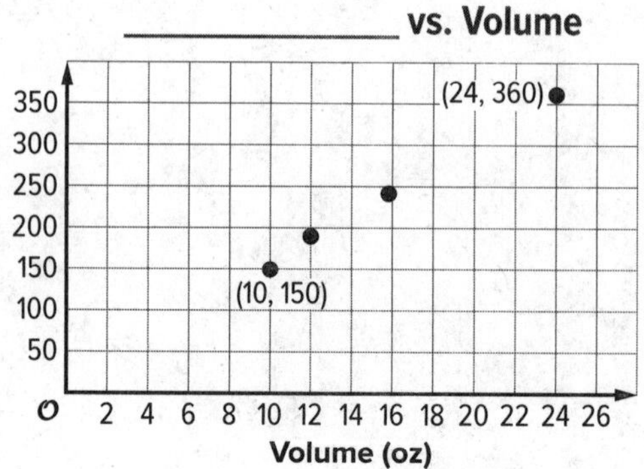

a. Which graph is which? Give them the correct titles.

b. Which quantities appear to be in a proportional relationship? Explain how you know.

c. For the proportional relationship, find the constant of proportionality. What does that number mean?

NAME ________________ DATE ________ PERIOD ________

2. Lin and Andre biked home from school at a steady pace. Lin biked 1.5 km and it took her 5 minutes. Andre biked 2 km and it took him 8 minutes.

a. Draw a graph with two lines that represent the bike rides of Lin and Andre.

b. For each line, highlight the point with coordinates $(1, k)$ and find k.

c. Who was biking faster?

3. Match each equation to its graph.

a. $y = 2x$

b. $y = \frac{4}{5}x$

c. $y = \frac{1}{4}x$

d. $y = \frac{2}{3}x$

e. $y = \frac{4}{3}x$

f. $y = \frac{3}{2}x$

Graph 1

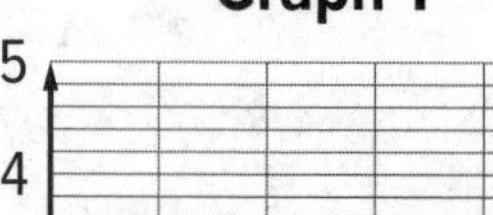

Graph 2

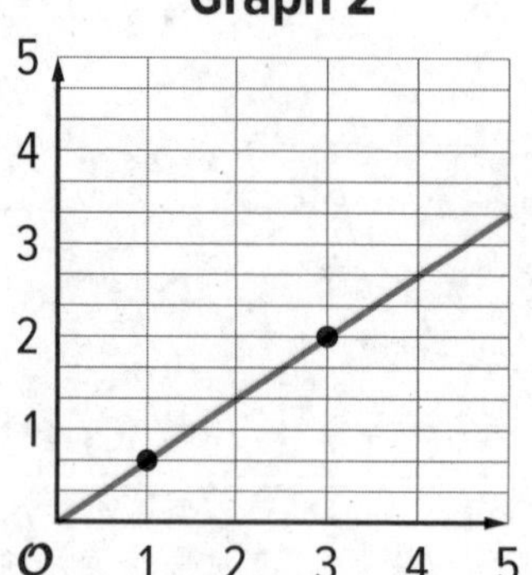

Graph 3

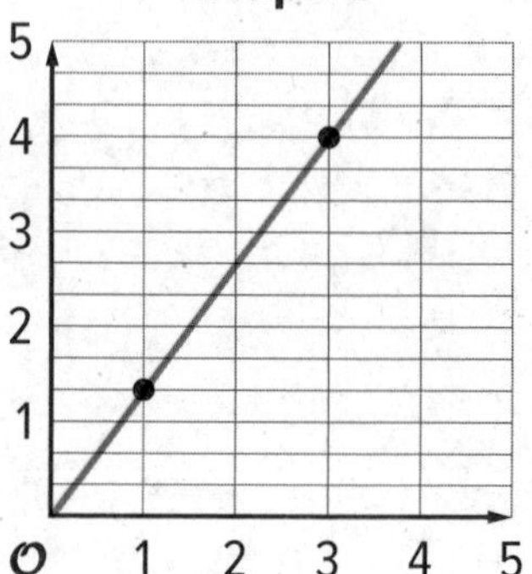

Graph 4

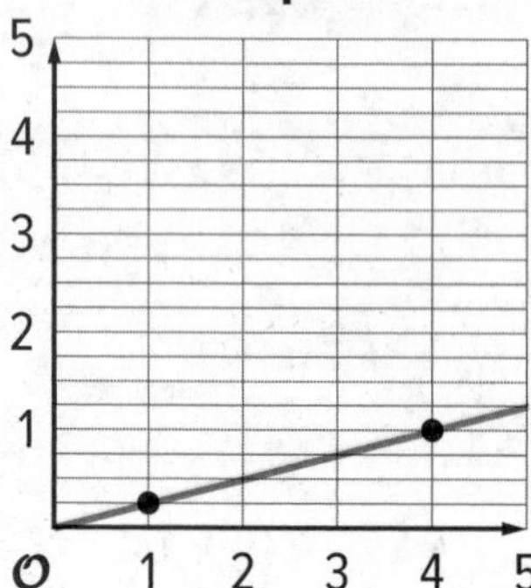

Graph 5

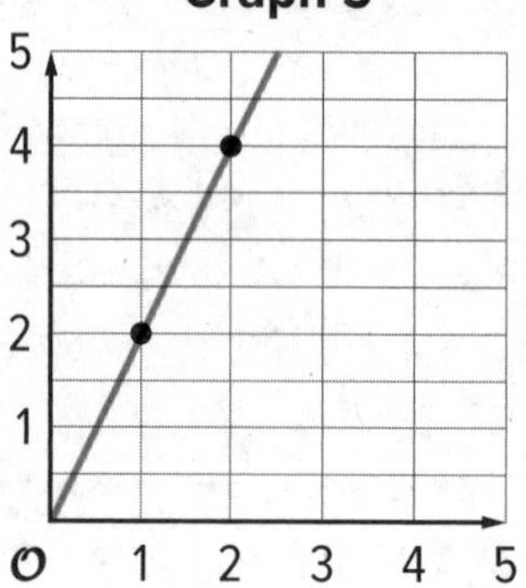

Graph 6

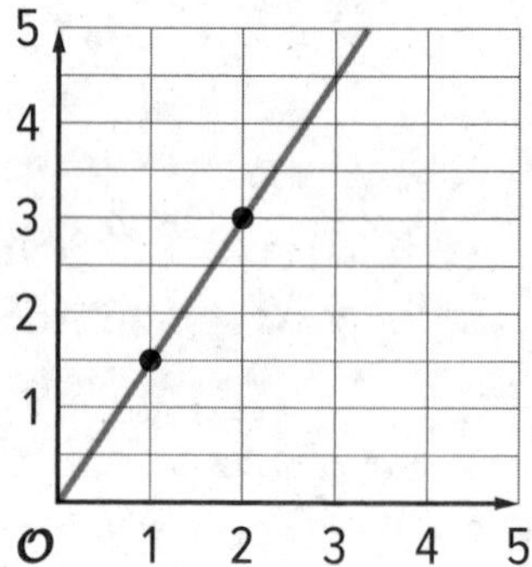

Lesson 2-13

Two Graphs for Each Relationship

NAME _______________ DATE _______________ PERIOD _______________

Learning Goal Let's use tables, equations, and graphs to answer questions about proportional relationships.

Warm Up

13.1 True or False: Fractions and Decimals

Decide whether each equation is true or false. Be prepared to explain your reasoning.

1. $\frac{3}{2} \cdot 16 = 3 \cdot 8$

2. $\frac{3}{4} \div \frac{1}{2} = \frac{6}{4} \div \frac{1}{4}$

3. $(2.8) \cdot (13) = (0.7) \cdot (52)$

Activity

13.2 Tables, Graphs, and Equations

Your teacher will assign you *one* of these three points:

$A = (10, 4)$, $B = (4, 5)$, $C = (8, 5)$.

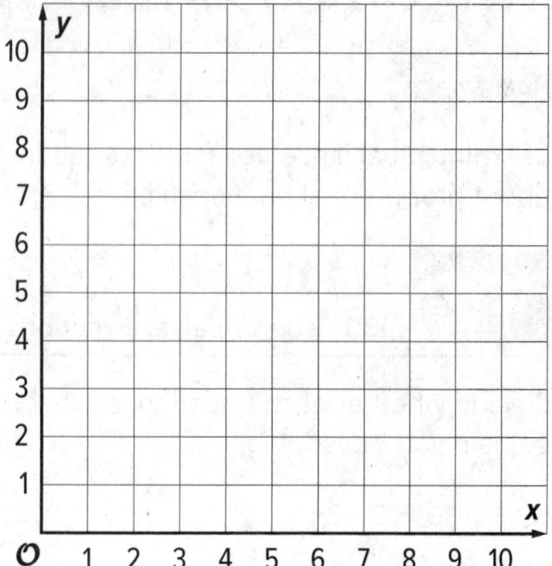

1. On the graph, plot and label *only* your assigned point.

2. Use a ruler to line up your point with the origin, (0, 0). Draw a line that starts at the origin, goes through your point, and continues to the edge of the graph.

3. Complete the table with the coordinates of points on your graph. Use a fraction to represent any value that is not a whole number.

4. Write an equation that represents the relationship between x and y defined by your point.

5. Compare your graph and table with the rest of your group. What is the same and what is different about:

   a. your tables?

   b. your equations?

   c. your graphs?

x	y	$\frac{y}{x}$
0		N/A
1		
2		
3		
4		
5		
6		
7		
8		
9		
10		

NAME ______________________ DATE ______________ PERIOD __________

6. What is the y-coordinate of your graph when the x-coordinate is 1? Plot and label this point on your graph. Where do you see this value in the table? Where do you see this value in your equation?

7. Describe any connections you see between the table, characteristics of the graph, and the equation.

Are you ready for more?

The graph of an equation of the form $y = kx$, where k is a positive number, is a line through (0, 0) and the point (1, k).

1. Name at least one line through (0, 0) that cannot be represented by an equation like this.

2. If you could draw the graphs of *all* of the equations of this form in the same coordinate plane, what would it look like?

Activity

13.3 Hot Dog Eating Contest

Andre and Jada were in a hot dog eating contest. Andre ate 10 hot dogs in 3 minutes. Jada ate 12 hot dogs in 5 minutes.

Here are two different graphs that both represent this situation.

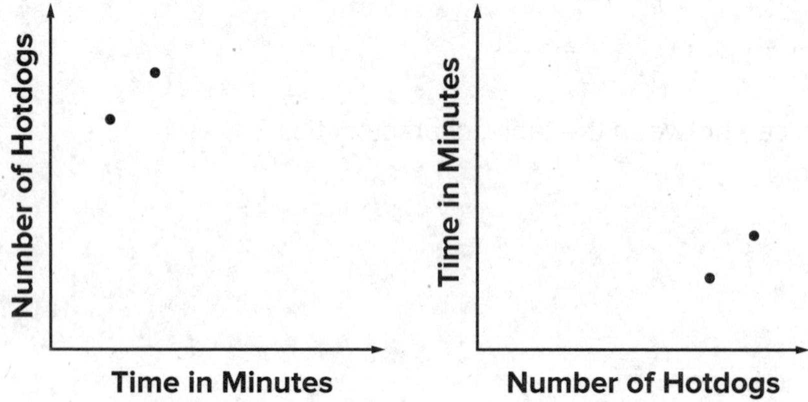

1. On the first graph, which point shows Andre's consumption and which shows Jada's consumption? Label them.

2. Draw two lines: one through the origin and Andre's point, and one through the origin and Jada's point.

3. Write an equation for each line. Use t to represent time in minutes and h to represent number of hot dogs.

   a. Andre:

   b. Jada:

4. For each equation, what does the constant of proportionality tell you?

5. Repeat the previous steps for the second graph.

   a. Andre:

   b. Jada:

NAME ______________________ DATE ______________ PERIOD ____________

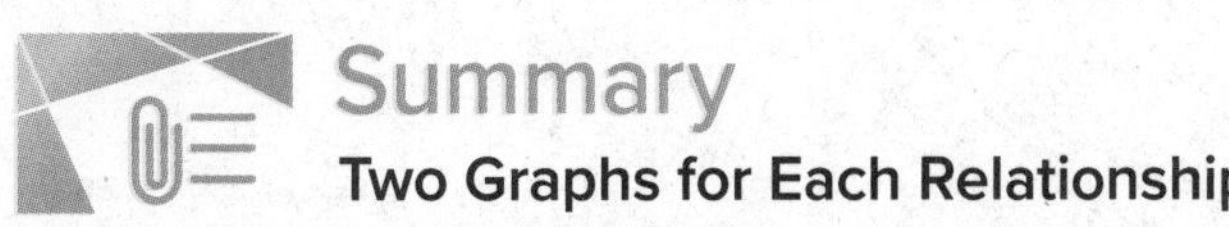

Summary
Two Graphs for Each Relationship

Imagine that a faucet is leaking at a constant rate and that every 2 minutes, 10 milliliters of water leak from the faucet. There is a proportional relationship between the volume of water and elapsed time.

- We could say that the elapsed time is proportional to the volume of water. The corresponding constant of proportionality tells us that the faucet is leaking at a rate of $\frac{1}{5}$ of a minute per milliliter.
- We could say that the volume of water is proportional to the elapsed time. The corresponding constant of proportionality tells us that the faucet is leaking at a rate of 5 milliliters per minute.

Let's use v to represent volume in milliliters and t to represent time in minutes. Here are graphs and equations that represent both ways of thinking about this relationship:

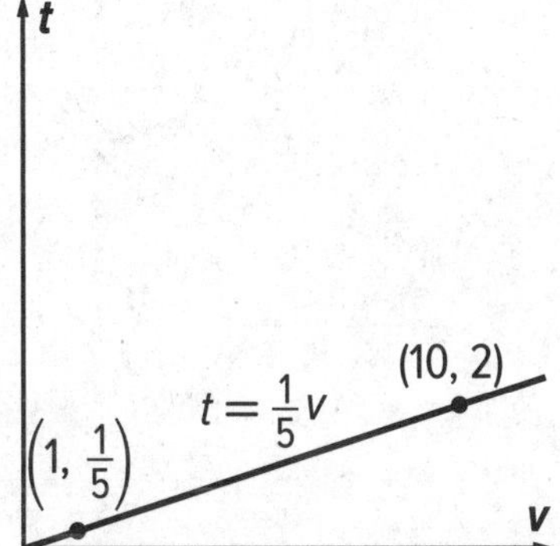

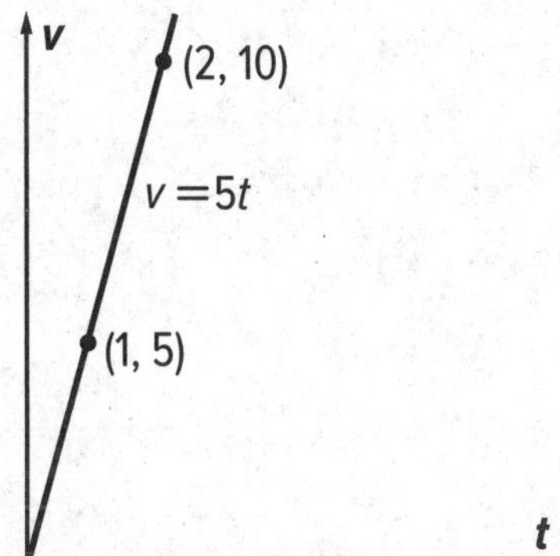

Even though the relationship between time and volume is the same, we are making a different choice in each case about which variable to view as the independent variable.

The graph on the left has v as the independent variable, and the graph on the right has t as the independent variable.

Practice

Two Graphs for Each Relationship

1. At the supermarket you can fill your own honey bear container. A customer buys 12 oz of honey for \$5.40.

   a. How much does honey cost per ounce?

   b. How much honey can you buy per dollar?

   c. Write two different equations that represent this situation. Use *h* for ounces of honey and *c* for cost in dollars.

   d. Choose one of your equations, and sketch its graph. Be sure to label the axes.

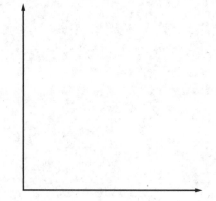

2. The point $\left(3, \frac{6}{5}\right)$ lies on the graph representing a proportional relationship. Which of the following points also lie on the same graph?

   Select **all** that apply.

   A. (1, 0.4)

   B. $\left(1.5, \frac{6}{10}\right)$

   C. $\left(\frac{6}{5}, 3\right)$

   D. $\left(4, \frac{11}{5}\right)$

   E. (15, 6)

NAME ________________ DATE ________ PERIOD ________

3. A trail mix recipe asks for 4 cups of raisins for every 6 cups of peanuts. There is a proportional relationship between the amount of raisins r (cups), and the amount of peanuts p (cups), in this recipe.

   a. Write the equation for the relationship that has constant of proportionality greater than 1. Graph the relationship.

   b. Write the equation for the relationship that has a constant of proportionality less than 1. Graph the relationship.

4. Here is a graph that represents a proportional relationship. **(Lesson 2-11)**

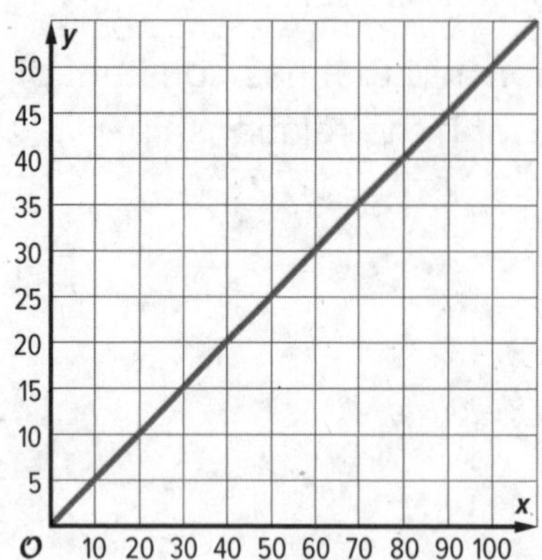

a. Come up with a situation that could be represented by this graph.

b. Label the axes with the quantities in your situation.

c. Give the graph a title.

d. Choose a point on the graph. What do the coordinates represent in your situation?

Lesson 2-14

Four Representations

NAME ____________________ DATE __________ PERIOD __________

Learning Goal Let's contrast relationships that are and are not proportional in four different ways.

Warm Up

14.1 Which Is the Bluest?

1. Which group of blocks is the bluest?

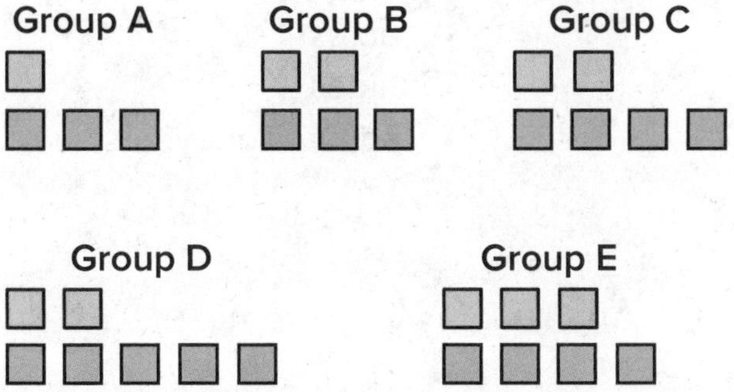

Group A   Group B   Group C

Group D   Group E

2. Order the groups of blocks from least blue to bluest.

Activity

14.2 One Scenario, Four Representations

1. Select two things from different lists. Make up a situation where there is a *proportional relationship* between quantities that involve these things.

creatures
- starfish
- centipedes
- earthworms
- dinosaurs

length
- centimeters
- cubits
- kilometers
- parsecs

time
- nanoseconds
- minutes
- years
- millennia

volume
- milliliters
- gallons
- bushels
- cubic miles

body parts
- legs
- eyes
- neurons
- digits

area
- square microns
- acres
- hides
- square light-years

weight
- nanograms
- ounces
- deben
- metric tonnes

substance
- helium
- oobleck
- pitch
- glue

2. Select two other things from the lists, and make up a situation where there is a relationship between quantities that involve these things, but the relationship is *not* proportional.

NAME ________________________________ DATE ______________ PERIOD ____________

3. Your teacher will give you two copies of the "One Scenario, Four Representations" sheet. For each of your situations, describe the relationships in detail. If you get stuck, consider asking your teacher for a copy of the sample response.

   a. Write one or more sentences describing the relationship between the things you chose.

   b. Make a table with titles in each column and at least 6 pairs of numbers relating the two things.

   c. Graph the situation and label the axes.

   d. Write an equation showing the relationship and explain in your own words what each number and letter in your equation means.

   e. Explain how you know whether each relationship is proportional or not proportional. Give as many reasons as you can.

Activity
14.3 Make a Poster

Create a visual display of your two situations that includes all the information from the previous activity.

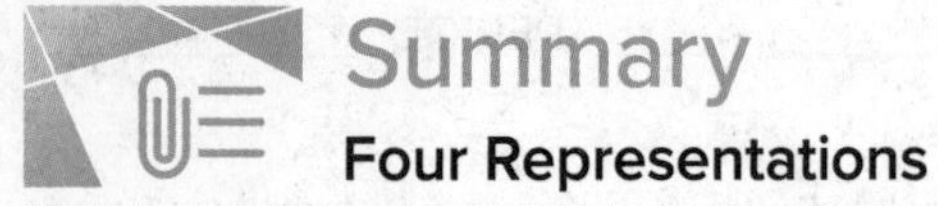

Summary

Four Representations

The constant of proportionality for a proportional relationship can often be easily identified in a graph, a table, and an equation that represents it. Here is an example of all three representations for the same relationship. The constant of proportionality is circled:

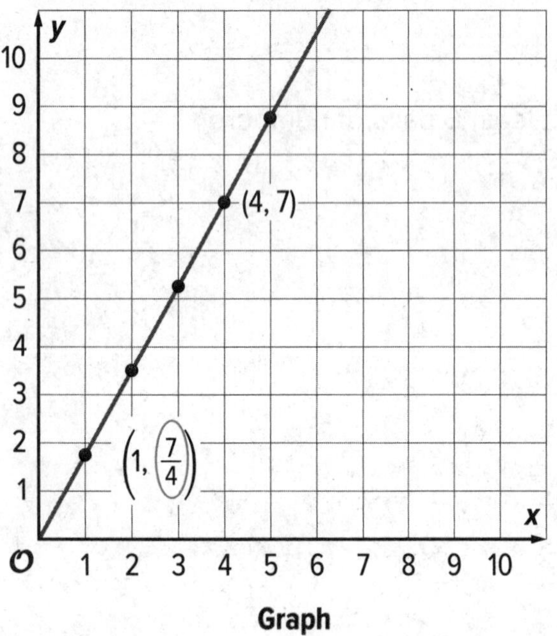

Graph

x	y
0	0
1	$\frac{7}{4}$
2	$\frac{7}{2}$
3	$\frac{21}{4}$
4	7

Table

$$y = \frac{7}{4}x$$

Equation

On the other hand, some relationships are not proportional...

- if the graph of a relationship is not a straight line through the origin,
- if the equation cannot be expressed in the form $y = kx$,
- or if the table does not have a constant of proportionality that you can multiply by any number in the first column to get the associated number in the second column,

then the relationship between the quantities is not a proportional relationship.

NAME ______________________ DATE ____________ PERIOD ________

Practice
Four Representations

1. The equation $c = 2.95g$ shows how much it costs to buy gas at a gas station on a certain day. In the equation, c represents the cost in dollars, and g represents how many gallons of gas were purchased.

   a. Write down at least four (gallons of gas, cost) pairs that fit this relationship.

   b. Create a graph of the relationship.

   c. What does 2.95 represent in this situation?

   d. Jada's mom remarks, "You can get about a third of a gallon of gas for a dollar." Is she correct? How did she come up with that?

2. There is a proportional relationship between a volume measured in cups and the same volume measured in tablespoons. Three cups is equivalent to 48 tablespoons, as shown in the graph.

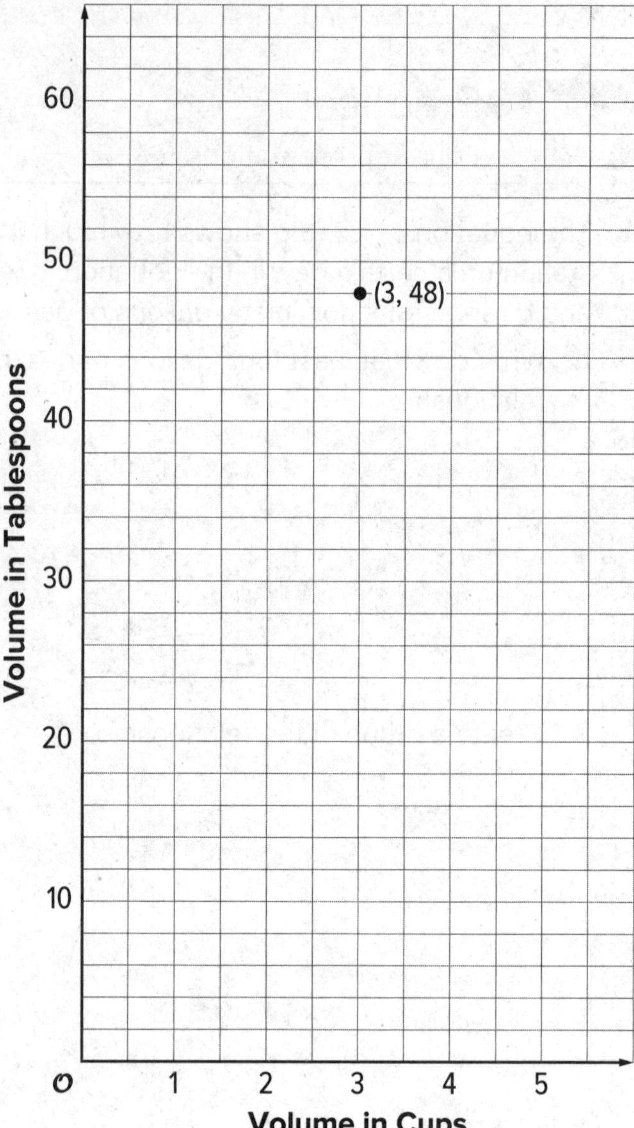

a. Plot and label at least two more points that represent the relationship.

b. Use a straightedge to draw a line that represents this proportional relationship.

c. For which value y is $(1, y)$ on the line you just drew?

d. What is the constant of proportionality for this relationship?

e. Write an equation representing this relationship. Use c for cups and t for tablespoons.

Lesson 2-15

Using Water Efficiently

NAME ______________________ DATE __________ PERIOD __________

Learning Goal Let's investigate saving water.

Warm Up

15.1 Comparing Baths and Showers

Some people say that it uses more water to take a bath than a shower. Others disagree.

1. What information would you collect in order to answer the question?

2. Estimate some reasonable values for the things you suggest.

Activity

15.2 Saving Water: Bath or Shower?

1. Describe a method for comparing the water usage for a bath and a shower.

2. Find out values for the measurements needed to use the method you described. You may ask your teacher or research them yourself.

3. Under what conditions does a bath use more water? Under what conditions does a shower use more water?

Activity

15.3 Representing Water Usage

1. Continue considering the problem from the previous activity. Name two quantities that are in a proportional relationship. Explain how you know they are in a proportional relationship.

2. What are two constants of proportionality for the proportional relationship? What do they tell us about the situation?

3. On graph paper, create a graph that shows how the two quantities are related. Make sure to label the axes.

4. Write two equations that relate the quantities in your graph. Make sure to record what each variable represents.

Learning Targets

Lesson	Learning Target(s)
2-1 One of These Things is Not Like the Others	• I can use equivalent ratios to describe scaled copies of shapes. • I know that two recipes will taste the same if the ingredients are in equivalent ratios.
2-2 Introducing Proportional Relationships with Tables	• I can use a table to reason about two quantities that are in a proportional relationship. • I understand the terms proportional relationship and constant of proportionality.
2-3 More about Constant of Proportionality	• I can find missing information in a proportional relationship using a table. • I can find the constant of proportionality from information given in a table.

(continued on the next page)

(continued from the previous page)

Lesson	Learning Target(s)
2-4 Proportional Relationships and Equations	• I can write an equation of the form $y = kx$ to represent a proportional relationship described by a table or a story. • I can write the constant of proportionality as an entry in a table.
2-5 Two Equations for Each Relationship	• I can find two constants of proportionality for a proportional relationship. • I can write two equations representing a proportional relationship described by a table or story.
2-6 Using Equations to Solve Problems	• I can find missing information in a proportional relationship using the constant of proportionality. • I can relate all parts of an equation like $y = kx$ to the situation it represents.
2-7 Comparing Relationships with Tables	• I can decide if a relationship represented by a table could be proportional and when it is definitely not proportional.

Lesson	Learning Target(s)
2-8 Comparing Relationships with Equations	• I can decide if a relationship represented by an equation is proportional or not.
2-9 Solving Problems about Proportional Relationships	• I can ask questions about a situation to determine whether two quantities are in a proportional relationship. • I can solve all kinds of problems involving proportional relationships.
2-10 Introducing Graphs of Proportional Relationships	• I know that the graph of a proportional relationship lies on a line through (0, 0).
2-11 Interpreting Graphs of Proportional Relationships	• I can draw the graph of a proportional relationship given a single point on the graph (other than the origin). • I can find the constant of proportionality from a graph. • I understand the information given by graphs of proportional relationships that are made up of points or a line.

(continued on the next page)

(continued from the previous page)

Lesson	Learning Target(s)
2-12 Using Graphs to Compare Relationships	• I can compare two, related proportional relationships based on their graphs. • I know that the steeper graph of two proportional relationships has a larger constant of proportionality.
2-13 Two Graphs for Each Relationship	• I can interpret a graph of a proportional relationship using the situation. • I can write an equation representing a proportional relationship from a graph.
2-14 Four Representations	• I can make connections between the graphs, tables, and equations of a proportional relationship. • I can use units to help me understand information about proportional relationships.
2-15 Using Water Efficiently	• I can answer a question by representing a situation using proportional relationships.

Notes:

Unit 3

Measuring Circles

At the end of this unit, you'll apply what you learned about measuring circles to design a stained-glass window.

Topics

- Circumference of a Circle
- Area of a Circle
- Let's Put It to Work

Unit 3

Measuring Circles

Circumference of a Circle

Area of a Circle

Let's Put It to Work

Lesson 3-1

How Well Can You Measure?

NAME ______________________ DATE ____________ PERIOD ____________

Learning Goal Let's see how accurately we can measure.

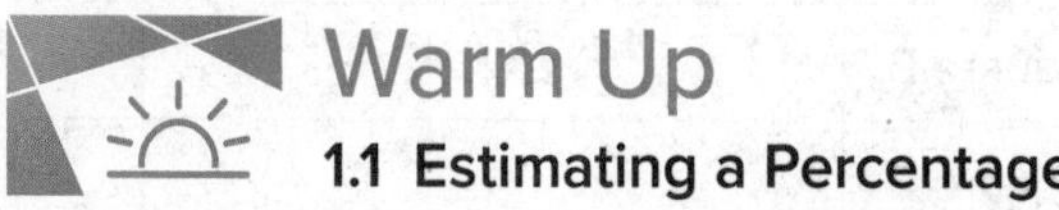

Warm Up

1.1 Estimating a Percentage

A student got 16 out of 21 questions correct on a quiz. Use mental estimation to answer these questions.

1. Did the student answer less than or more than 80% of the questions correctly?

2. Did the student answer less than or more than 75% of the questions correctly?

Activity

1.2 Perimeter of a Square

Your teacher will give you a picture of 9 different squares and will assign your group 3 of these squares to examine more closely.

1. For each of your assigned squares, measure the length of the diagonal and the perimeter of the square in centimeters.

   Check your measurements with your group. After you come to an agreement, record your measurements in the table.

	Diagonal (cm)	Perimeter (cm)
Square A		
Square B		
Square C		
Square D		
Square E		
Square F		
Square G		
Square H		
Square I		

2. Plot the diagonal and perimeter values from the table on the coordinate plane.

3. What do you notice about the points on the graph?

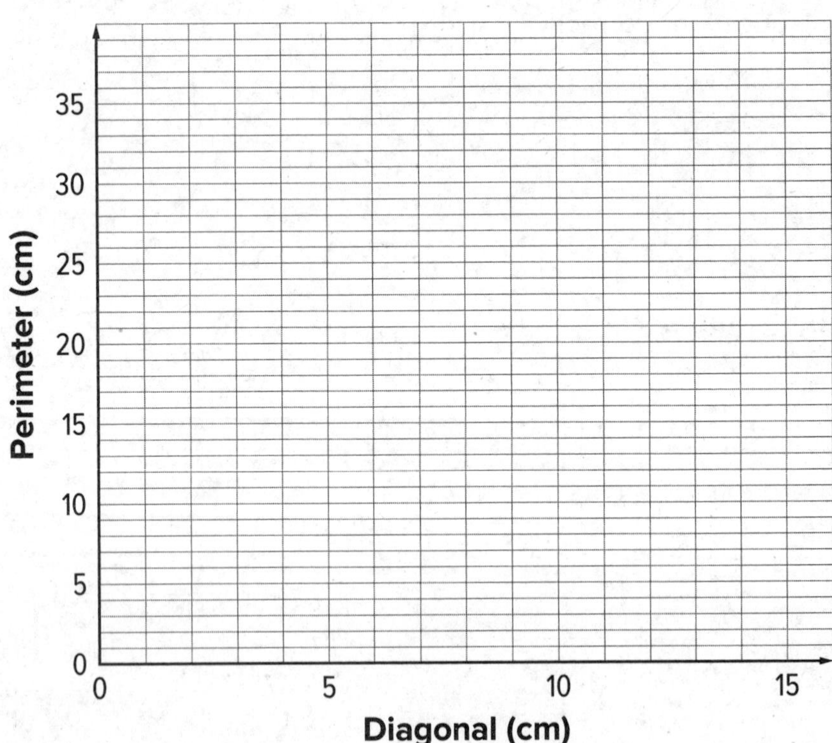

Pause here so your teacher can review your work.

4. Record measurements of the other squares to complete your table.

NAME ______________________ DATE ____________ PERIOD ____________

Activity

1.3 Area of a Square

1. In the table, record the length of the diagonal for each of your assigned squares from the previous activity. Next, calculate the area of each of your squares.

	Diagonal (cm)	Area (cm^2)
Square A		
Square B		
Square C		
Square D		
Square E		
Square F		
Square G		
Square H		
Square I		

Pause here so your teacher can review your work. Be prepared to share your values with the class.

2. Examine the class graph of these values. What do you notice?

3. How is the relationship between the diagonal and area of a square the same as the relationship between the diagonal and perimeter of a square from the previous activity? How is it different?

Are you ready for more?

Here is a rough map of a neighborhood.

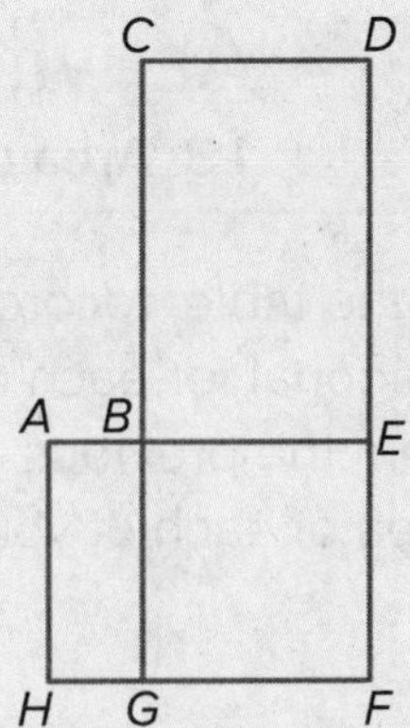

There are 4 mail routes during the week.

- On Monday, the mail truck follows the route *A-B-E-F-G-H-A*, which is 14 miles long.
- On Tuesday, the mail truck follows the route *B-C-D-E-F-G-B*, which is 22 miles long.
- On Wednesday, the mail truck follows the route *A-B-C-D-E-F-G-H-A*, which is 24 miles long.
- On Thursday, the mail truck follows the route *B-E-F-G-B*.

How long is the route on Thursdays?

NAME _______________ DATE _______________ PERIOD _______________

Summary

How Well Can You Measure?

When we measure the values for two related quantities, plotting the measurements in the coordinate plane can help us decide if it makes sense to model them with a proportional relationship. If the points are close to a line through (0, 0), then a proportional relationship is a good model.

For example, here is a graph of the values for the height, measured in millimeters, of different numbers of pennies placed in a stack.

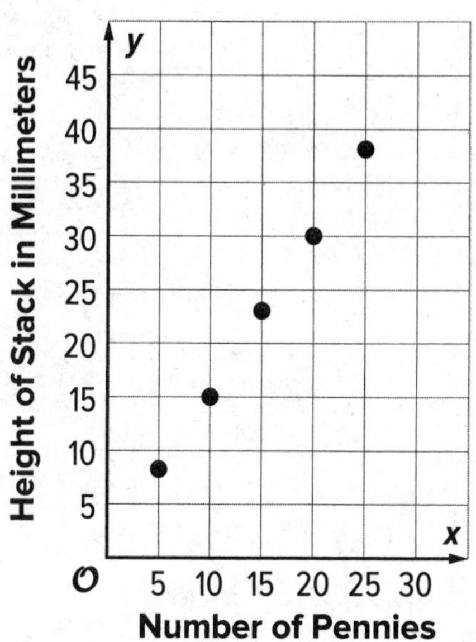

Because the points are close to a line through (0, 0), the height of the stack of pennies appears to be proportional to the number of pennies in a stack. This makes sense because we can see that the heights of the pennies only vary a little bit.

An additional way to investigate whether or not a relationship is proportional is by making a table. Here is some data for the weight of different numbers of pennies in grams, along with the corresponding number of grams per penny.

Number of Pennies	Grams	Grams per Penny
1	3.1	3.1
2	5.6	2.8
5	13.1	2.6
10	25.6	2.6

Though we might expect this relationship to be proportional, the quotients are not very close to one another. In fact, the metal in pennies changed in 1982, and older pennies are heavier. This explains why the weight per penny for different numbers of pennies are so different!

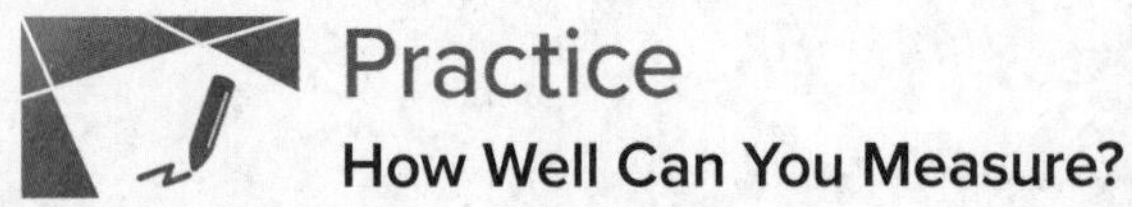

Practice

How Well Can You Measure?

1. Estimate the side length of a square that has a 9 cm long diagonal.

2. Select **all** quantities that are proportional to the diagonal length of a square.

   A. area of a square

   B. perimeter of a square

   C. side length of a square

3. Diego made a graph of two quantities that he measured and said, "The points all lie on a line except one, which is a little bit above the line. This means that the quantities can't be proportional." Do you agree with Diego? Explain.

4. The graph shows that while it was being filled, the amount of water in gallons in a swimming pool was approximately proportional to the time that has passed in minutes.

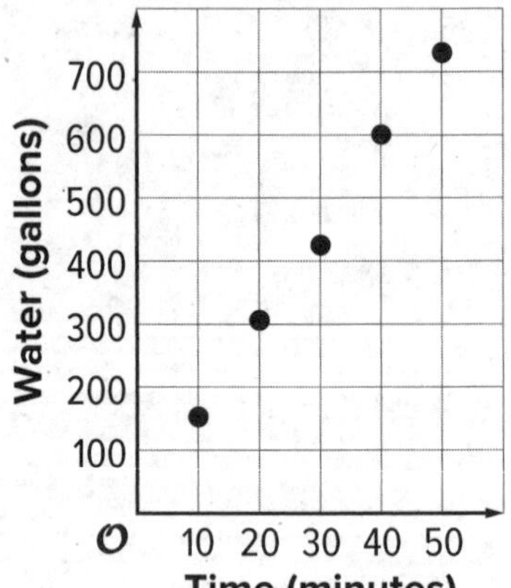

   a. About how much water was in the pool after 25 minutes?

   b. Approximately when were there 500 gallons of water in the pool?

   c. Estimate the constant of proportionality for the number of gallons of water per minute going into the pool.

Lesson 3-2

Exploring Circles

NAME ______________________ DATE ____________ PERIOD ____________

Learning Goal Let's explore circles.

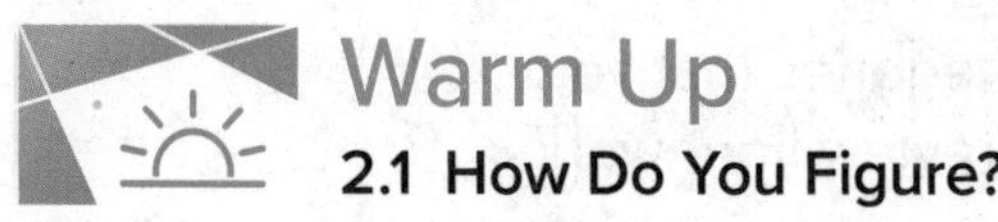

Warm Up

2.1 How Do You Figure?

Here are two figures.

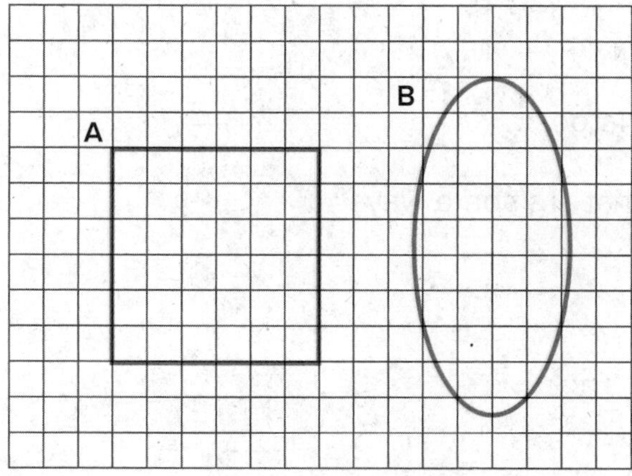

Figure C looks more like Figure A than like Figure B. Sketch what Figure C might look like. Explain your reasoning.

Activity

2.2 Sorting Round Objects

Your teacher will give you some pictures of different objects.

1. How could you sort these pictures into two groups? Be prepared to share your reasoning.

2. Work with your partner to sort the pictures into the categories that your class has agreed on. Pause here so your teacher can review your work.

3. What are some characteristics that all **circles** have in common?

4. Put the circular objects in order from smallest to largest.

5. Select one of the pictures of a circular object. What are some ways you could measure the actual size of your circle?

Are you ready for more?

On January 3rd, Earth is 147,500,000 kilometers away from the Sun. On July 4th, Earth is 152,500,000 kilometers away from the Sun. The Sun has a radius of about 865,000 kilometers.

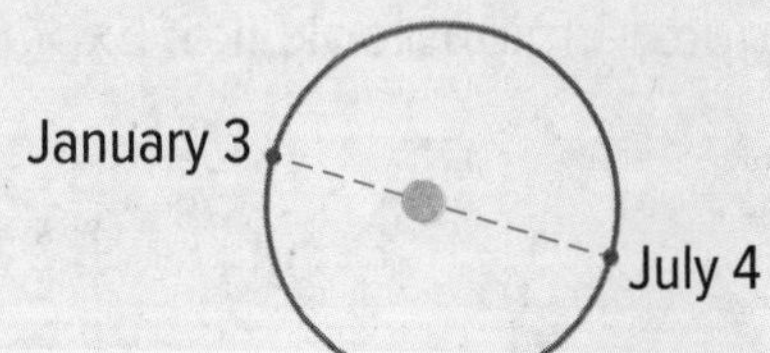

Could Earth's orbit be a circle with some point in the Sun as its center? Explain your reasoning.

NAME ______________________ DATE ____________ PERIOD ________

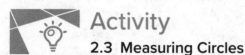

Activity
2.3 Measuring Circles

Priya, Han, and Mai each measured one of the circular objects from earlier.

- Priya says that the bike wheel is 24 inches.
- Han says that the yo-yo trick is 24 inches.
- Mai says that the glow necklace is 24 inches.

1. Do you think that all these circles are the same size?

2. What part of the circle did each person measure? Explain your reasoning.

Activity
2.4 Drawing Circles

Draw and label each circle.

1. Circle A, with a **diameter** of 6 cm.

2. Circle B, with a **radius** of 5 cm. Pause here so your teacher can review your work.

3. Circle C, with a radius that is equal to Circle A's diameter.

NAME ______________________________ DATE ______________ PERIOD __________

4. Circle D, with a diameter that is equal to Circle B's radius.

5. Use a compass to recreate one of these designs.

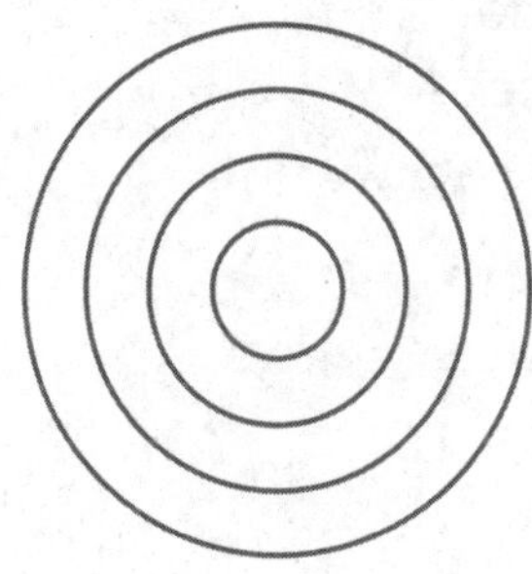

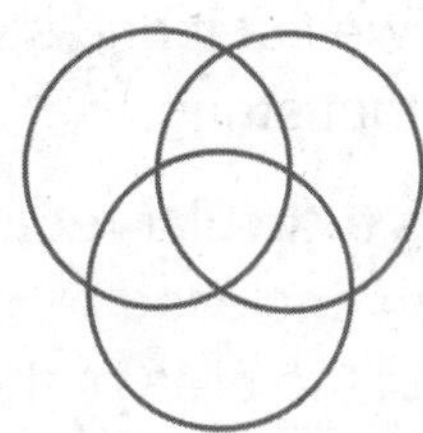

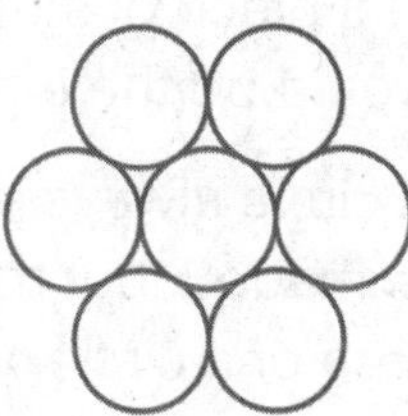

Summary

Exploring Circles

A **circle** consists of all of the points that are the same distance away from a particular point called the *center* of the circle.

A segment that connects the center with any point on the circle is called a **radius**.

For example, segments *QG*, *QH*, *QI*, and *QJ* are all radii of circle 2. (We say one radius and two radii.) The length of any radius is always the same for a given circle. For this reason, people also refer to this distance as the *radius* of the circle.

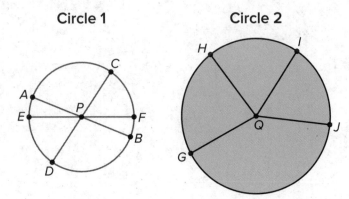

A segment that connects two opposite points on a circle (passing through the circle's center) is called a **diameter**.

For example, segments *AB*, *CD*, and *EF* are all diameters of circle 1.

All diameters in a given circle have the same length because they are composed of two radii. For this reason, people also refer to the length of such a segment as the *diameter* of the circle.

The **circumference** of a circle is the distance around it. If a circle was made of a piece of string and we cut it and straightened it out, the circumference would be the length of that string.

A circle always encloses a circular region. The region enclosed by circle 2 is shaded, but the region enclosed by circle 1 is not. When we refer to the area of a circle, we mean the area of the enclosed circular region.

Glossary

circle
circumference
diameter
radius

NAME _______________ DATE _______________ PERIOD _______________

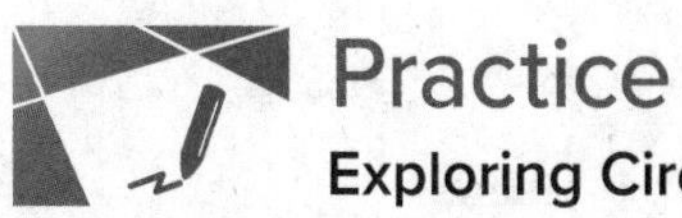

Practice
Exploring Circles

1. Use a geometric tool to draw a circle. Draw and measure a radius and a diameter of the circle.

2. Here is a circle with center *H* and some line segments and curves joining points on the circle.

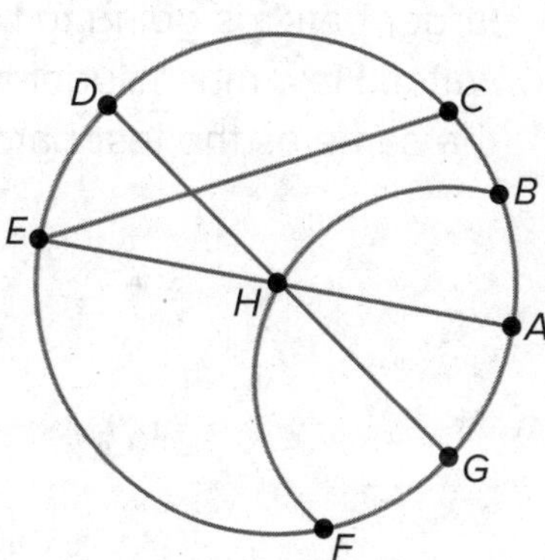

Identify examples of the following. Explain your reasoning.

a. Diameter

b. Radius

3. Lin measured the diameter of a circle in two different directions. Measuring vertically, she got 3.5 cm, and measuring horizontally, she got 3.6 cm. Explain some possible reasons why these measurements differ.

4. A small, test batch of lemonade used $\frac{1}{4}$ cup of sugar added to 1 cup of water and $\frac{1}{4}$ cup of lemon juice. After confirming it tasted good, a larger batch is going to be made with the same ratios using 10 cups of water. How much sugar should be added so that the large batch tastes the same as the test batch? **(Lesson 2-1)**

5. The graph of a proportional relationship contains the point with coordinates (3, 12). What is the constant of proportionality of the relationship? **(Lesson 2-13)**

Lesson 3-3

Exploring Circumference

NAME ________________ DATE ________ PERIOD ________

Learning Goal Let's explore the circumference of circles.

Warm Up

3.1 Which Is Greater?

Clare wonders if the height of the toilet paper tube or the distance around the tube is greater. What information would she need in order to solve the problem? How could she find this out?

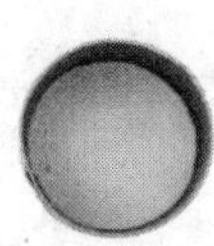

Activity

3.2 Measuring Circumference and Diameter

Your teacher will give you several circular objects.

1. Measure the diameter and the circumference of the circle in each object to the nearest tenth of a centimeter. Record your measurements in the table.

Object	Diameter (cm)	Circumference (cm)

2. Plot the diameter and circumference values from the table on the coordinate plane. What do you notice?

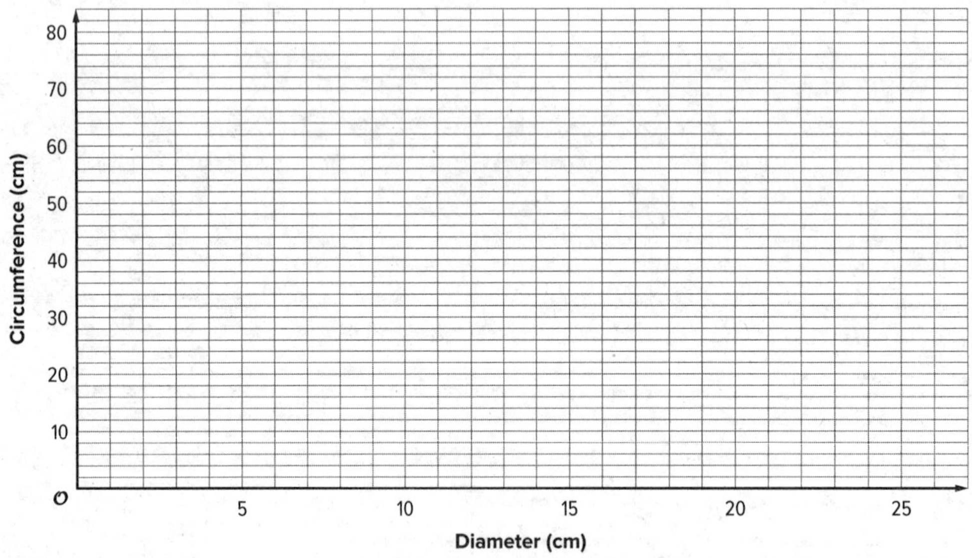

3. Plot the points from two other groups on the same coordinate plane. Do you see the same pattern that you noticed earlier?

NAME ______________________ DATE ____________ PERIOD ____________

Activity

3.3 Calculating Circumference and Diameter

Here are five circles.
One measurement for each circle is given in the table.

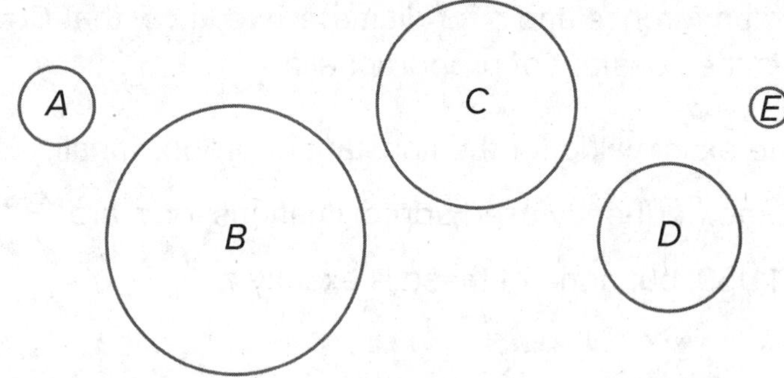

Use the constant of proportionality estimated in the previous activity to complete the table.

	Diameter (cm)	Circumference (cm)
Circle A	3	
Circle B	10	
Circle C		24
Circle D		18
Circle E	1	

Are you ready for more?

The circumference of Earth is approximately 40,000 km. If you made a circle of wire around the globe that is only 10 meters (0.01 km) longer than the circumference of the globe, could a flea, a mouse, or even a person creep under it?

Summary

Exploring Circumference

There is a proportional relationship between the diameter and circumference of any circle. That means that if we write C for circumference and d for diameter, we know that $C = kd$, where k is the constant of proportionality.

The exact value for the constant of proportionality is called π. Some frequently used approximations for π are $\frac{22}{7}$, 3.14, and 3.14159, but none of these is exactly π.

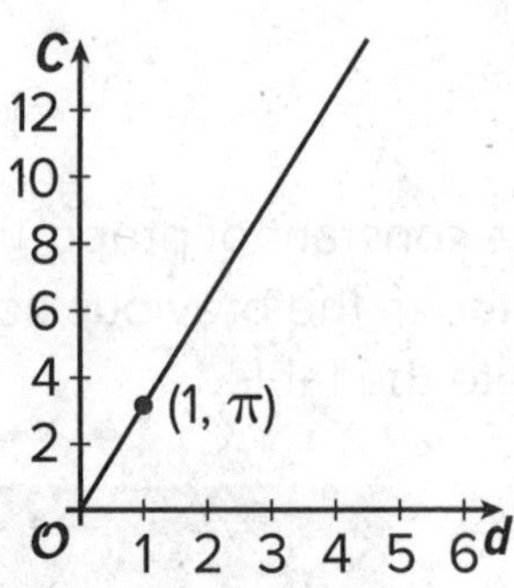

We can use this to estimate the circumference if we know the diameter, and vice versa. For example, using 3.1 as an approximation for π, if a circle has a diameter of 4 cm, then the circumference is about $(3.1) \cdot 4 = 12.4$ or 12.4 cm.

The relationship between the circumference and the diameter can be written as $C = \pi d$.

Glossary

pi (π)

NAME ______________________ DATE ____________ PERIOD ________

Practice

Exploring Circumference

1. Diego measured the diameter and circumference of several circular objects and recorded his measurements in the table.

Object	Diameter (cm)	Circumference (cm)
Half Dollar Coin	3	10
Flying Disc	23	28
Jar Lid	8	25
Flower Pot	15	48

One of his measurements is inaccurate. Which measurement is it? Explain how you know.

2. Complete the table. Use one of the approximate values for π discussed in class (for example 3.14, $\frac{22}{7}$, 3.1416). Explain or show your reasoning.

Object	Diameter	Circumference
Hula Hoop	35 in	
Circular Pond		556 ft
Magnifying Glass	5.2 cm	
Car Tire		71.6 in

3. *A* is the center of the circle, and the length of *CD* is 15 centimeters. **(Lesson 3-2)**

a. Name a segment that is a radius. How long is it?

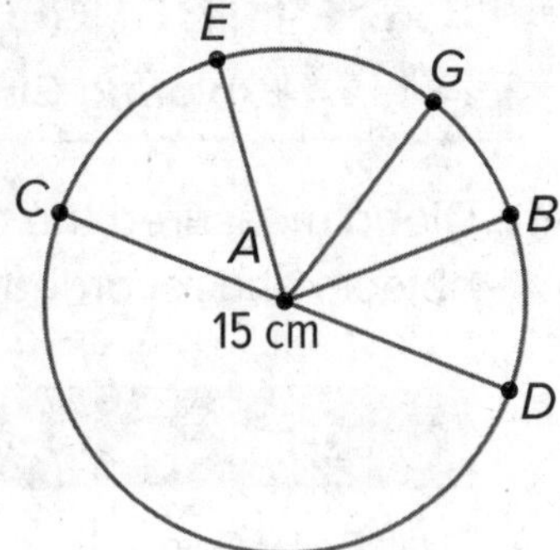

b. Name a segment that is a diameter. How long is it?

4. Respond to each of the following. **(Lesson 2-10)**

a. Consider the equation $y = 1.5x + 2$. Find four pairs of x and y values that make the equation true. Plot the points (x, y) on the coordinate plane.

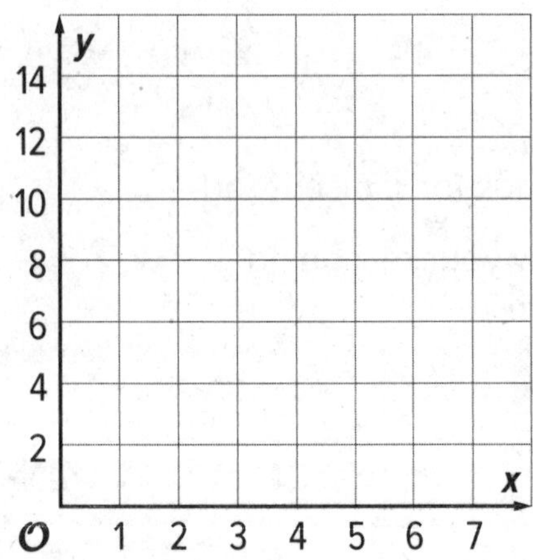

b. Based on the graph, can this be a proportional relationship? Why or why not?

Lesson 3-4

Applying Circumference

NAME ____________________ DATE ____________ PERIOD ____________

Learning Goal Let's use π to solve problems.

Warm Up

4.1 What Do We Know? What Can We Estimate?

Here are some pictures of circular objects, with measurement tools shown.

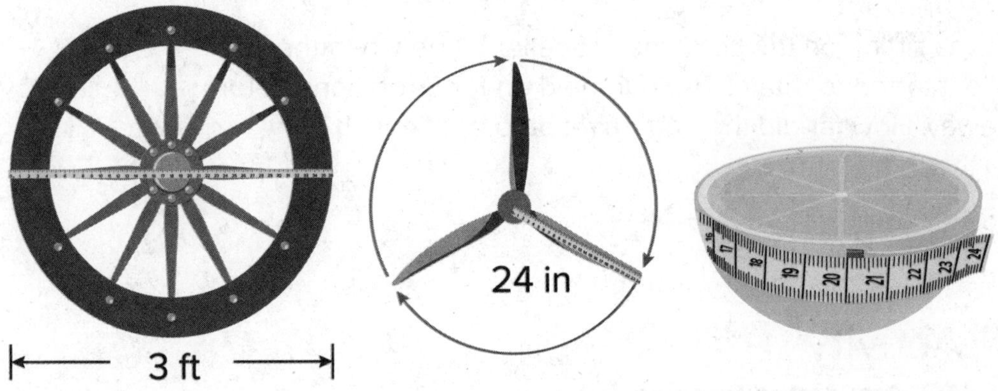

The measurement tool on each picture reads as follows:

- Wagon wheel: 3 feet
- Plane propeller: 24 inches
- Sliced orange: 20 centimeters

1. For each picture, which measurement is shown?

2. Based on this information, what measurement(s) could you estimate for each picture?

Activity
4.2 Using π

In the previous activity, we looked at pictures of circular objects. One measurement for each object is listed in the table.

Your teacher will assign you an approximation of π to use for this activity.

1. Complete the table.

Object	Radius	Diameter	Circumference
Wagon Wheel		3 ft	
Airplane Propeller	24 in		
Orange Slice			20 cm

2. A bug was sitting on the tip of the propeller blade when the propeller started to rotate. The bug held on for 5 rotations before flying away. How far did the bug travel before it flew off?

Activity
4.3 Around the Running Track

The field inside a running track is made up of a rectangle that is 84.39 m long and 73 m wide, together with a half-circle at each end.

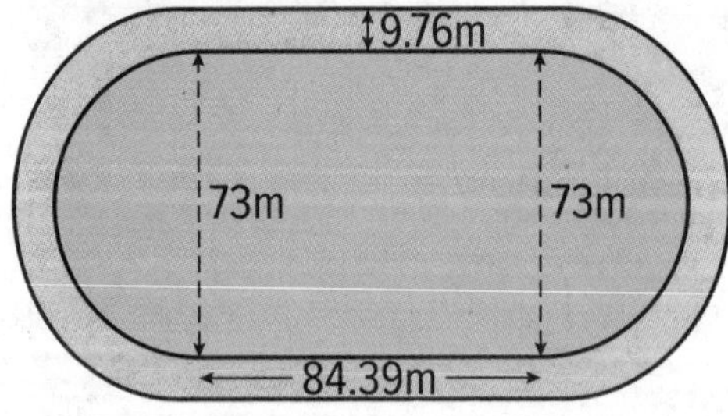

1. What is the distance around the inside of the track? Explain or show your reasoning.

NAME ______________________ DATE ______________ PERIOD ____________

2. The track is 9.76 m wide all the way around. What is the distance around the outside of the track? Explain or show your reasoning.

Are you ready for more?

This size running track is usually called a 400-meter track. However, if a person ran as close to the "inside" as possible on the track, they would run less than 400 meters in one lap. How far away from the inside border would someone have to run to make one lap equal exactly 400 meters?

Activity
4.4 Measuring a Picture Frame

Kiran bent some wire around a rectangle to make a picture frame. The rectangle is 8 inches by 10 inches.

1. Find the perimeter of the wire picture frame. Explain or show your reasoning.

2. If the wire picture frame were stretched out to make one complete circle, what would its radius be?

Summary
Applying Circumference

The circumference of a circle, C, is π times the diameter, d.

The diameter is twice the radius, r.

So if we know any one of these measurements for a particular circle, we can find the others. We can write the relationships between these different measures using equations:

$d = 2r$ $\quad\quad$ $C = \pi d$ $\quad\quad$ $C = 2\pi r$

- If the diameter of a car tire is 60 cm, that means the radius is 30 cm and the circumference is $60 \cdot \pi$ or about 188 cm.
- If the radius of a clock is 5 in, that means the diameter is 10 in, and the circumference is $10 \cdot \pi$ or about 31 in.
- If a ring has a circumference of 44 mm, that means the diameter is $44 \div \pi$, which is about 14 mm, and the radius is about 7 mm.

NAME ____________ DATE ________ PERIOD ________

Practice

Applying Circumference

1. Here is a picture of a Ferris wheel. It has a diameter of 80 meters.

Vadim Petrakov/Shutterstock

   a. On the picture, draw and label a diameter.

   b. How far does a rider travel in one complete rotation around the Ferris wheel?

2. Identify each measurement as the diameter, radius, or circumference of the circular object. Then, estimate the other two measurements for the circle.

   a. The length of the minute hand on a clock is 5 in.

   b. The distance across a sink drain is 3.8 cm.

   c. The tires on a mining truck are 14 ft tall.

d. The fence around a circular pool is 75 ft long.

e. The distance from the tip of a slice of pizza to the crust is 7 in.

f. Breaking a cookie in half creates a straight side 10 cm long.

g. The length of the metal rim around a glass lens is 190 mm.

h. From the center to the edge of a DVD measures 60 mm.

3. A half circle is joined to an equilateral triangle with side lengths of 12 units. What is the perimeter of the resulting shape?

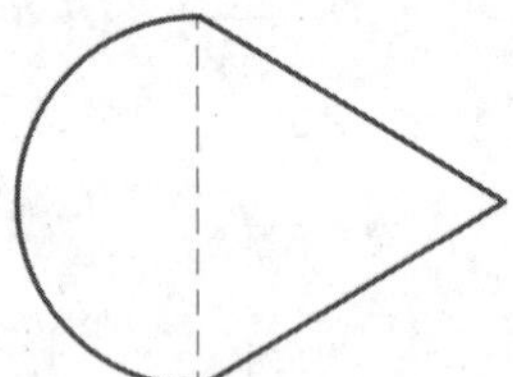

4. Circle A has a diameter of 1 foot. Circle B has a circumference of 1 meter. Which circle is bigger? Explain your reasoning. (1 inch ≈ 2.54 centimeters)

5. The circumference of Tyler's bike tire is 72 inches. What is the diameter of the tire? **(Lesson 3-3)**

Lesson 3-5

Circumference and Wheels

NAME ______________________ DATE ____________ PERIOD ________

Learning Goal Let's explore how far different wheels roll.

Warm Up

5.1 A Rope and a Wheel

Han says that you can wrap a 5-foot rope around a wheel with a 2-foot diameter because $\frac{5}{2}$ is less than pi. Do you agree with Han? Explain your reasoning.

Activity

5.2 Rolling, Rolling, Rolling

Your teacher will give you a circular object.

1. Follow these instructions to create the drawing:
   - On a separate piece of paper, use a ruler to draw a line all the way across the page.
   - Roll your object along the line and mark where it completes one rotation.
   - Use your object to draw tick marks along the line that are spaced as far apart as the diameter of your object.
2. What do you notice?

3. Use your ruler to measure each distance. Record these values in the first row of the table.

   a. the diameter of your object

   b. how far your object rolled in one complete rotation

   c. the quotient of how far your object rolled divided by the diameter of your object

Object	Diameter	Distance Traveled in One Rotation	Distance ÷ Diameter

4. If you wanted to trace two complete rotations of your object, how long of a line would you need?

5. Share your results with your group and record their measurements in the table.

6. If each person in your group rolled their object along the entire length of the classroom, which object would complete the most rotations? Explain or show your reasoning.

NAME ____________________ DATE ____________ PERIOD ____________

Activity

5.3 Rotations and Distance

1. A car wheel has a diameter of 20.8 inches.

   a. About how far does the car wheel travel in 1 rotation? 5 rotations? 30 rotations?

   b. Write an equation relating the distance the car travels in inches, c, to the number of wheel rotations, x.

   c. About how many rotations does the car wheel make when the car travels 1 mile? Explain or show your reasoning.

2. A bike wheel has a radius of 13 inches.

   a. About how far does the bike wheel travel in 1 rotation? 5 rotations? 30 rotations?

   b. Write an equation relating the distance the bike travels in inches, b, to the number of wheel rotations, x.

c. About how many rotations does the bike wheel make when the bike travels 1 mile? Explain or show your reasoning.

Are you ready for more?

Here are some photos of a spring toy.

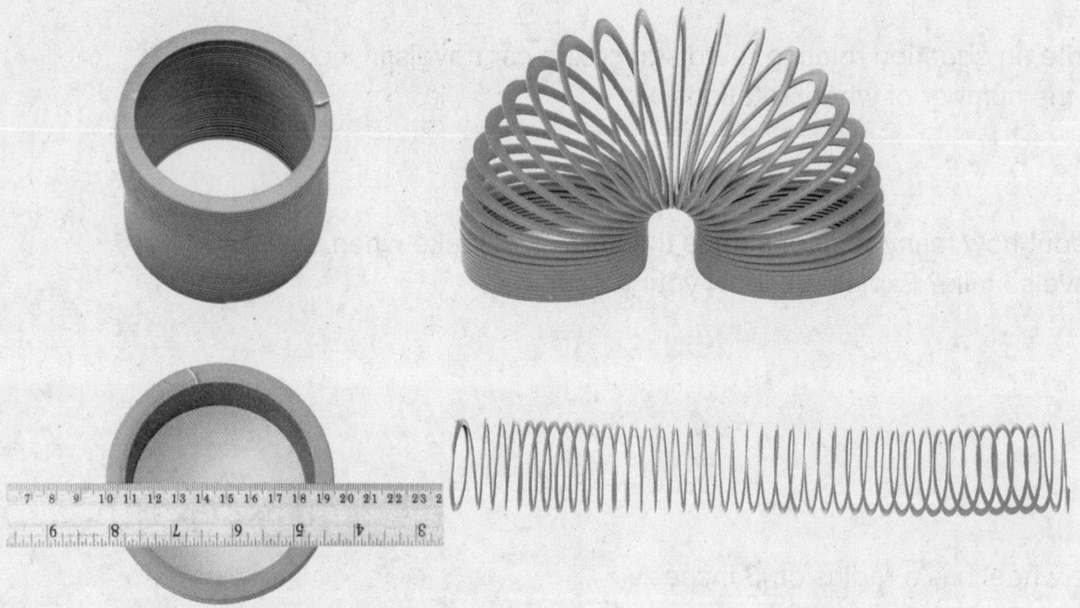

If you could stretch out the spring completely straight, how long would it be? Explain or show your reasoning.

NAME ______________________ DATE ____________ PERIOD ________

Activity

5.4 Rotations and Speed

The circumference of a car wheel is about 65 inches.

1. If the car wheel rotates once per second, how far does the car travel in one minute?

2. If the car wheel rotates once per second, about how many miles does the car travel in one hour?

3. If the car wheel rotates 5 times per second, about how many miles does the car travel in one hour?

4. If the car is traveling 65 miles per hour, about how many times per second does the wheel rotate?

Summary

Circumference and Wheels

The circumference of a circle is the distance around the circle. This is also how far the circle rolls on flat ground in one rotation.

For example, a bicycle wheel with a diameter of 24 inches has a circumference of 24π inches and will roll 24π inches (or 2π feet) in one complete rotation.

There is an equation relating the number of rotations of the wheel to the distance it has traveled.

To see why, let's look at a table showing how far the bike travels when the wheel makes different numbers of rotations.

Number of Rotations	Distance Traveled (feet)
1	2π
2	4π
3	6π
10	20π
50	100π
x	?

In the table, we see that the relationship between the distance traveled and the number of wheel rotations is a proportional relationship. The constant of proportionality is 2π.

To find the missing value in the last row of the table, note that each rotation of the wheel contributes 2π feet of distance traveled. So after x rotations the bike will travel $2\pi x$ feet.

If d is the distance, in feet, traveled when this wheel makes x rotations, we have the relationship: $d = 2\pi x$.

NAME ______________________ DATE ____________ PERIOD ________

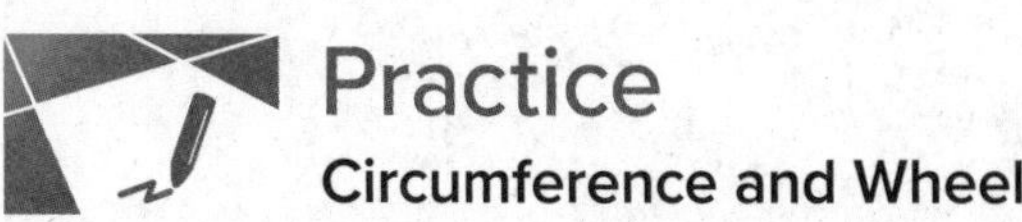

Practice
Circumference and Wheels

1. The diameter of a bike wheel is 27 inches. If the wheel makes 15 complete rotations, how far does the bike travel?

2. The wheels on Kiran's bike are 64 inches in circumference. How many times do the wheels rotate if Kiran rides 300 yards?

3. The numbers are measurements of radius, diameter, and circumference of Circles A and B. Circle A is smaller than Circle B. Which number belongs to which quantity? 2.5, 5, 7.6, 15.2, 15.7, 47.7 **(Lesson 3-4)**

4. Circle A has a circumference of $2\frac{2}{3}$ m. Circle B has a diameter that is $1\frac{1}{2}$ times as long as Circle A's diameter. What is the circumference of Circle B? **(Lesson 3-3)**

5. The length of segment *AE* is 5 centimeters. **(Lesson 3-2)**

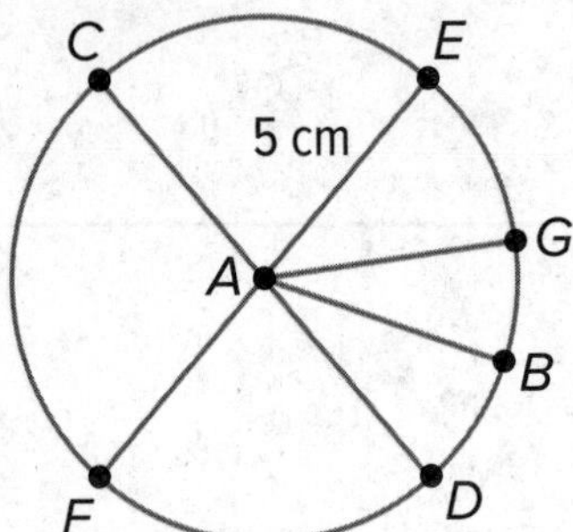

a. What is the length of segment *CD*?

b. What is the length of segment *AB*?

c. Name a segment that has the same length as segment *AB*.

Lesson 3-6

Estimating Areas

NAME ____________________ DATE ____________ PERIOD ________

Learning Goal Let's estimate the areas of weird shapes.

Warm Up

6.1 Mental Calculations

Find a strategy to make each calculation mentally.

1. $599 + 87$
2. $254 - 88$
3. $99 \cdot 75$

Activity

6.2 House Floorplan

Here is a floor plan of a house. Approximate lengths of the walls are given.

What is the approximate area of the home, including the balcony? Explain or show your reasoning.

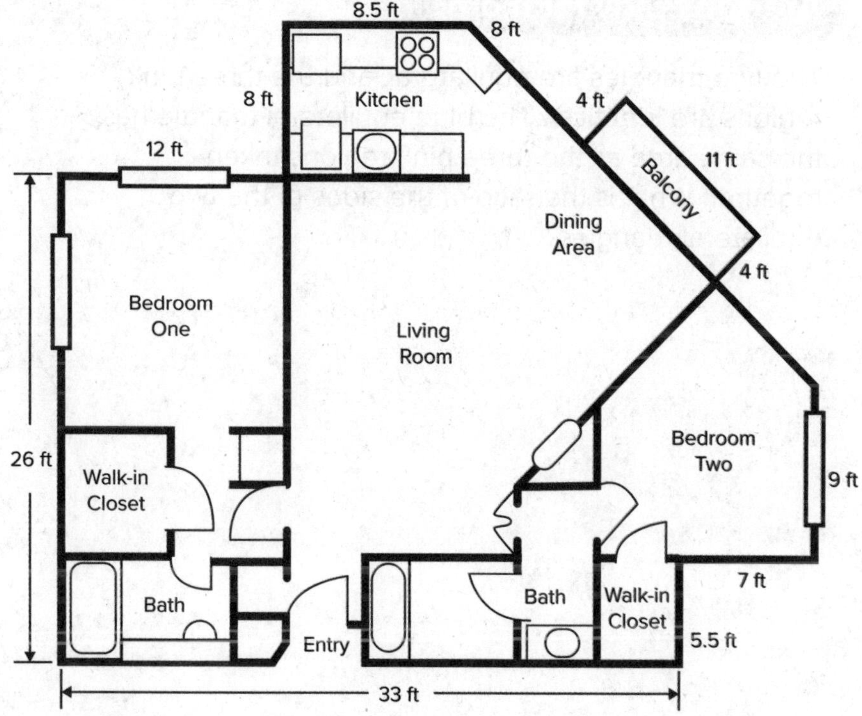

Activity

6.3 Area of Nevada

Estimate the area of Nevada in square miles. Explain or show your reasoning.

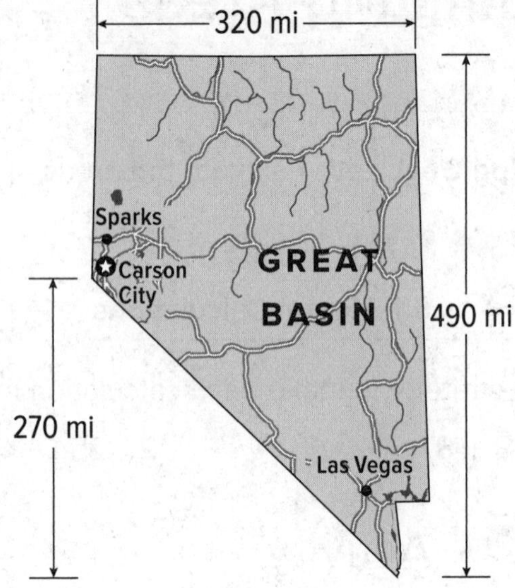

Are you ready for more?

The two triangles are equilateral, and the three pink regions are identical. The blue equilateral triangle has the same area as the three pink regions taken together. What is the ratio of the sides of the two equilateral triangles?

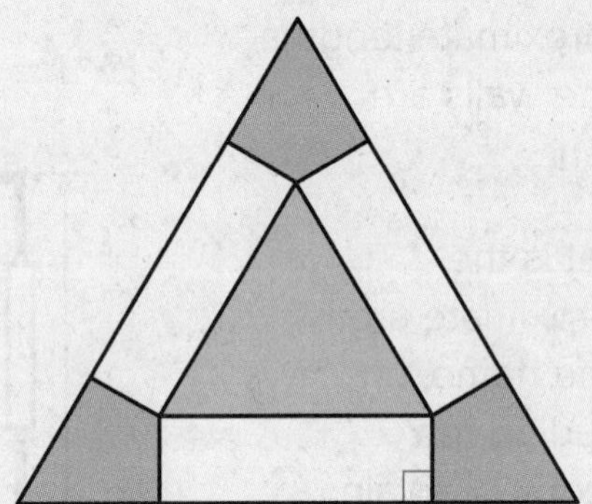

NAME ____________________ DATE ______________ PERIOD ______________

Summary

Estimating Areas

We can find the area of some complex polygons by surrounding them with a simple polygon like a rectangle. For example, this octagon is contained in a rectangle.

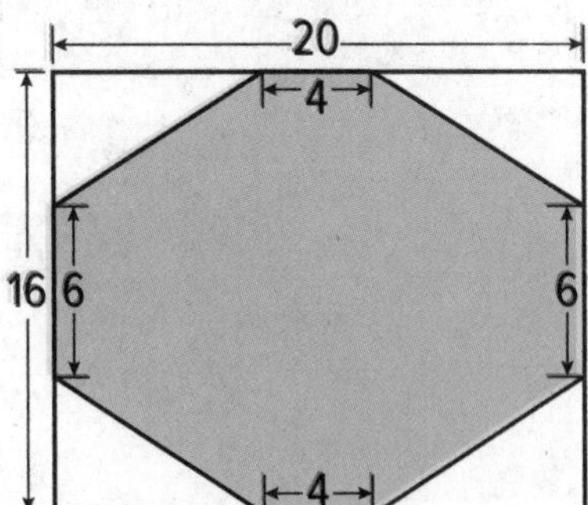

The rectangle is 20 units long and 16 units wide, so its area is 320 square units.

- To get the area of the octagon, we need to subtract the areas of the four right triangles in the corners.
- These triangles are each 8 units long and 5 units wide, so they each have an area of 20 square units.

The area of the octagon is 320 – (4 • 20) or 240 square units.

We can estimate the area of irregular shapes by approximating them with a polygon and finding the area of the polygon.

For example, here is an illustration of Lake Tahoe with some one-dimensional measurements around the lake.

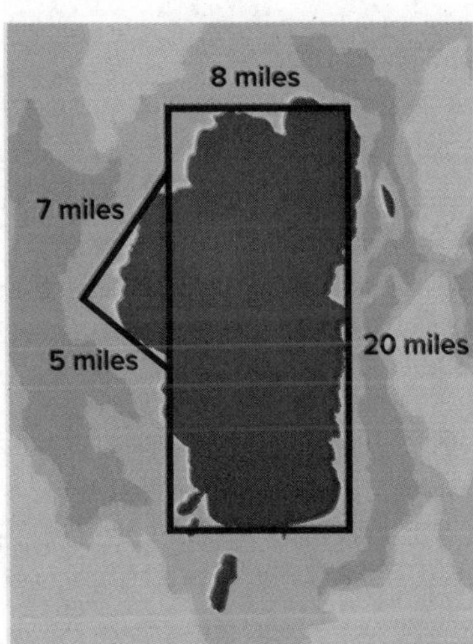

The area of the rectangle is 160 square miles, and the area of the triangle is 17.5 square miles for a total of 177.5 square miles. We recognize that this is an approximation, and not likely the exact area of the lake.

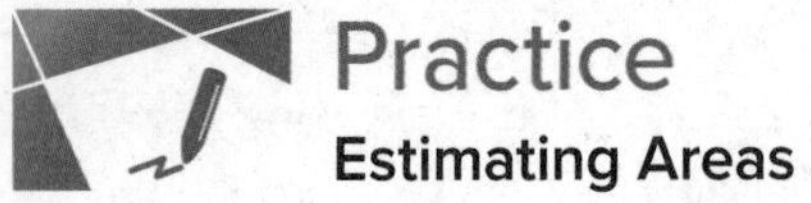

Practice

Estimating Areas

1. Find the area of the polygon.

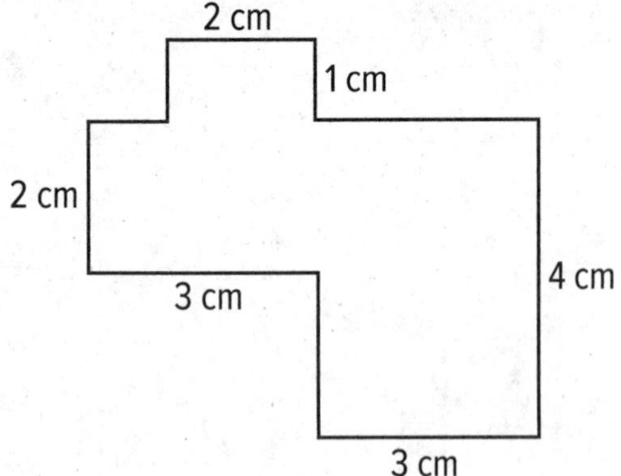

2. a. Draw polygons on the map that could be used to approximate the area of Virginia.

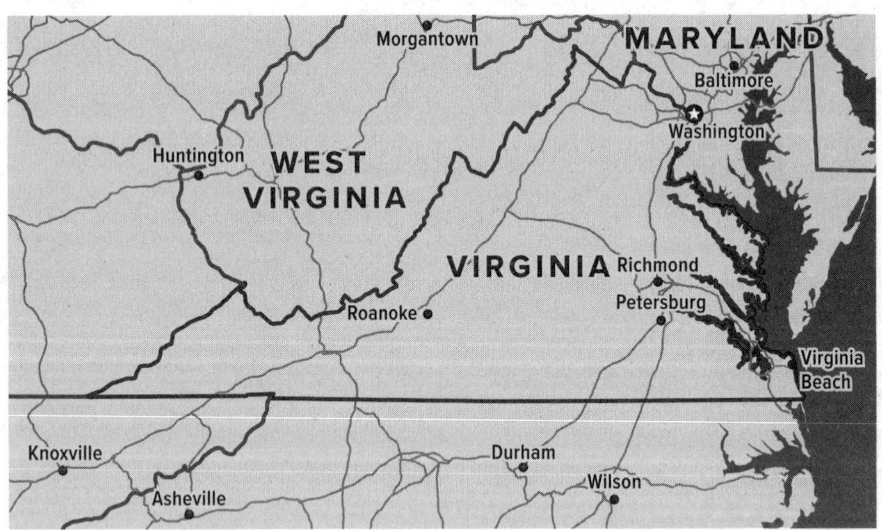

NAME ____________________ DATE ____________ PERIOD ________

b. Which measurements would you need to know in order to calculate an approximation of the area of Virginia? Label the sides of the polygons whose measurements you would need. (Note: You aren't being asked to calculate anything.)

3. Jada's bike wheels have a diameter of 20 inches. How far does she travel if the wheels rotate 37 times? (Lesson 3-5)

4. The radius of Earth is approximately 6,400 km. The equator is the circle around Earth dividing it into the northern and southern hemispheres. (The center of the earth is also the center of the equator.) What is the length of the equator? (Lesson 3-4)

5. Here are several recipes for sparkling lemonade. For each recipe describe how many tablespoons of lemonade mix it takes per cup of sparkling water. **(Lesson 2-1)**

Recipe 1: 4 tablespoons of lemonade mix and 12 cups of sparkling water

Recipe 2: 4 tablespoons of lemonade mix and 6 cups of sparkling water

Recipe 3: 3 tablespoons of lemonade mix and 5 cups of sparkling water

Recipe 4: $\frac{1}{2}$ tablespoon of lemonade mix and $\frac{3}{4}$ cups of sparkling water

Lesson 3-7

Exploring the Area of a Circle

NAME ________________ DATE ________ PERIOD ________

Learning Goal Let's investigate the areas of circles.

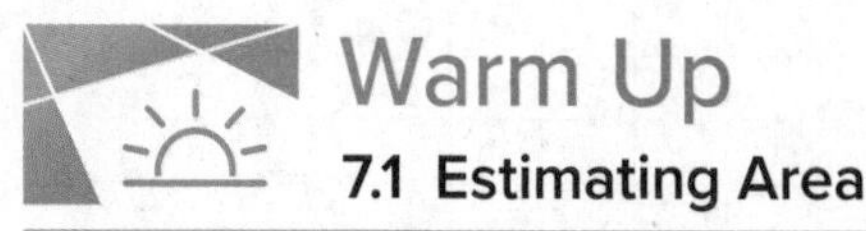

Warm Up

7.1 Estimating Areas

Your teacher will show you some figures. Decide which figure has the largest area. Be prepared to explain your reasoning.

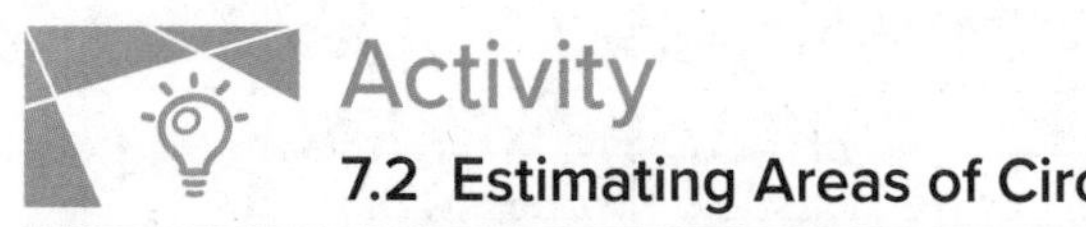

Activity

7.2 Estimating Areas of Circles

Your teacher will give your group two circles of different sizes.

1. For each circle, use the squares on the graph paper to measure the diameter and estimate the **area of the circle**. Record your measurements in the table.

Diameter (cm)	Estimated Area (cm^2)

2. Plot the values from the table on the class coordinate plane. Then plot the class's data points on your coordinate plane.

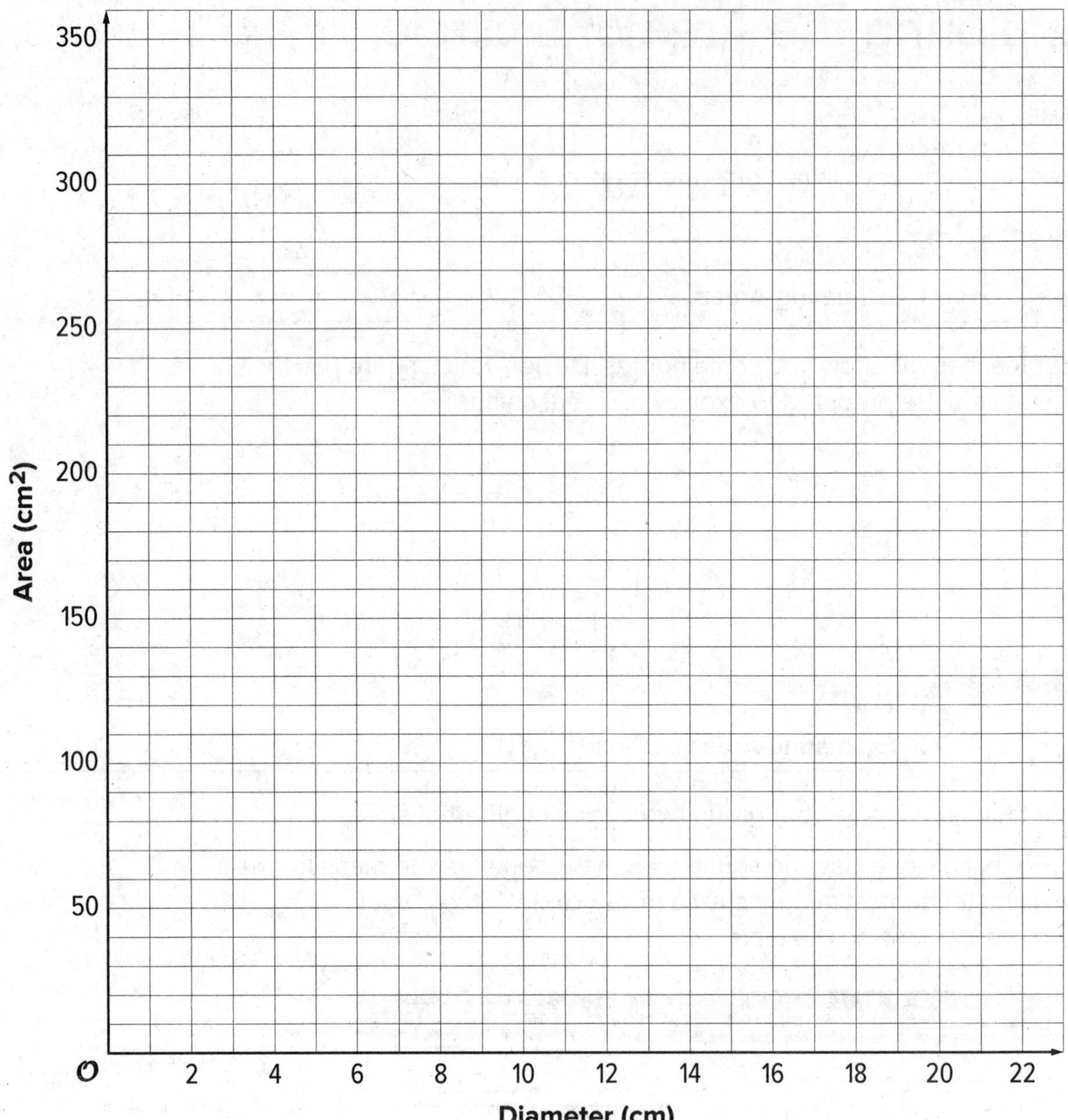

3. In a previous lesson, you graphed the relationship between the diameter and circumference of a circle. How is this graph the same? How is it different?

NAME ______________________ DATE ______________ PERIOD ____________

Are you ready for more?

How many circles of radius 1 unit can you fit inside each of the following so that they do not overlap?

1. a circle of radius 2 units?

2. a circle of radius 3 units?

3. a circle of radius 4 units?

If you get stuck, consider using coins or other circular objects.

Activity

7.3 Covering a Circle

Here is a square whose side length is the same as the radius of the circle.

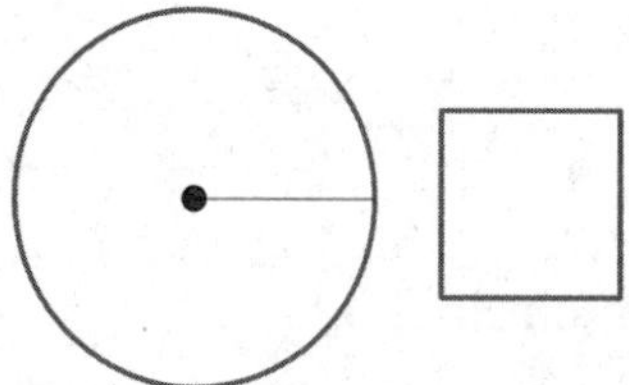

How many of these squares do you think it would take to cover the circle exactly?

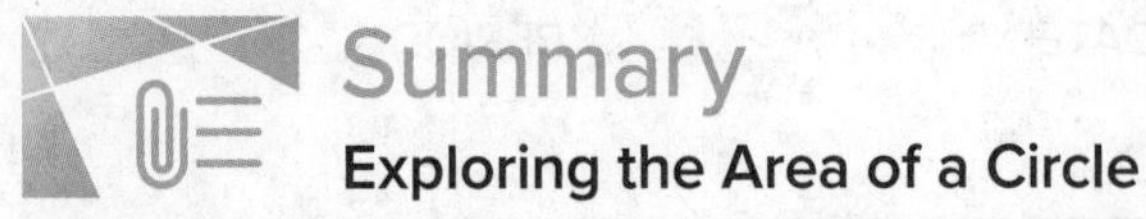

Summary

Exploring the Area of a Circle

The circumference C of a circle is proportional to the diameter d, and we can write this relationship as $C = \pi d$.

The circumference is also proportional to the radius of the circle, and the constant of proportionality is $2 \cdot \pi$ because the diameter is twice as long as the radius.

However, the **area of a circle** is *not* proportional to the diameter (or the radius).

The area of a circle with radius r is a little more than 3 times the area of a square with side r so the area of a circle of radius r is approximately $3r^2$.

We saw earlier that the circumference of a circle of radius r is $2\pi r$. If we write C for the circumference of a circle, this proportional relationship can be written $C = 2\pi r$.

The area A of a circle with radius r is approximately $3r^2$. Unlike the circumference, the area is not proportional to the radius because $3r^2$ cannot be written in the form kr for a number k.

We will investigate and refine the relationship between the area and the radius of a circle in future lessons.

Glossary

area of a circle

NAME ______________________________ DATE ______________ PERIOD __________

Practice

Exploring the Area of a Circle

1. The *x*-axis of each graph has the diameter of a circle in meters. Label the *y*-axis on each graph with the appropriate measurement of a circle: radius (m), circumference (m), or area (m²).

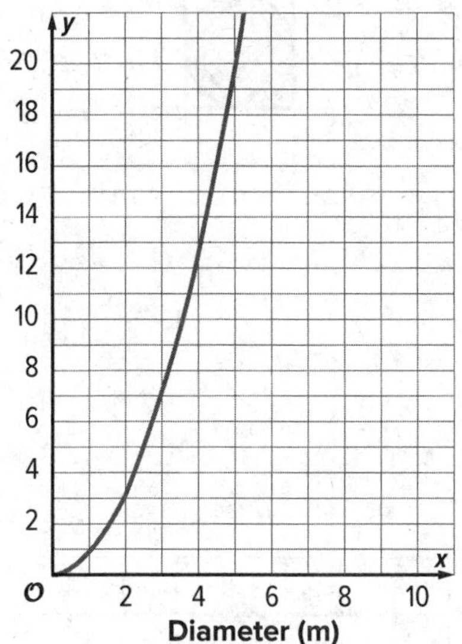

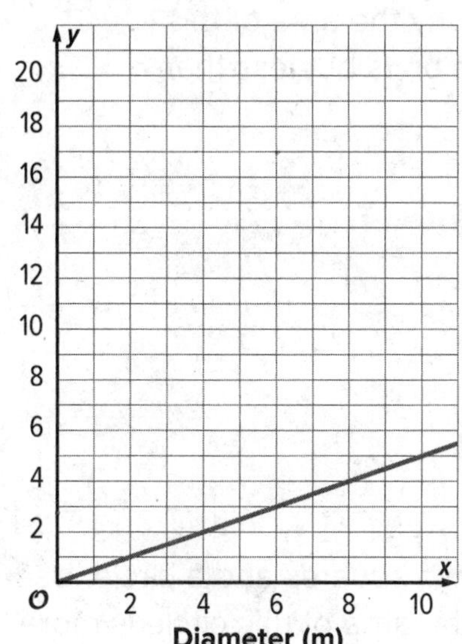

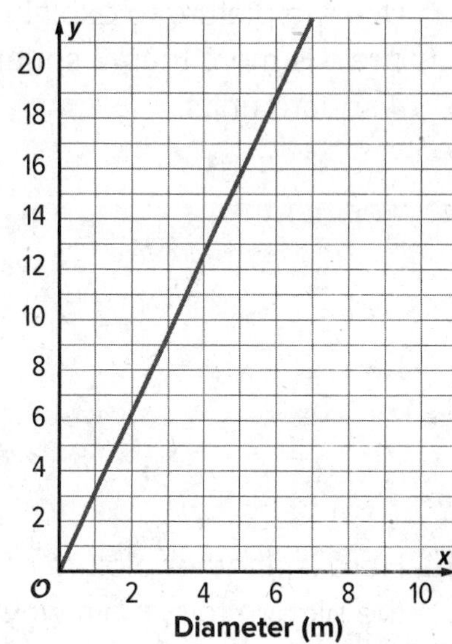

2. Circle A has area 500 in². The diameter of circle B is three times the diameter of circle A. Estimate the area of circle B.

3. Lin's bike travels 100 meters when her wheels rotate 55 times. What is the circumference of her wheels? **(Lesson 3-5)**

4. Priya drew a circle whose circumference is 25 cm. Clare drew a circle whose diameter is 3 times the diameter of Priya's circle. What is the circumference of Clare's circle? **(Lesson 3-3)**

5. Respond to each of the following.

   a. Here is a picture of two squares and a circle. Use the picture to explain why the area of this circle is more than 2 square units but less than 4 square units.

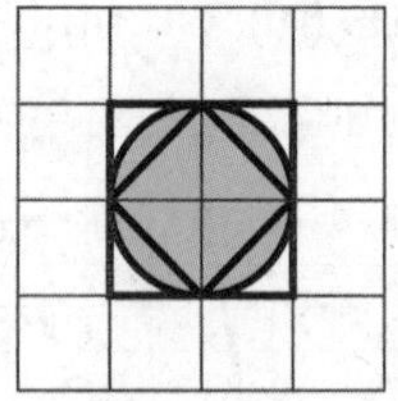

   b. Here is another picture of two squares and a circle. Use the picture to explain why the area of this circle is more than 18 square units and less than 36 square units.

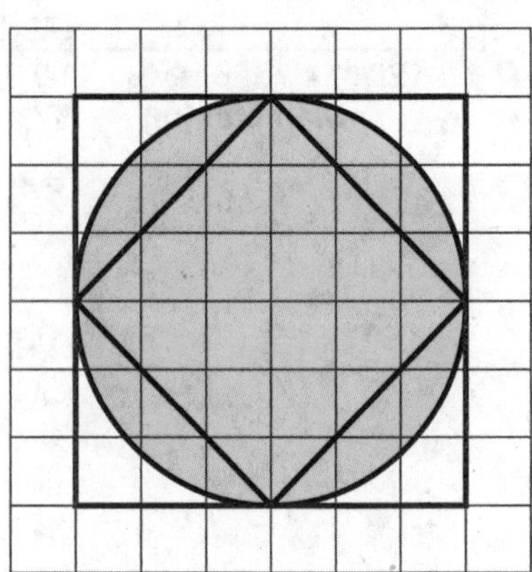

6. Find the circumference of this circle. **(Lesson 3-3)**

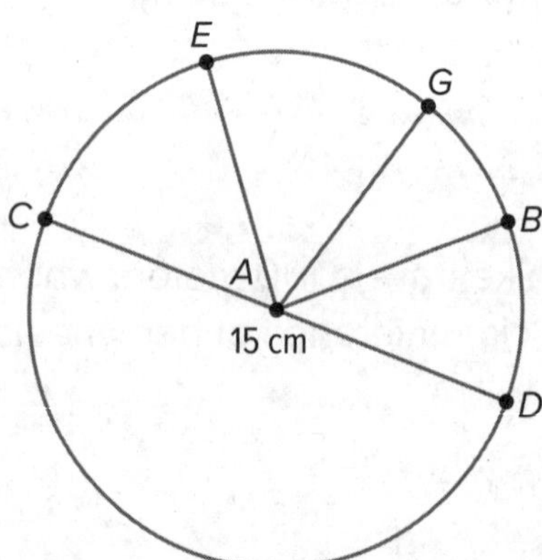

Lesson 3-8

Relating Area to Circumference

NAME ______________________ DATE ____________ PERIOD ____________

Learning Goal Let's rearrange circles to calculate their areas.

Warm Up

8.1 Irrigating a Field

A circular field is set into a square with an 800 m side length. Estimate the field's area.

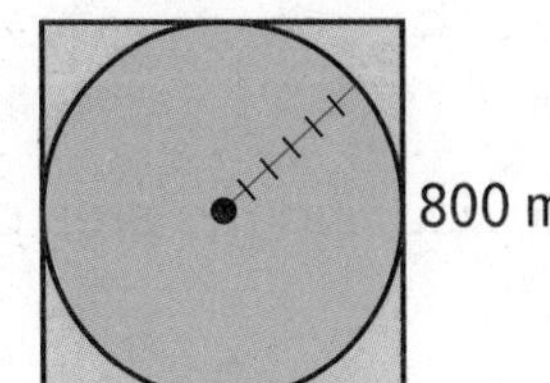

A. about 5,000 m^2

B. about 50,000 m^2

C. about 500,000 m^2

D. about 5,000,000 m^2

E. about 50,000,000 m^2

Activity

8.2 Making a Polygon out of a Circle

Your teacher will give you a circular object, a marker, and two pieces of paper of different colors.

Follow these instructions to create a visual display.

1. Using a thick marker, trace your circle in two separate places on the same piece of paper.
2. Cut out both circles, cutting around the marker line.
3. Fold and cut one of the circles into fourths.
4. Arrange the fourths so that straight sides are next to each other, but the curved edges are alternately on top and on bottom. Pause here so your teacher can review your work.
5. Fold and cut the fourths in half to make eighths. Arrange the eighths next to each other, like you did with the fourths.

6. If your pieces are still large enough, repeat the previous step to make sixteenths.
7. Glue the remaining circle and the new shape onto a piece of paper that is a different color.

After you finish gluing your shapes, answer the following questions.

1. How do the areas of the two shapes compare?
2. What polygon does the shape made of the circle pieces most resemble?
3. How could you find the area of this polygon?

Activity
8.3 Making Another Polygon out of a Circle

Imagine a circle made of rings that can bend, but not stretch.

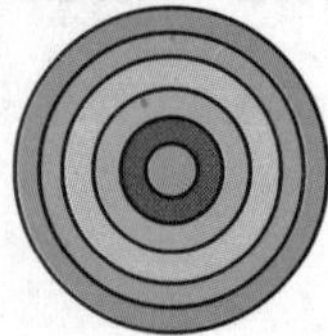

A circle is made of rings.

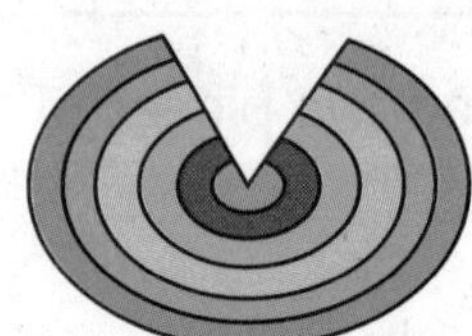

The rings are unrolled.

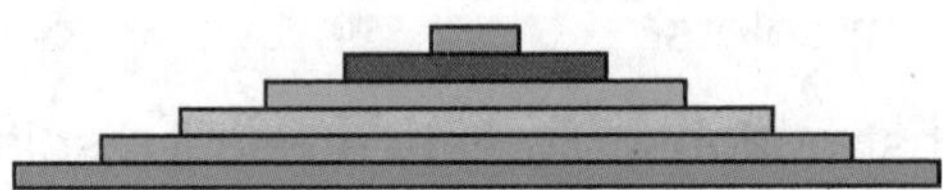

The circle has been made into a new shape.

1. What polygon does the new shape resemble?
2. How does the area of the polygon compare to the area of the circle?

NAME ______________________ DATE ____________ PERIOD ____________

3. How can you find the area of the polygon?

4. Show, in detailed steps, how you could find the polygon's area in terms of the circle's measurements. Show your thinking. Organize it so it can be followed by others.

5. After you finish, trade papers with a partner and check each other's work. If you disagree, work to reach an agreement. Discuss:
   - Do you agree or disagree with each step?
   - Is there a way to make the explanation clearer?

6. Return your partner's work, and revise your explanation based on the feedback you received.

Activity
8.4 Tiling a Table

Elena wants to tile the top of a circular table. The diameter of the table top is 28 inches. What is its area?

Are you ready for more?

A box contains 20 square tiles that are 2 inches on each side. How many boxes of tiles will Elena need to tile the table?

Summary

Relating Area to Circumference

If C is a circle's circumference and r is its radius, then $C = 2\pi r$.

The area of a circle can be found by taking the product of half the circumference and the radius.

If A is the area of the circle, this gives the equation $A = \frac{1}{2}(2\pi r) \cdot r$.

This equation can be rewritten as $A = \pi r^2$.

Remember that when we have $r \cdot r$ we can write r^2 and we can say "r squared".

This means that if we know the radius, we can find the area. For example, if a circle has radius 10 cm, then the area is about $(3.14) \cdot 100$ which is 314 cm^2.

If we know the diameter, we can figure out the radius, and then we can find the area. For example, if a circle has a diameter of 30 ft, then the radius is 15 ft, and the area is about $(3.14) \cdot 225$ which is approximately 707 ft^2.

Glossary

squared

NAME ______________________ DATE ____________ PERIOD ________

Practice

Relating Area to Circumference

1. The picture shows a circle divided into 8 equal wedges which are rearranged.

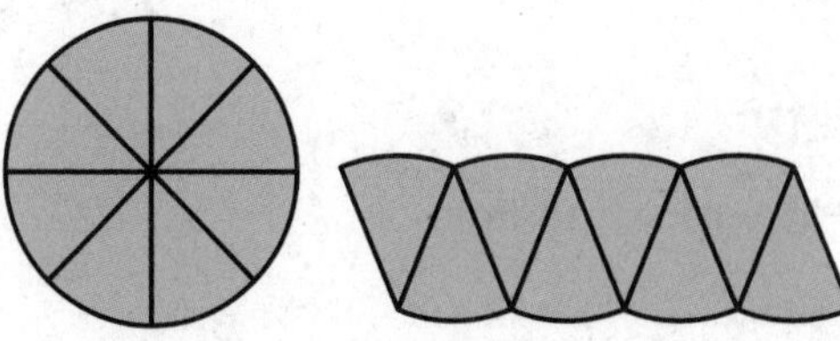

The radius of the circle is r and its circumference is $2\pi r$. How does the picture help to explain why the area of the circle is πr^2?

2. A circle's circumference is approximately 76 cm. Estimate the radius, diameter, and area of the circle.

3. Jada paints a circular table that has a diameter of 37 inches. What is the area of the table?

4. The Carousel on the National Mall has 4 rings of horses. Kiran is riding on the inner ring, which has a radius of 9 feet. Mai is riding on the outer ring, which is 8 feet farther out from the center than the inner ring is. **(Lesson 3-4)**

   a. In one rotation of the carousel, how much farther does Mai travel than Kiran?

   b. One rotation of the carousel takes 12 seconds. How much faster does Mai travel than Kiran?

5. Here are the diameters of four coins:

Coin	Penny	Nickel	Dime	Quarter
Diameter	1.9 cm	2.1 cm	1.8 cm	2.4 cm

   a. A coin rolls a distance of 33 cm in 5 rotations. Which coin is it?

   b. A quarter makes 8 rotations. How far did it roll?

   c. A dime rolls 41.8 cm. How many rotations did it make? **(Lesson 3-5)**

Lesson 3-9

Applying Area of Circles

NAME ______________________ DATE ______________ PERIOD ____________

Learning Goal Let's find the areas of shapes made up of circles.

Warm Up

9.1 Still Irrigating the Field

The area of this field is about 500,000 m^2. What is the field's area to the nearest square meter? Assume that the side lengths of the square are exactly 800 m.

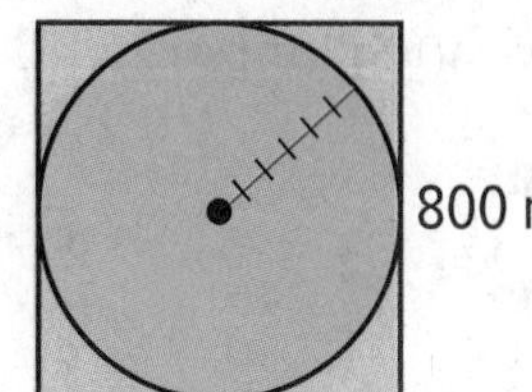

(A.) 502,400 m^2

(B.) 502,640 m^2

(C.) 502,655 m^2

(D.) 502,656 m^2

(E.) 502,857 m^2

Activity

9.2 Comparing Areas Made of Circles

1. Each square has a side length of 12 units. Compare the areas of the shaded regions in the 3 figures. Which figure has the largest shaded region? Explain or show your reasoning.

Figure A

Figure B

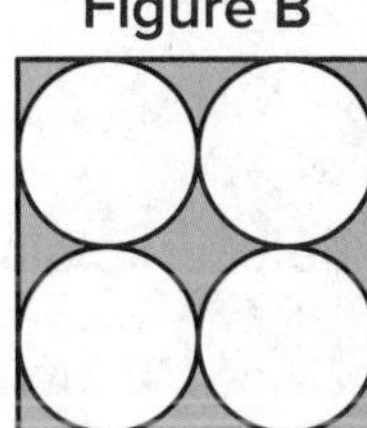

Figure C

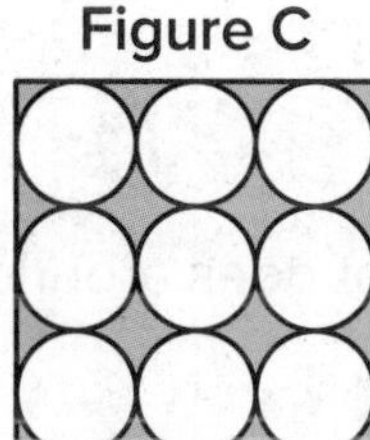

2. Each square in Figures D and E has a side length of 1 unit. Compare the area of the two figures. Which figure has more area? How much more? Explain or show your reasoning.

Figure D

Figure E

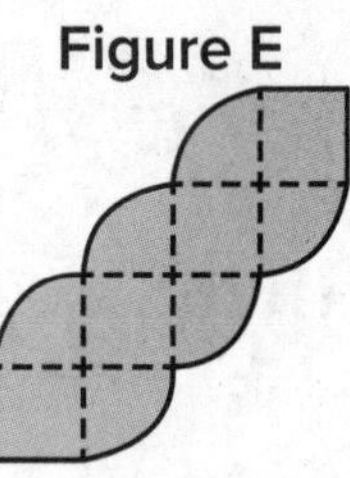

Are you ready for more?

Which figure has a longer perimeter, Figure D or Figure E? How much longer?

Activity

9.3 The Running Track Revisited

The field inside a running track is made up of a rectangle 84.39 m long and 73 m wide, together with a half-circle at each end. The running lanes are 9.76 m wide all the way around.

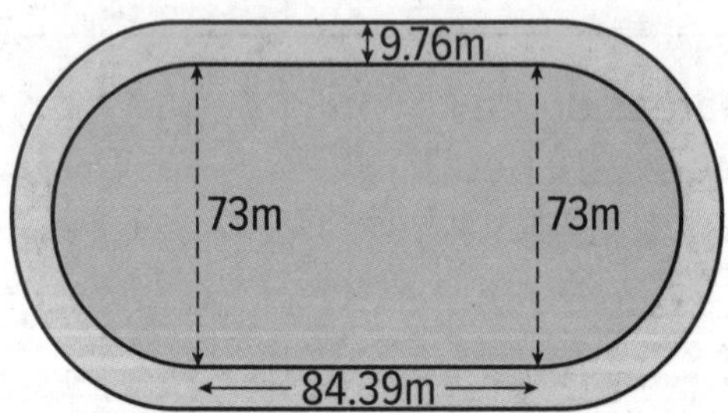

What is the area of the running track that goes around the field? Explain or show your reasoning.

NAME ______________________ DATE ______________ PERIOD ____________

Summary
Applying Area of Circles

The relationship between A, the area of a circle, and r, its radius, is $A = \pi r^2$. We can use this to find the area of a circle if we know the radius.

For example, if a circle has a radius of 10 cm, then the area is $\pi \cdot 10^2$ or 100π cm^2.

We can also use the formula to find the radius of a circle if we know the area. For example, if a circle has an area of 49π m^2 then its radius is 7 m and its diameter is 14 m.

Sometimes instead of leaving π in expressions for the area, a numerical approximation can be helpful.

For the examples above, a circle of radius 10 cm has area about 314 cm^2. In a similar way, a circle with an area of 154 m^2 has a radius of about 7 m.

We can also figure out the area of a fraction of a circle.

For example, the figure shows a circle divided into 3 pieces of equal area. The shaded part has an area of $\frac{1}{3}\pi r^2$.

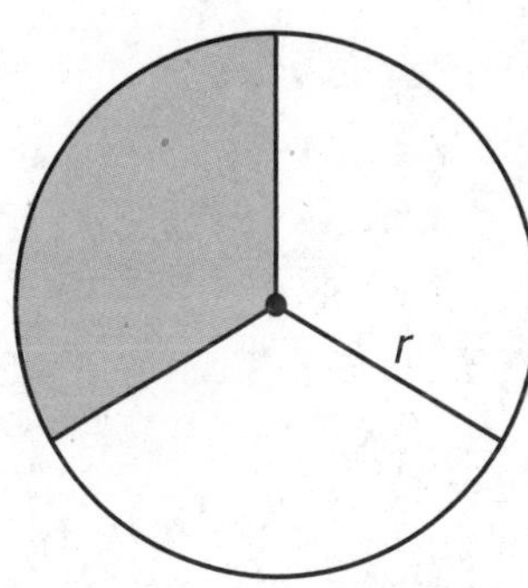

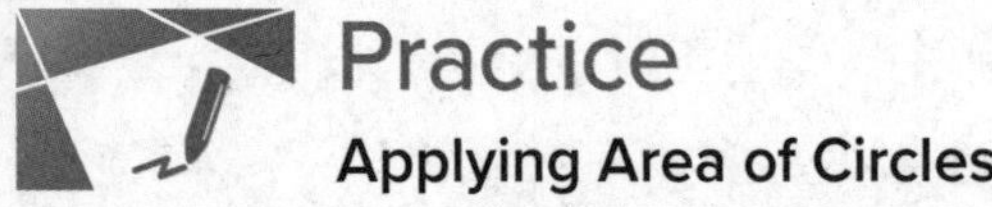

Practice

Applying Area of Circles

1. A circle with a 12-inch diameter is folded in half and then folded in half again. What is the area of the resulting shape?

2. Find the area of the shaded region. Express your answer in terms of π.

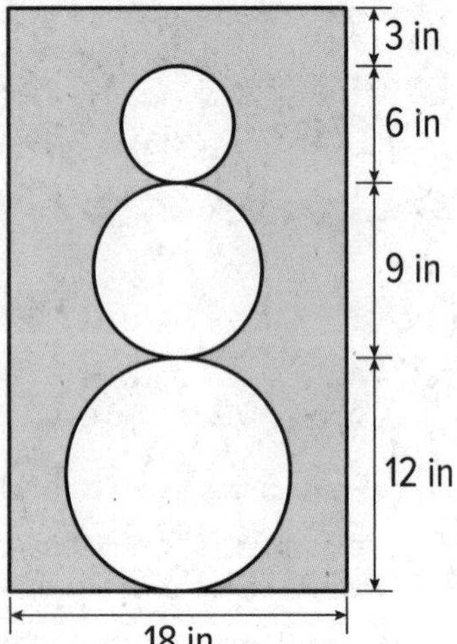

NAME ____________________ DATE ____________ PERIOD ____________

3. The face of a clock has a circumference of 63 in. What is the area of the face of the clock? (Lesson 3-8)

4. Which of these pairs of quantities are proportional to each other? For the quantities that are proportional, what is the constant of proportionality? (Lesson 3-7)

a. radius and diameter of a circle

b. radius and circumference of a circle

c. radius and area of a circle

d. diameter and circumference of a circle

e. diameter and area of a circle

5. Find the area of this shape in two different ways. **(Lesson 3-6)**

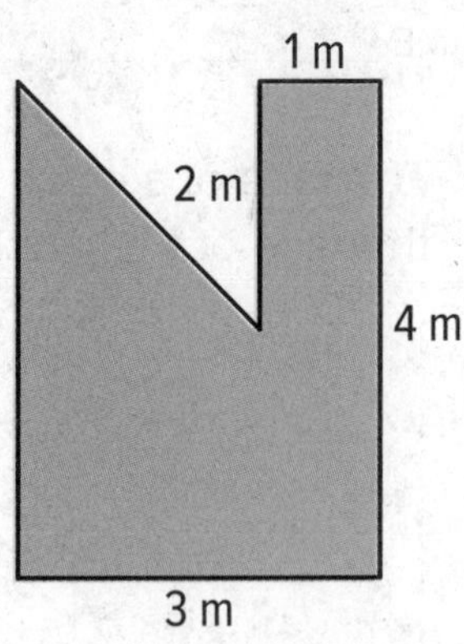

6. Elena and Jada both read at a constant rate, but Elena reads more slowly. For every 4 pages that Elena can read, Jada can read 5. **(Lesson 2-5)**

   a. Complete the table.

Pages Read by Elena	Pages Read by Jada
4	5
1	
9	
e	
	15
	j

   b. Here is an equation for the table: $j = 1.25e$. What does the 1.25 mean?

   c. Write an equation for this relationship that starts $e =$ ...

Lesson 3-10

Distinguishing Circumference and Area

NAME ______________________ DATE ______________ PERIOD ______________

Learning Goal Let's contrast circumference and area.

Warm Up

10.1 Filling the Plate

About how many mints can fit on the plate in a single layer? Be prepared to explain your reasoning.

Activity

10.2 Card Sort: Circle Problems

Your teacher will give you cards with questions about circles.

1. Sort the cards into two groups based on whether you would use the circumference or the area of the circle to answer the question. Pause here so your teacher can review your work.

2. Your teacher will assign you a card to examine more closely. What additional information would you need in order to answer the question on your card?

3. Estimate measurements for the circle on your card.

4. Use your estimates to calculate the answer to the question.

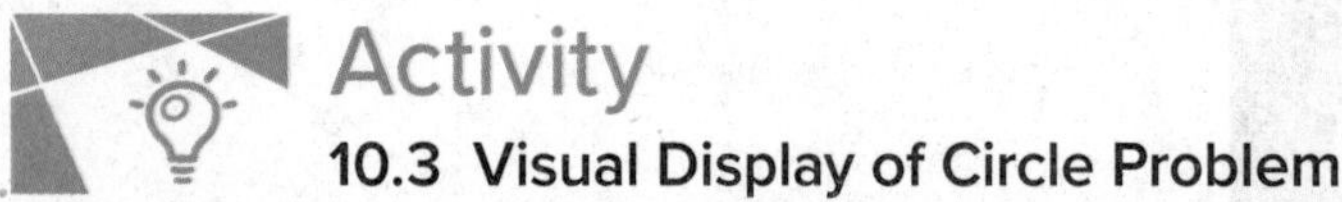

Activity
10.3 Visual Display of Circle Problem

In the previous activity, you estimated the answer to a question about circles.

Create a visual display that includes:

- the question you were answering.
- a diagram of a circle labeled with your estimated measurements.
- your thinking, organized so that others can follow it.
- your answer, expressed in terms of π and also expressed as a decimal approximation.

NAME ______________________ DATE ______________ PERIOD __________

Activity
10.4 Analyzing Circle Claims

Here are two students' answers for each question. Do you agree with either of them? Explain or show your reasoning.

1. How many feet are traveled by a person riding once around the merry-go-round?

   - Clare says, "The radius of the merry-go-round is about 4 feet, so the distance around the edge is about 8π feet."
   - Andre says, "The diameter of the merry-go-round is about 4 feet, so the distance around the edge is about 4π feet."

2. How much room is there to spread frosting on the cookie?

   - Clare says, "The radius of the cookie is about 3 centimeters, so the space for frosting is about 6π cm^2."
   - Andre says, "The diameter of the cookie is about 3 inches, so the space for frosting is about 2.25π in^2."

3. How far does the unicycle move when the wheel makes 5 full rotations?

- Clare says, "The diameter of the unicycle wheel is about 0.5 meters. In 5 complete rotations it will go about $\frac{5}{2}\pi$ m^2."
- Andre says, "I agree with Clare's estimate of the diameter, but that means the unicycle will go about $\frac{5}{4}\pi$ m."

Are you ready for more?

A goat (point G) is tied with a 6-foot rope to the corner of a shed. The floor of the shed is a square whose sides are each 3 feet long. The shed is closed and the goat can't go inside. The space all around the shed is flat, grassy, and the goat can't reach any other structures or objects. What is the area over which the goat can roam?

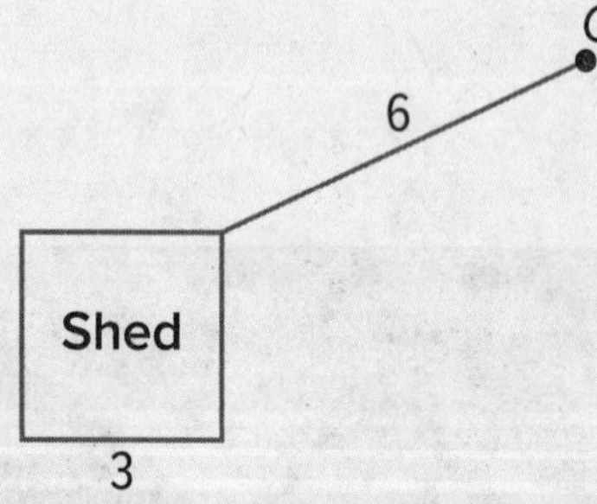

NAME ______________________ DATE ______________ PERIOD ____________

Summary

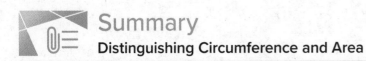

Distinguishing Circumference and Area

Sometimes we need to find the circumference of a circle, and sometimes we need to find the area.

Here are some examples of quantities related to the circumference of a circle.

- the length of a circular path
- the distance a wheel will travel after one complete rotation
- the length of a piece of rope coiled in a circle

Here are some examples of quantities related to the area of a circle.

- the amount of land that is cultivated on a circular field
- the amount of frosting needed to cover the top of a round cake
- the number of tiles needed to cover a round table

In both cases, the radius (or diameter) of the circle is all that is needed to make the calculation.

The circumference of a circle with radius r is $2\pi r$ while its area is πr^2.

The circumference is measured in linear units (such as cm, in, km) while the area is measured in square units (such as cm^2, in^2, km^2).

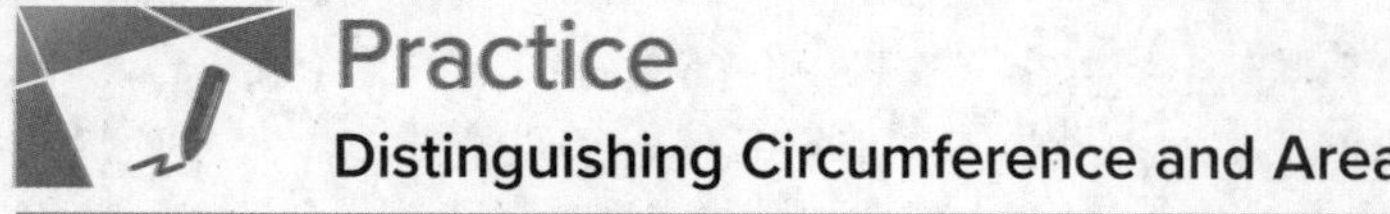

Practice

Distinguishing Circumference and Area

1. For each problem, decide whether the circumference of the circle or the area of the circle is most useful for finding a solution. Explain your reasoning.

   a. A car's wheels spin at 1000 revolutions per minute. The diameter of the wheels is 23 inches. You want to know how fast the car is traveling.

   b. A circular kitchen table has a diameter of 60 inches. You want to know how much fabric is needed to cover the table top.

   c. A circular puzzle is 20 inches in diameter. All of the pieces are about the same size. You want to know about how many pieces there are in the puzzle.

   d. You want to know about how long it takes to walk around a circular pond.

NAME ________________________ DATE ____________ PERIOD ____________

2. The city of Paris, France is completely contained within an almost circular road that goes around the edge. Use the map with its scale to:

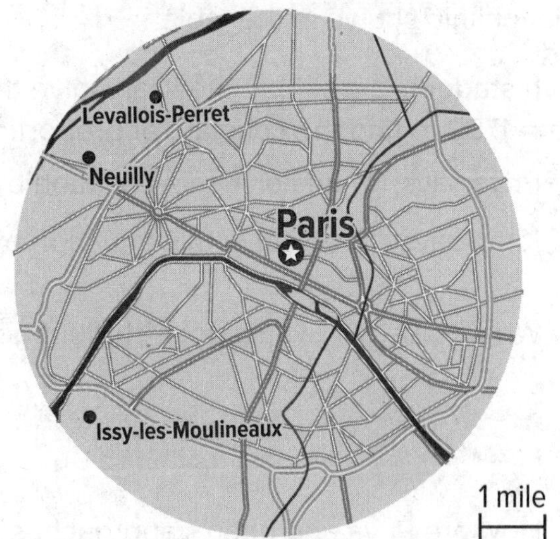

   a. Estimate the circumference of Paris.

   b. Estimate the area of Paris.

3. Here is a diagram of a softball field.

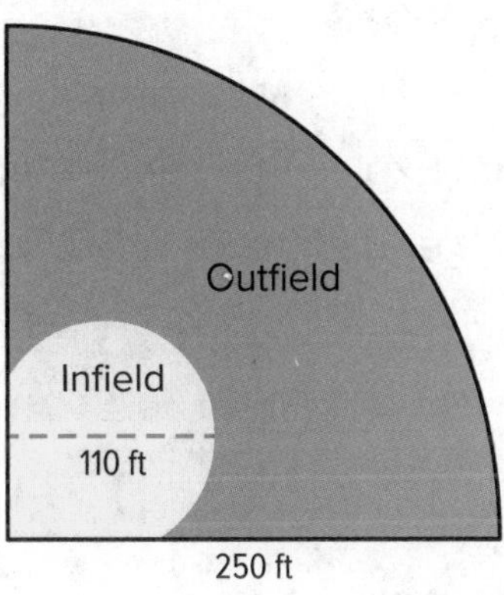

   a. About how long is the fence around the field?

   b. About how big is the outfield?

4. While in math class, Priya and Kiran come up with two ways of thinking about the proportional relationship shown in the table.

x	y
2	?
5	1750

Both students agree that they can solve the equation $5k = 1750$ to find the constant of proportionality.

- Priya says, "I can solve this equation by dividing 1750 by 5."
- Kiran says, "I can solve this equation by multiplying 1750 by $\frac{1}{5}$."

a. What value of k would each student get using their own method?

b. How are Priya and Kiran's approaches related?

c. Explain how each student might approach solving the equation $\frac{2}{3}k = 50$. **(Lesson 2-5)**

Lesson 3-11

Stained-Glass Windows

NAME ________________________ DATE ______________ PERIOD __________

Learning Goal Let's use circumference and area to design stained-glass windows.

Activity

11.1 Cost of a Stained-Glass Window

The students in art class are designing a stained-glass window to hang in the school entryway. The window will be 3 feet tall and 4 feet wide. Here is their design.

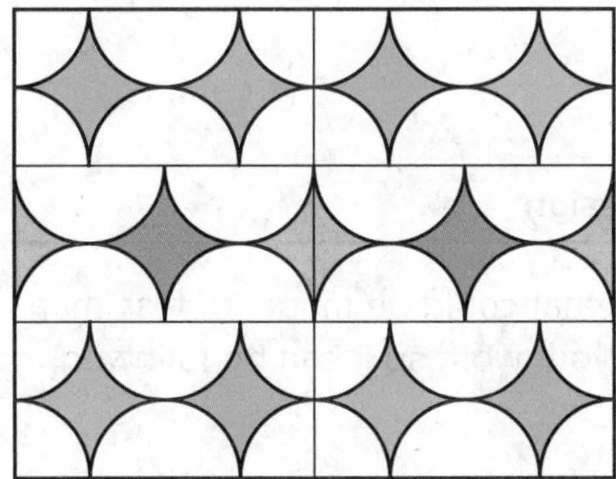

They have raised \$100 for the project. The colored glass costs \$5 per square foot and the clear glass costs \$2 per square foot. The material they need to join the pieces of glass together costs 10 cents per foot and the frame around the window costs \$4 per foot.

Do they have enough money to cover the cost of making the window?

Activity
11.2 A Bigger Window

A local community member sees the school's stained-glass window and really likes the design. They ask the students to create a larger copy of the window using a scale factor of 3. Would $450 be enough to buy the materials for the larger window? Explain or show your reasoning.

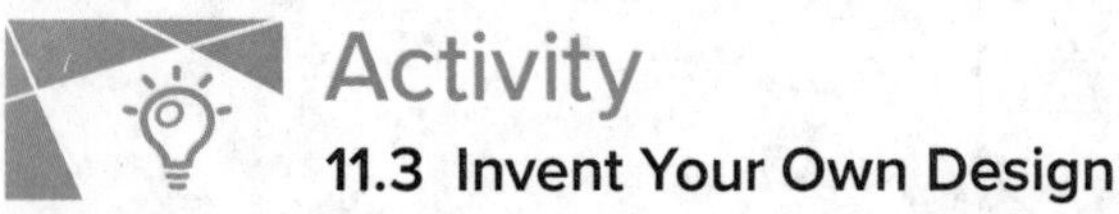

Activity
11.3 Invent Your Own Design

Draw a stained-glass window design that could be made for less than $450. Show your thinking. Organize your work so it can be followed by others.

Learning Targets

Lesson		Learning Target(s)
3-1	How Well Can You Measure?	• I can examine quotients and use a graph to decide whether two associated quantities are in a proportional relationship. • I understand that it can be difficult to measure the quantities in a proportional relationship accurately.
3-2	Exploring Circles	• I can describe the characteristics that make a shape a circle. • I can identify the diameter, center, radius, and circumference of a circle.
3-3	Exploring Circumference	• I can describe the relationship between circumference and diameter of any circle. • I can explain what π means.
3-4	Applying Circumference	• I can choose an approximation for π based on the situation or problem. • If I know the radius, diameter, or circumference of a circle, I can find the other two.

(continued on the next page)

(continued from the previous page)

Lesson		Learning Target(s)
3-5	Circumference and Wheels	• If I know the radius or diameter of a wheel, I can find the distance the wheel travels in some number of revolutions.
3-6	Estimating Areas	• I can calculate the area of a complicated shape by breaking it into shapes whose area I know how to calculate.
3-7	Exploring the Area of a Circle	• If I know a circle's radius or diameter, I can find an approximation for its area. • I know whether or not the relationship between the diameter and area of a circle is proportional and can explain how I know.
3-8	Relating Area to Circumference	• I can explain how the area of a circle and its circumference are related to each other. • I know the formula for the area of a circle.

Lesson	Learning Target(s)
3-9 Applying Area of Circles	• I can calculate the area of more complicated shapes that include fractions of circles. • I can write exact answers in terms of π.
3-10 Distinguishing Circumference and Area	• I can decide whether a situation about a circle has to do with area or circumference. • I can use formulas for circumference and area of a circle to help solve problems.
3-11 Stained-Glass Windows	• I can apply my understanding of area and circumference of circles to solve more complicated problems.

Notes:

(continued on the next page)

(continued from the previous page)

Unit 4

Proportional Relationships and Percentages

Green sea turtles live in the ocean, but lay their eggs on shore. You'll learn more about how proportional relationships can help protect these turtles' eggs in this unit.

Topics

- Proportional Relationships with Fractions
- Percent Increase and Decrease
- Applying Percentages
- Let's Put It to Work

Unit 4

Proportional Relationships and Percentages

Proportional Relationships with Fractions

Percent Increase and Decrease

Applying Percentages

Let's Put It to Work

Lesson 4-1

Lots of Flags

NAME ______________________ DATE ____________ PERIOD ________

Learning Goal Let's explore the U.S. flag.

Warm Up

1.1 Scaled or Not?

1. Which of the geometric objects are scaled versions of each other?

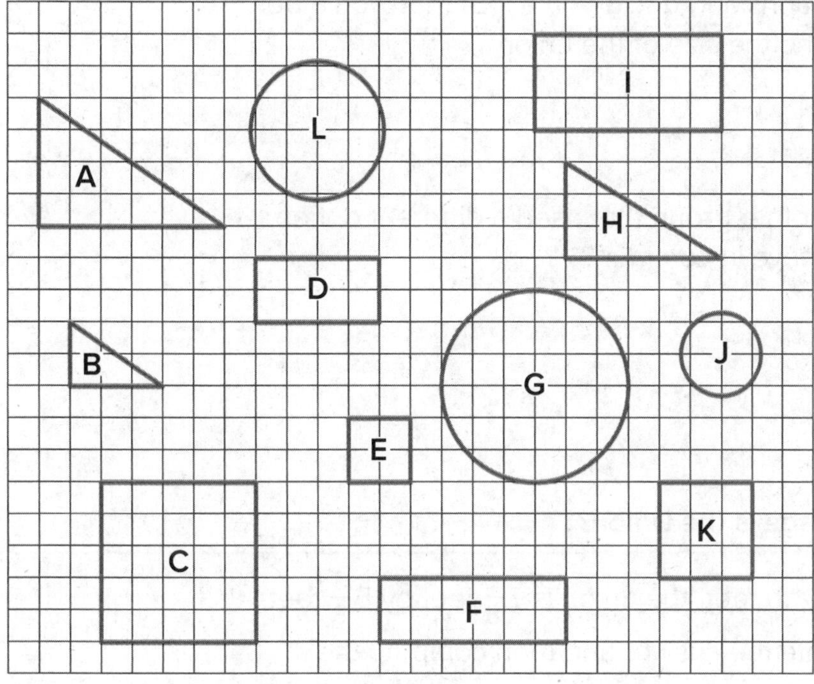

2. Pick two of the objects that are scaled copies and find the scale factor.

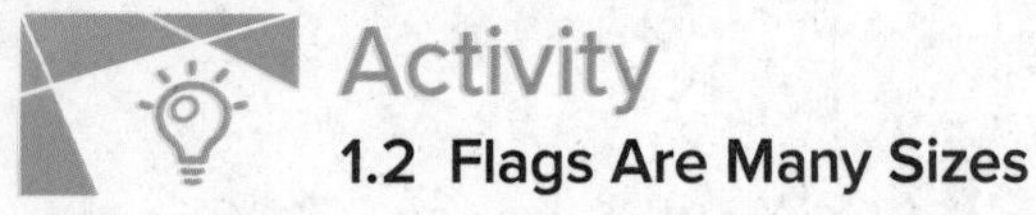

Activity

1.2 Flags Are Many Sizes

One standard size for the United States flag is 19 feet by 10 feet. On a flag of this size, the union (the blue rectangle in the top-left corner) is $7\frac{5}{8}$ feet by $5\frac{3}{8}$ feet.

There are many places that display flags of different sizes.

- Many classrooms display a U.S. flag.
- Flags are often displayed on stamps.
- There was a flag on the space shuttle.
- Astronauts on the Apollo missions had a flag on a shoulder patch.

1. Choose one of the four options and decide on a size that would be appropriate for this flag. Find the size of the union.

2. Share your answer with another group that used a different option. What do your dimensions have in common?

Activity

1.3 What Percentage Is the Union?

On a U.S. flag that is 19 feet by 10 feet, the union is $7\frac{5}{8}$ feet by $5\frac{3}{8}$ feet. For each question, first estimate the answer and then compute the actual percentage.

1. What percentage of the flag is taken up by the union?

2. What percentage of the flag is red? Be prepared to share your reasoning.

NAME ____________________ DATE ____________ PERIOD ____________

Are you ready for more?

The largest U.S. flag in the world is 225 feet by 505 feet.

1. Is the ratio of the length to the width equivalent to 1 : 1.9, the ratio for official government flags?

2. If a square yard of the flag weighs about 3.8 ounces, how much does the entire flag weigh in pounds?

Summary

Lots of Flags

Imagine you have a painting that is 15 feet wide and 5 feet high. To sketch a scaled copy of the painting, the ratio of the width and height of a scaled copy must be equivalent to 15 : 5.

Width	Height
15	5
2	h

What is the height of a scaled copy that is 2 feet across?

We know that the height is $\frac{1}{3}$ the width, so $h = \frac{1}{3} \cdot 2$ or $\frac{2}{3}$.

Sometimes ratios include fractions and decimals. We will be working with these kinds of ratios in the next few lessons.

Glossary

percentage

Practice

Lots of Flags

1. A rectangle has a height to width ratio of 3 : 4.5. Give two examples of dimensions for rectangles that could be scaled versions of this rectangle.

2. One rectangle measures 2 units by 7 units. A second rectangle measures 11 units by 37 units. Are these two figures scaled versions of each other? If so, find the scale factor. If not, briefly explain why.

3. Ants have 6 legs. Elena and Andre write equations showing the proportional relationship between the number of ants, a, to the number of ant legs l. Elena writes $a = 6 \cdot l$ and Andre writes $l = \frac{1}{6} \cdot a$. Do you agree with either of the equations? Explain your reasoning. **(Lesson 2-5)**

NAME _______________ DATE _______________ PERIOD _______________

4. On the grid, draw a scaled copy of quadrilateral *ABCD* with a scale factor $\frac{2}{3}$. **(Lesson 1-4)**

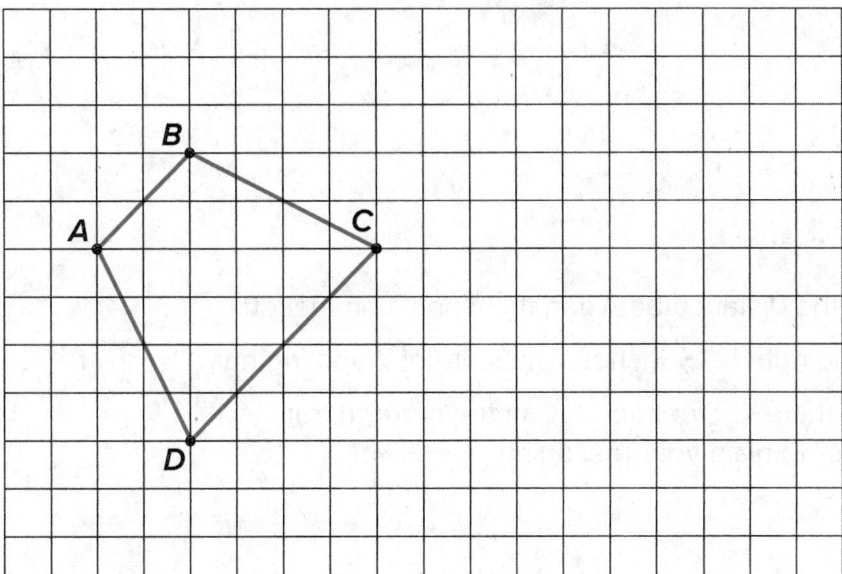

5. Solve each equation mentally. **(Lesson 1-5)**

a. $\frac{5}{2} \cdot x = 1$

b. $x \cdot \frac{7}{3} = 1$

c. $1 \div \frac{11}{2} = x$

6. Lin has a scale model of a modern train. The model is created at a scale of 1 to 48.

   a. The height of the model train is 102 millimeters. What is the actual height of the train in meters? Explain your reasoning.

   b. On the scale model, the distance between the wheels on the left and the wheels on the right is $1\frac{1}{4}$ inches. The state of Wyoming has old railroad tracks that are 4.5 feet apart. Can the modern train travel on those tracks? Explain your reasoning. **(Lesson 1-11)**

Lesson 4-2

Ratios and Rates with Fractions

NAME ________________ DATE ________ PERIOD ________

Learning Goal Let's calculate some rates with fractions.

Warm Up

2.1 Number Talk: Division

Find each quotient mentally.

1. $5 \div \frac{1}{3}$

2. $2 \div \frac{1}{3}$

3. $\frac{1}{2} \div \frac{1}{3}$

4. $2\frac{1}{2} \div \frac{1}{3}$

Steve Boyko/Shutterstock

Activity

2.2 A Train is Traveling at . . .

A train is traveling at a constant speed and goes 7.5 kilometers in 6 minutes. At that rate:

1. How far does the train go in 1 minute?

2. How far does the train go in 100 minutes?

Activity

2.3 Comparing Running Speeds

Lin ran $2\frac{3}{4}$ miles in $\frac{2}{5}$ of an hour. Noah ran $8\frac{2}{3}$ miles in $\frac{4}{3}$ of an hour.

1. Pick one of the questions that was displayed, but don't tell anyone which question you picked. Find the answer to the question.

2. When you and your partner are both done, share the answer you got (do not share the question) and ask your partner to guess which question you answered. If your partner can't guess, explain the process you used to answer the question.

3. Switch with your partner and take a turn guessing the question that your partner answered.

Are you ready for more?

Nothing can go faster than the speed of light, which is 299,792,458 meters per second. Which of these are possible?

1. Traveling a billion meters in 5 seconds.

2. Traveling a meter in 2.5 nanoseconds. (A nanosecond is a billionth of a second.)

3. Traveling a parsec in a year. (A parsec is about 3.26 light years and a light year is the distance light can travel in a year.)

NAME ______________________ DATE ______________ PERIOD ____________

Activity

2.4 Scaling the Mona Lisa

In real life, the Mona Lisa measures $2\frac{1}{2}$ feet by $1\frac{3}{4}$ feet. A company that makes office supplies wants to print a scaled copy of the Mona Lisa on the cover of a notebook that measures 11 inches by 9 inches.

1. What size should they use for the scaled copy of the Mona Lisa on the notebook cover?

2. What is the scale factor from the real painting to its copy on the notebook cover?

3. Discuss your thinking with your partner. Did you use the same scale factor? If not, is one more reasonable than the other?

Summary

Ratios and Rates with Fractions

There are 12 inches in a foot, so we can say that for every 1 foot, there are 12 inches, or the ratio of feet to inches is 1 : 12. We can find the **unit rates** by dividing the numbers in the ratio:

$1 \div 12 = \frac{1}{12}$, so there is $\frac{1}{12}$ foot per inch.

$12 \div 1 = 12$, so there are 12 inches per foot.

The numbers in a ratio can be fractions, and we calculate the unit rates the same way: by dividing the numbers in the ratio. For example, if someone runs $\frac{3}{4}$ mile in $\frac{11}{2}$ minutes, the ratio of minutes to miles is $\frac{11}{2} : \frac{3}{4}$.

$\frac{11}{2} \div \frac{3}{4} = \frac{22}{3}$, so the person's pace is $\frac{22}{3}$ minutes per mile.

$\frac{3}{4} \div \frac{11}{2} = \frac{3}{22}$, so the person's speed is $\frac{3}{22}$ miles per minute.

Glossary

unit rate

Practice

Ratios and Rates with Fractions

1. A cyclist rode 3.75 miles in 0.3 hours.

   a. How fast was she going in miles per hour?

   b. At that rate, how long will it take her to go 4.5 miles?

2. A recipe for sparkling grape juice calls for $1\frac{1}{2}$ quarts of sparkling water and $\frac{3}{4}$ quart of grape juice.

   a. How much sparkling water would you need to mix with 9 quarts of grape juice?

   b. How much grape juice would you need to mix with $\frac{15}{4}$ quarts of sparkling water?

   c. How much of each ingredient would you need to make 100 quarts of punch?

NAME ______________________ DATE ____________ PERIOD ____________

3. At a deli counter...

- someone bought $1\frac{3}{4}$ pounds of ham for \$14.50.
- someone bought $2\frac{1}{2}$ pounds of turkey for \$26.25.
- someone bought $\frac{3}{8}$ pound of roast beef for \$5.50.

Which meat is the least expensive per pound? Which meat is the most expensive per pound? Explain how you know.

4. Respond to the following questions. **(Lesson 3-10)**

a. Draw a scaled copy of the circle using a scale factor of 2.

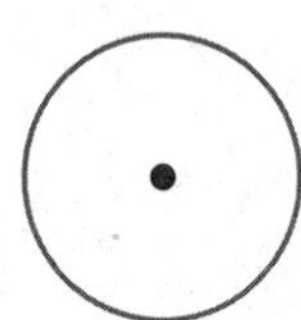

b. How does the circumference of the scaled copy compare to the circumference of the original circle?

c. How does the area of the scaled copy compare to the area of the original circle?

5. Jada has a scale map of Kansas that fits on a page in her book. The page is 5 inches by 8 inches. Kansas is about 210 miles by 410 miles. Select **all** scales that could be a scale of the map. (There are 2.54 centimeters in an inch.) **(Lesson 1-11)**

A. 1 in to 1 mi

B. 1 cm to 1 km

C. 1 in to 10 mi

D. 1 ft to 100 mi

E. 1 cm to 200 km

F. 1 in to 100 mi

G. 1 cm to 1000 km

Lesson 4-3

Revisiting Proportional Relationships

NAME ______________________ DATE ______________ PERIOD ____________

Learning Goal Let's use constants of proportionality to solve more problems.

Warm Up

3.1 Recipe Ratios

A recipe calls for $\frac{1}{2}$ cup sugar and 1 cup flour. Complete the table to show how much sugar and flour to use in different numbers of batches of the recipe.

Sugar (cups)	Flour (cups)
$\frac{1}{2}$	1
$\frac{3}{4}$	
	$1\frac{3}{4}$
1	
	$2\frac{1}{2}$

Activity

3.2 The Price of Rope

Two students are solving the same problem: At a hardware store, they can cut a length of rope off of a big roll, so you can buy any length you like. The cost for 6 feet of rope is $7.50. How much would you pay for 50 feet of rope, at this rate?

1. Kiran knows he can solve the problem this way.

   What would be Kiran's answer?

	Length of Rope (feet)	Price of Rope (dollars)	
$\cdot \frac{1}{6}$	6	7.50	$\cdot \frac{1}{6}$
$\cdot 50$	1	1.25	$\cdot 50$
	50		

2. Kiran wants to know if there is a more efficient way of solving the problem. Priya says she can solve the problem with only 2 rows in the table.

What do you think Priya's method is?

Length of Rope (feet)	Price of Rope (dollars)
6	7.50
50	

Activity

3.3 Swimming, Manufacturing, and Painting

1. Tyler swims at a constant speed, 5 meters every 4 seconds. How long does it take him to swim 114 meters?

Distance (meters)	Time (seconds)
5	4
114	

2. A factory produces 3 bottles of sparkling water for every 8 bottles of plain water. How many bottles of sparkling water does the company produce when it produces 600 bottles of plain water?

Number of Bottles of Sparkling Water	Number of Bottles of Plain Water

3. A certain shade of light blue paint is made by mixing $1\frac{1}{2}$ quarts of blue paint with 5 quarts of white paint. How much white paint would you need to mix with 4 quarts of blue paint?

NAME ______________________ DATE ______________ PERIOD __________

4. For each of the previous three situations, write an equation to represent the proportional relationship.

Are you ready for more?

Different nerve signals travel at different speeds.

- Pressure and touch signals travel about 250 feet per second.
- Dull pain signals travel about 2 feet per second.

1. How long does it take you to feel an ant crawling on your foot?

2. How much longer does it take to feel a dull ache in your foot?

Activity

3.4 Finishing the Race and More Orange Juice

1. Lin runs $2\frac{3}{4}$ miles in $\frac{2}{5}$ of an hour. Tyler runs $8\frac{2}{3}$ miles in $\frac{4}{3}$ of an hour. How long does it take each of them to run 10 miles at that rate?

2. Priya mixes $2\frac{1}{2}$ cups of water with $\frac{1}{3}$ cup of orange juice concentrate. Diego mixes $1\frac{2}{3}$ cups of water with $\frac{1}{4}$ cup of orange juice concentrate. How much concentrate should each of them mix with 100 cups of water to make juice that tastes the same as their original recipe? Explain or show your reasoning.

Summary

Revisiting Proportional Relationships

If we identify two quantities in a problem and one is proportional to the other, then we can calculate the constant of proportionality and use it to answer other questions about the situation.

For example, Andre runs at a constant speed, 5 meters every 2 seconds. How long does it take him to run 91 meters at this rate?

In this problem there are two quantities, time (in seconds) and distance (in meters). Since Andre is running at a constant speed, time is proportional to distance. We can make a table with distance and time as column headers and fill in the given information.

Distance (meters)	Time (seconds)
5	2
91	

To find the value in the right column, we multiply the value in the left column by $\frac{2}{5}$ because $\frac{2}{5} \cdot 5 = 2$. This means that it takes Andre $\frac{2}{5}$ seconds to run one meter.

At this rate, it would take Andre $\frac{2}{5} \cdot 91 = \frac{182}{5}$, or 36.4 seconds to walk 91 meters. In general, if t is the time it takes to walk d meters at that pace, then $t = \frac{2}{5}d$.

NAME ______________________ DATE ____________ PERIOD ____________

Practice

Revisiting Proportional Relationships

1. It takes an ant farm 3 days to consume $\frac{1}{2}$ of an apple. At that rate, in how many days will the ant farm consume 3 apples?

2. To make a shade of paint called jasper green, mix 4 quarts of green paint with $\frac{2}{3}$ cups of black paint. How much green paint should be mixed with 4 cups of black paint to make jasper green?

3. An airplane is flying from New York City to Los Angeles. The distance it travels in miles, d, is related to the time in seconds, t, by the equation $d = 0.15t$.

   a. How fast is it flying? Be sure to include the units.

   b. How far will it travel in 30 seconds?

   c. How long will it take to go 12.75 miles?

4. A grocer can buy strawberries for $1.38 per pound.

   a. Write an equation relating c, the cost, and p, the pounds of strawberries.

   b. A strawberry order cost $241.50. How many pounds did the grocer order?

5. Crater Lake in Oregon is shaped like a circle with a diameter of about 5.5 miles. **(Lesson 3-10)**

   a. How far is it around the perimeter of Crater Lake?

   b. What is the area of the surface of Crater Lake?

6. A 50-centimeter piece of wire is bent into a circle. What is the area of this circle? **(Lesson 3-8)**

7. Suppose Quadrilaterals A and B are both squares. Are A and B necessarily scaled copies of one another? Explain. **(Lesson 1-2)**

Lesson 4-4

Half as Much Again

NAME _______________ DATE _______________ PERIOD _______________

Learning Goal Let's use fractions to describe increases and decreases.

Warm Up

4.1 Notice and Wonder: Tape Diagrams

What do you notice? What do you wonder?

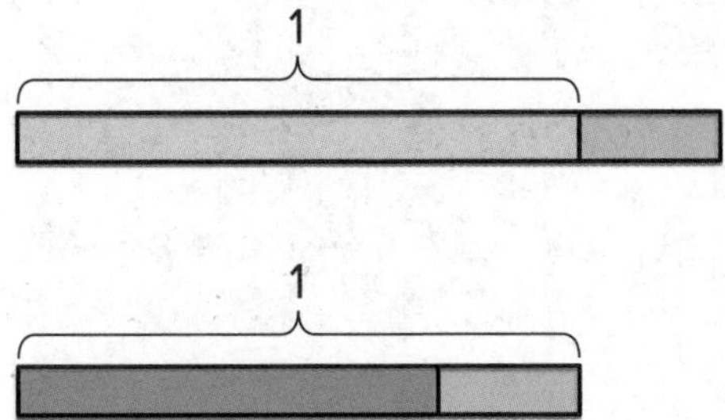

Activity

4.2 Walking Half as Much Again

1. Complete the table to show the total distance walked in each case.
   a. Jada's pet turtle walked 10 feet, and then half that length again.
   b. Jada's baby brother walked 3 feet, and then half that length again.
   c. Jada's hamster walked 4.5 feet, and then half that length again.
   d. Jada's robot walked 1 foot, and then half that length again.
   e. A person walked x feet, and then half that length again.

Initial Distance	Total Distance
10	
3	
4.5	
1	
x	

2. Explain how you computed the total distance in each case.

3. Two students each wrote an equation to represent the relationship between the initial distance walked (x) and the total distance walked (y).

- Mai wrote $y = x + \frac{1}{2}x$.
- Kiran wrote $y = \frac{3}{2}x$.

Do you agree with either of them? Explain your reasoning.

Are you ready for more?

Zeno jumped 8 meters. Then he jumped half as far again (4 meters). Then he jumped half as far again (2 meters). So, after 3 jumps, he was $8 + 4 + 2 = 14$ meters from his starting place.

1. Zeno kept jumping half as far again. How far would he be after 4 jumps? 5 jumps? 6 jumps?

2. Before he started jumping, Zeno put a mark on the floor that was exactly 16 meters from his starting place. How close can Zeno get to the mark if he keeps jumping half as far again?

3. If you enjoyed thinking about this problem, consider researching Zeno's Paradox.

NAME ______________________ DATE ______________ PERIOD ________

Activity

4.3 More and Less

1. Match each situation with a diagram. A diagram may not have a match.

   a. Han ate x ounces of blueberries. Mai ate $\frac{1}{3}$ less than that.

   b. Mai biked x miles. Han biked $\frac{2}{3}$ more than that.

   c. Han bought x pounds of apples. Mai bought $\frac{2}{3}$ of that.

A

x

y

B

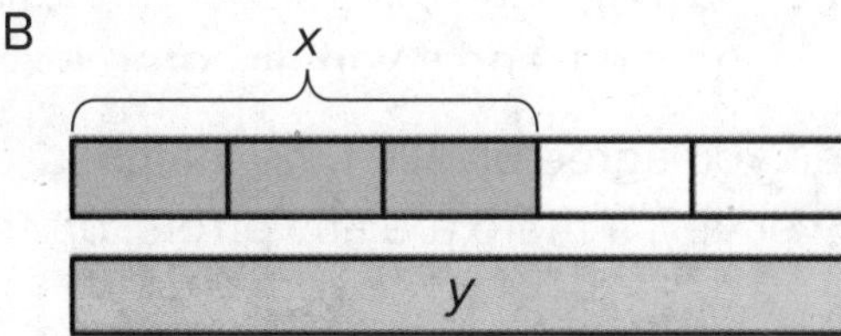

C

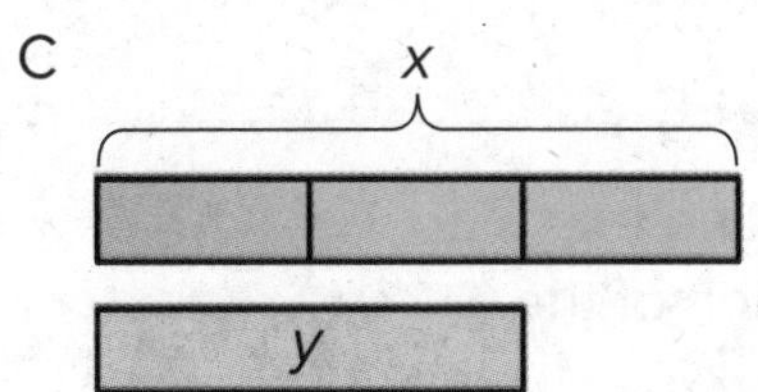

D

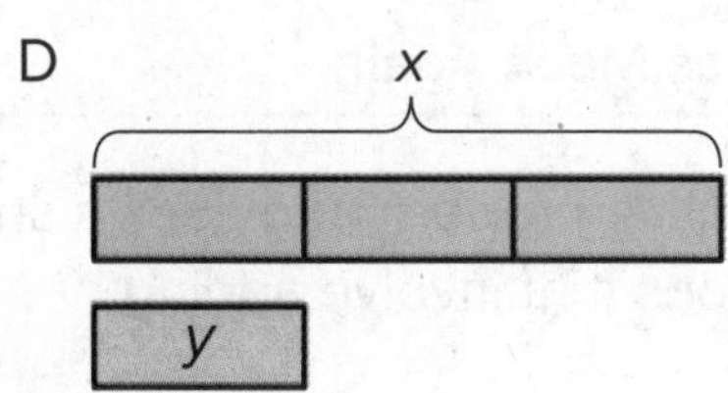

2. For each diagram, write an equation that represents the relationship between x and y.

   a. Diagram A:

   b. Diagram B:

   c. Diagram C:

   d. Diagram D:

3. Write a story for one of the diagrams that doesn't have a match.

Activity

4.4 Card Sort: Representations of Proportional Relationships

Your teacher will give you a set of cards that have proportional relationships represented 3 different ways: as descriptions, equations, and tables. Mix up the cards and place them all face-up.

1. Take turns with a partner to match a description with an equation and a table.
   a. For each match you find, explain to your partner how you know it's a match.
   b. For each match your partner finds, listen carefully to their explanation, and if you disagree, explain your thinking.
2. When you agree on all of the matches, check your answers with the answer key. If there are any errors, discuss why and revise your matches.

Summary

Half as Much Again

Using the distributive property provides a shortcut for calculating the final amount in situations that involve adding or subtracting a fraction of the original amount.

For example, one day Clare runs 4 miles. The next day, she plans to run that same distance plus half as much again. How far does she plan to run the next day?

Tomorrow she will run 4 miles plus $\frac{1}{2}$ of 4 miles.
We can use the distributive property to find this in one step: $1 \cdot 4 + \frac{1}{2} \cdot 4 = \left(1 + \frac{1}{2}\right) \cdot 4$

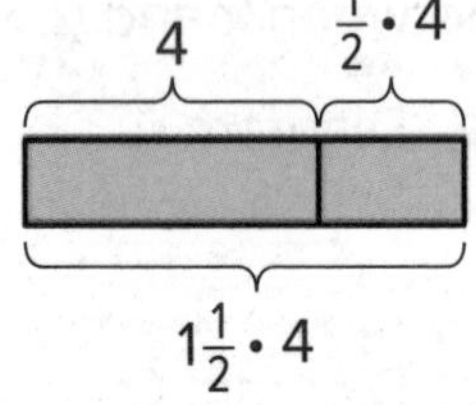

Clare plans to run $1\frac{1}{2} \cdot 4$, or 6 miles.

This works when we decrease by a fraction, too. If Tyler spent x dollars on a new shirt, and Noah spent $\frac{1}{3}$ less than Tyler, then Noah spent $\frac{2}{3}x$ dollars since $x - \frac{1}{3}x = \frac{2}{3}x$.

Glossary

tape diagram

NAME ____________________ DATE ____________ PERIOD ____________

Practice
Half as Much Again

1. Match each situation with a diagram.

   a. Diego drank x ounces of juice. Lin drank $\frac{1}{4}$ less than that.

   b. Lin ran x miles. Diego ran $\frac{3}{4}$ more than that.

   c. Diego bought x pounds of almonds. Lin bought $\frac{1}{4}$ of that.

A

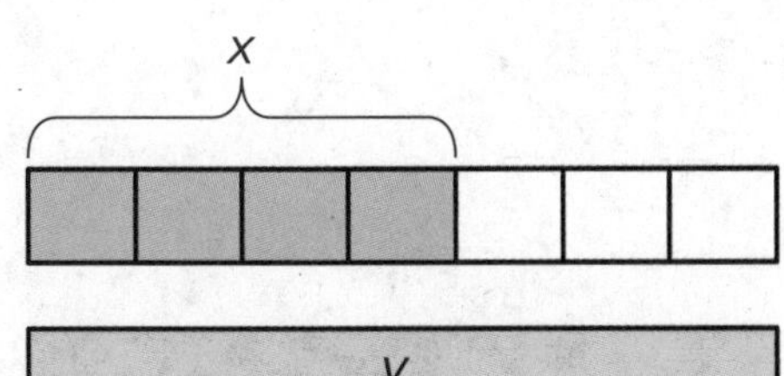

B

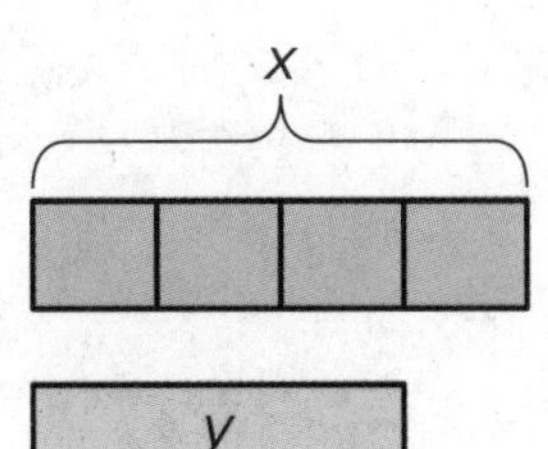

C

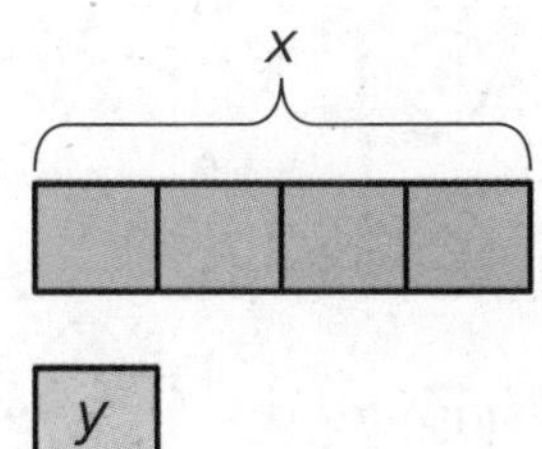

2. Elena walked 12 miles. Then she walked $\frac{1}{4}$ that distance. How far did she walk all together? Select **all** that apply.

   A. $12 + \frac{1}{4}$

   B. $12 \cdot \frac{1}{4}$

   C. $12 + \frac{1}{4} \cdot 12$

   D. $12\left(1 + \frac{1}{4}\right)$

   E. $12 \cdot \frac{3}{4}$

   F. $12 \cdot \frac{5}{4}$

3. Write a story that can be represented by the equation $y = x + \frac{1}{4}x$.

4. Select **all** ratios that are equivalent to 4 : 5. **(Lesson 4-1)**

(A.) 2 : 2.5

(B.) 2 : 3

(C.) 3 : 3.75

(D.) 7 : 8

(E.) 8 : 10

(F.) 14 : 27.5

5. Jada is making circular birthday invitations for her friends. The diameter of the circle is 12 cm. She bought 180 cm of ribbon to glue around the edge of each invitation. How many invitations can she make? **(Lesson 3-10)**

Lesson 4-5

Say It with Decimals

NAME ______________________ DATE ____________ PERIOD ____________

Learning Goal Let's use decimals to describe increases and decreases.

Warm Up

5.1 Notice and Wonder: Fractions to Decimals

A calculator gives the following decimal representations for some unit fractions:

$\frac{1}{2} = 0.5$

$\frac{1}{3} = 0.3333333$

$\frac{1}{4} = 0.25$

$\frac{1}{5} = 0.2$

$\frac{1}{6} = 0.1666667$

$\frac{1}{7} = 0.142857143$

$\frac{1}{8} = 0.125$

$\frac{1}{9} = 0.1111111$

$\frac{1}{10} = 0.1$

$\frac{1}{11} = 0.0909091$

What do you notice? What do you wonder?

Activity

5.2 Repeating Decimals

1. Use long division to express each fraction as a decimal.

$\frac{9}{25}$ $\frac{11}{30}$ $\frac{4}{11}$

2. What is similar about your answers to the previous question? What is different?

3. Use the decimal representations to decide which of these fractions has the greatest value. Explain your reasoning.

Are you ready for more?

One common approximation for π is $\frac{22}{7}$. Express this fraction as a decimal. How does this approximation compare to 3.14?

NAME ______________________ DATE ____________ PERIOD ____________

Activity

5.3 More and Less with Decimals

1. Match each diagram with a description and an equation.

Diagrams

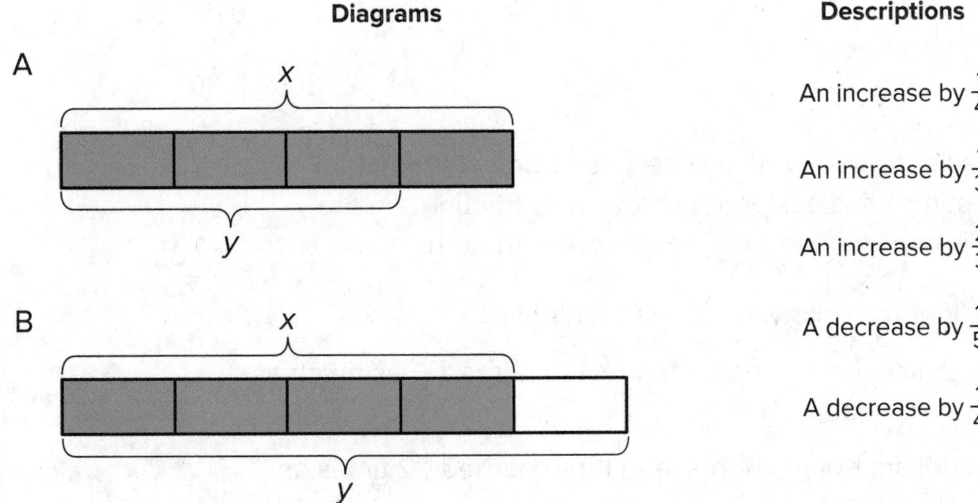

Descriptions	Equations
An increase by $\frac{1}{4}$	$y = 1.\overline{6}x$
An increase by $\frac{1}{3}$	$y = 1.\overline{3}x$
An increase by $\frac{2}{3}$	$y = 0.75x$
A decrease by $\frac{1}{5}$	$y = 0.4x$
A decrease by $\frac{1}{4}$	$y = 1.25x$

2. Draw a diagram for one of the unmatched equations.

Activity

5.4 Card Sort: More Representations

Your teacher will give you a set of cards that have proportional relationships represented 2 different ways: as descriptions and equations. Mix up the cards and place them all face-up.

Take turns with a partner to match a description with an equation.

1. For each match you find, explain to your partner how you know it's a match.

2. For each match your partner finds, listen carefully to their explanation, and if you disagree, explain your thinking.

3. When you have agreed on all of the matches, check your answers with the answer key. If there are any errors, discuss why and revise your matches.

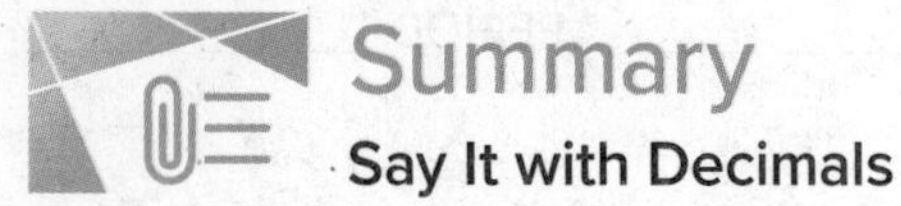

Summary

Say It with Decimals

Long division gives us a way of finding decimal representations for fractions.

For example, to find a decimal representation for $\frac{9}{8}$, we can divide 9 by 8.

$$\begin{array}{r} 1.125 \\ 8\overline{)9.000} \\ \underline{8} \\ 10 \\ \underline{8} \\ 20 \\ \underline{16} \\ 40 \\ \underline{40} \\ 0 \end{array}$$

So $\frac{9}{8} = 1.125$.

Sometimes it is easier to work with the decimal representation of a number, and sometimes it is easier to work with its fraction representation. It is important to be able to work with both.

For example, consider the following pair of problems:

- Priya earned x dollars doing chores, and Kiran earned $\frac{6}{5}$ as much as Priya. How much did Kiran earn?
- Priya earned x dollars doing chores, and Kiran earned 1.2 times as much as Priya. How much did Kiran earn?

Since $\frac{6}{5} = 1.2$, these are both exactly the same problem, and the answer is $\frac{6}{5}x$ or $1.2x$.

When we work with percentages in later lessons, the decimal representation will come in especially handy.

Glossary

long division
repeating decimal

NAME ______________________ DATE ______________ PERIOD __________

Practice
Say It with Decimals

1. Respond to the following.

   a. Match each diagram with a description and an equation.

Descriptions	Equations
An increase by $\frac{2}{3}$	$y = 1.8\overline{3}x$
An increase by $\frac{5}{6}$	$y = 1.\overline{6}x$
A decrease by $\frac{2}{5}$	$y = 0.6x$
A decrease by $\frac{5}{11}$	$y = 0.4x$

Diagrams

A

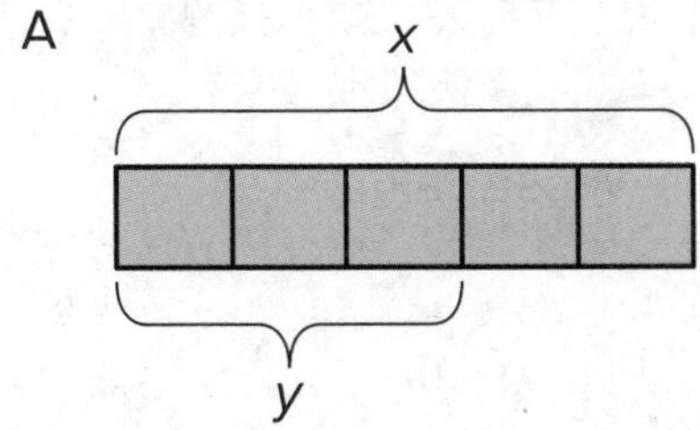

B

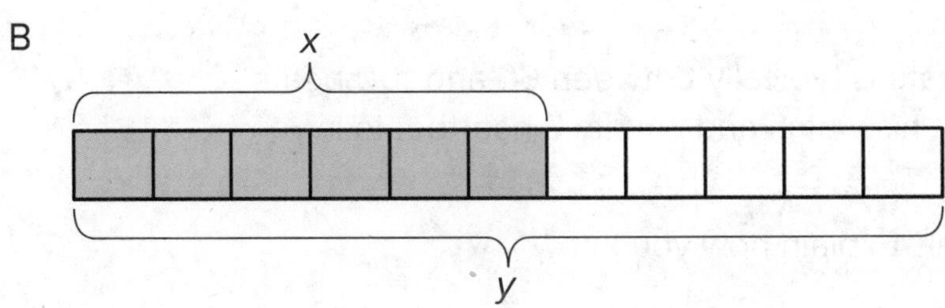

   b. Draw a diagram for one of the unmatched equations.

2. At the beginning of the month, there were 80 ounces of peanut butter in the pantry. Since then, the family ate 0.3 of the peanut butter. How many ounces of peanut butter are in the pantry now?

A. $0.7 \cdot 80$

B. $0.3 \cdot 80$

C. $80 - 0.3$

D. $(1 + 0.3) \cdot 80$

3. Respond to each of the following questions. **(Lesson 4-4)**

a. On a hot day, a football team drank an entire 50-gallon cooler of water and half as much again. How much water did they drink?

b. Jada has 12 library books checked out and Han has $\frac{1}{3}$ less than that. How many books does Han have checked out?

4. If x represents a positive number, select **all** expressions whose value is greater than x. **(Lesson 4-4)**

A. $\left(1 - \frac{1}{4}\right)x$

B. $\left(1 + \frac{1}{4}\right)x$

C. $\frac{7}{8}x$

D. $\frac{9}{8}x$

5. A person's resting heart rate is typically between 60 and 100 beats per minute. Noah looks at his watch and counts 8 heartbeats in 10 seconds. **(Lesson 2-6)**

a. Is his heart rate typical? Explain how you know.

b. Write an equation for h, the number of times Noah's heart beats (at this rate) in m minutes.

Lesson 4-6

Increasing and Decreasing

NAME ______________________ DATE ____________ PERIOD ____________

Learning Goal Let's use percentages to describe increases and decreases.

Warm Up

6.1 Improving Their Game

Here are the scores from 3 different sports teams from their last 2 games.

Sports Team	Total Points in Game 1	Total Points in Game 2
Football Team	22	30
Basketball Team	100	108
Baseball Team	4	12

1. What do you notice about the teams' scores? What do you wonder?

2. Which team improved the most? Explain your reasoning.

Activity

6.2 More Cereal and a Discounted Shirt

1. A cereal box says that now it contains 20% more. Originally, it came with 18.5 ounces of cereal. How much cereal does the box come with now?

2. The price of a shirt is $18.50, but you have a coupon that lowers the price by 20%. What is the price of the shirt after using the coupon?

Activity

6.3 Using Tape Diagrams

1. Match each situation to a diagram. Be prepared to explain your reasoning.

   a. Compared with last year's strawberry harvest, this year's strawberry harvest is a 25% increase.

   b. This year's blueberry harvest is 75% of last year's.

   c. Compared with last year, this year's peach harvest decreased 25%.

   d. This year's plum harvest is 125% of last year's plum harvest.

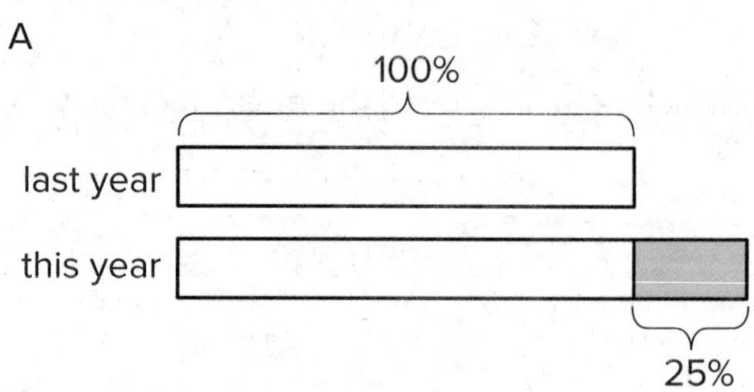

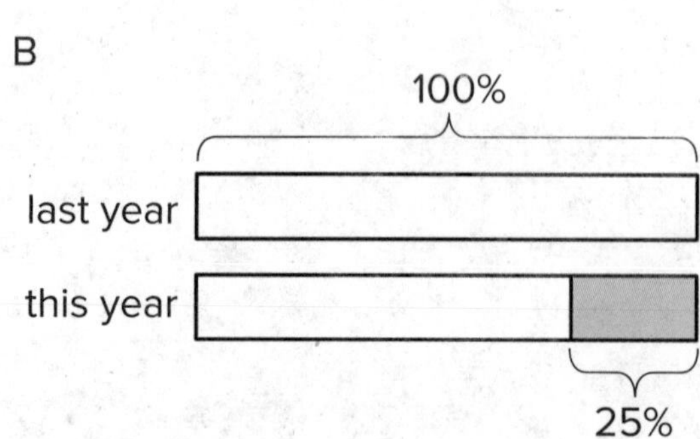

NAME ______________________ DATE ______________ PERIOD ____________

2. Draw a diagram to represent these situations.

   a. The number of ducks living at the pond increased by 40%.

   b. The number of mosquitoes decreased by 80%.

Are you ready for more?

What could it mean to say there is a 100% decrease in a quantity? Give an example of a quantity where this makes sense.

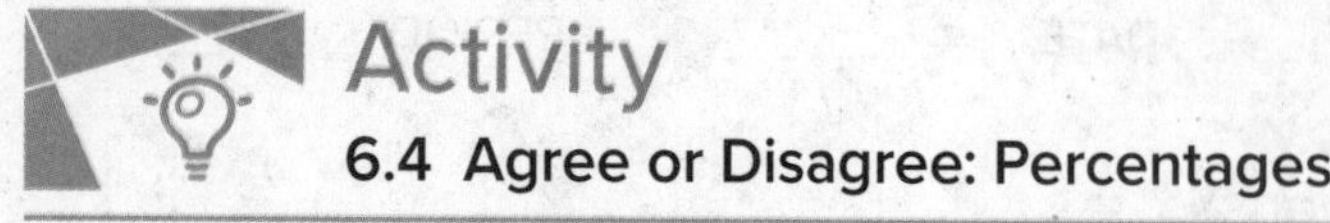

Activity

6.4 Agree or Disagree: Percentages

Do you agree or disagree with each statement? Explain your reasoning.

1. Employee A gets a pay raise of 50%. Employee B gets a pay raise of 45%. So Employee A gets the bigger pay raise.

2. Shirts are on sale for 20% off. You buy two of them. As you pay, the cashier says, "20% off of each shirt means 40% off of the total price."

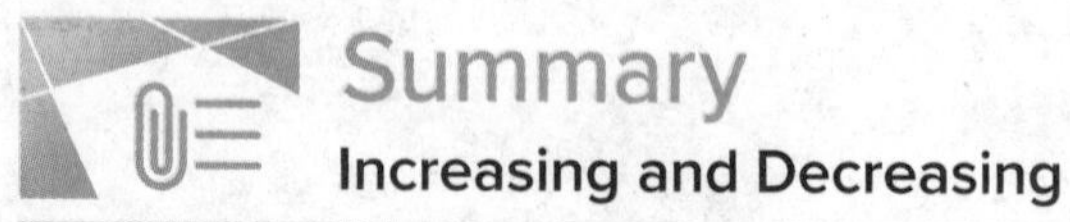

Summary

Increasing and Decreasing

Imagine that it takes Andre $\frac{3}{4}$ more than the time it takes Jada to get to school. Then we know that Andre's time is $1\frac{3}{4}$ or 1.75 times Jada's time. We can also describe this in terms of percentages:

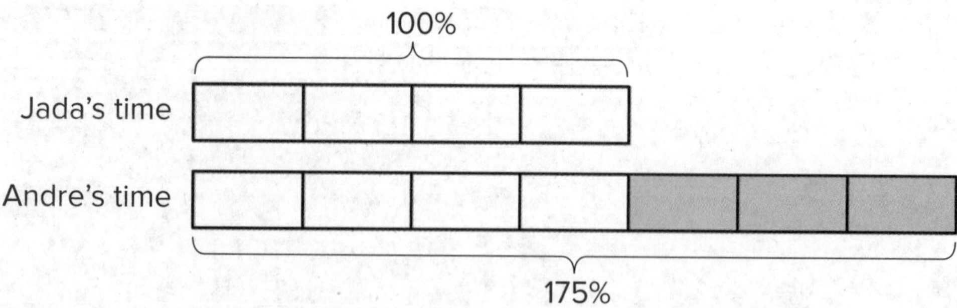

We say that Andre's time is 75% more than Jada's time. We can also see that Andre's time is 175% of Jada's time.

NAME ______________________________ DATE ______________ PERIOD ____________

In general, the terms **percent increase** and **percent decrease** describe an increase or decrease in a quantity as a percentage of the starting amount.

For example, if there were 500 grams of cereal in the original package, then "20% more" means that 20% of 500 grams has been added to the initial amount, $500 + (0.2) \cdot 500 = 600$, so there are 600 grams of cereal in the new package.

We can see that the new amount is 120% of the initial amount because $500 + (0.2) \cdot 500 = (1 + 0.2)\, 500$.

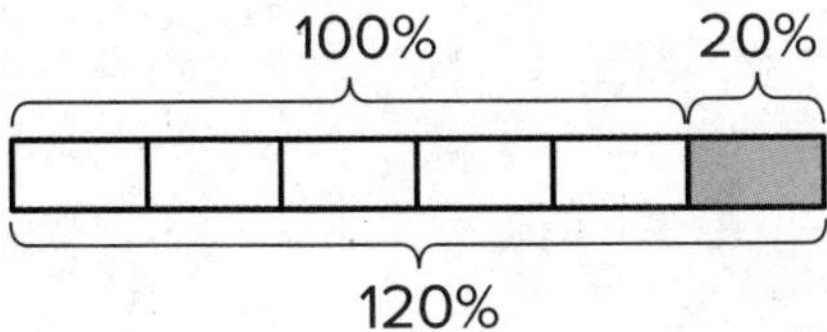

Glossary

percent decrease
percent increase

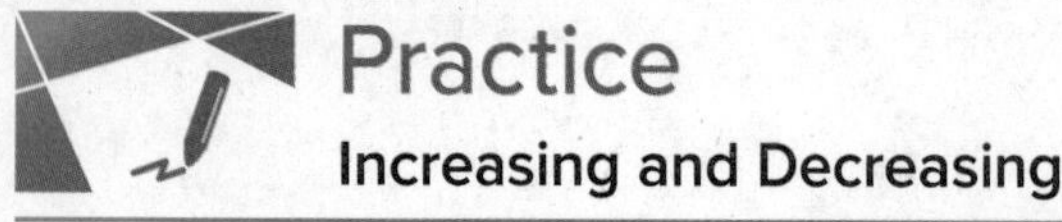

Practice

Increasing and Decreasing

1. For each diagram, decide if y is an increase or a decrease relative to x. Then determine the percent increase or decrease.

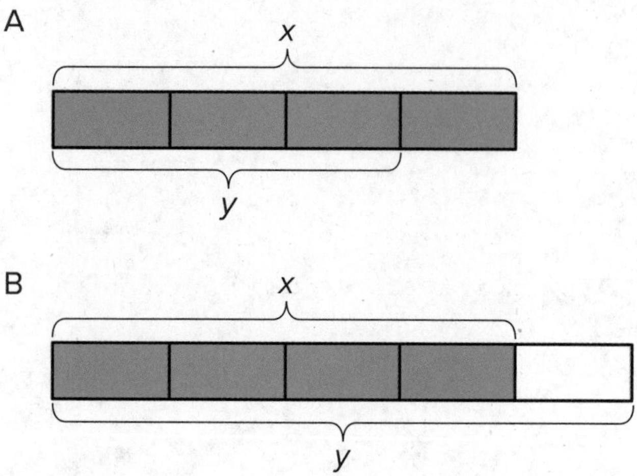

2. Draw diagrams to represent the following situations.

   a. The amount of flour that the bakery used this month was a 50% increase relative to last month.

   b. The amount of milk that the bakery used this month was a 75% decrease relative to last month.

NAME ______________________ DATE ____________ PERIOD ________

3. Write each percent increase or decrease as a percentage of the initial amount. The first one is done for you.

   a. This year, there was 40% more snow than last year.
   The amount of snow this year is 140% of the amount of snow last year.

   b. This year, there were 25% fewer sunny days than last year.

   c. Compared to last month, there was a 50% increase in the number of houses sold this month.

   d. The runner's time to complete the marathon was 10% less than the time to complete the last marathon.

4. The graph shows the relationship between the diameter and the circumference of a circle with the point $(1, \pi)$ shown. Find 3 more points that are on the line. **(Lesson 3-3)**

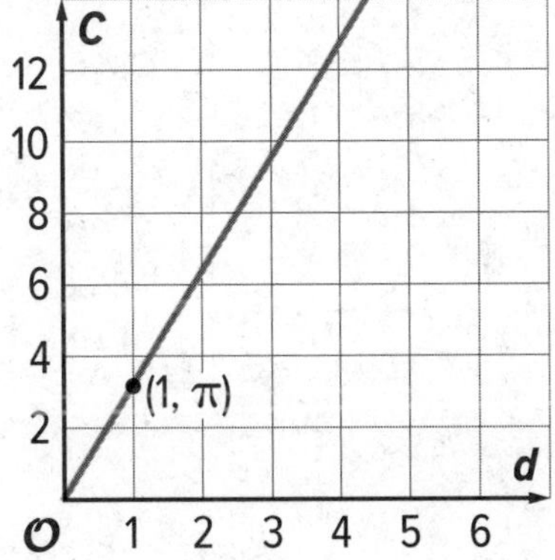

5. Priya bought x grams of flour. Clare bought $\frac{3}{8}$ more than that. Select **all** equations that represent the relationship between the amount of flour that Priya bought, x, and the amount of flour that Clare bought, y. **(Lesson 4-4)**

(A.) $y = \frac{3}{8}x$

(B.) $y = \frac{5}{8}x$

(C.) $y = x + \frac{3}{8}x$

(D.) $y = x - \frac{3}{8}x$

(E.) $y = \frac{11}{8}x$

Lesson 4-7

One Hundred Percent

NAME ______________________ DATE ____________ PERIOD ____________

Learning Goal Let's solve more problems about percent increase and percent decrease.

Warm Up

7.1 Notice and Wonder: Double Number Line

What do you notice? What do you wonder?

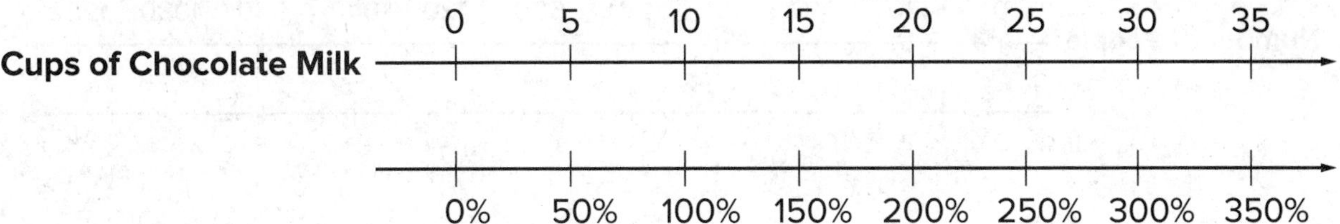

Activity

7.2 Double Number Lines

For each problem, complete the double number line diagram to show the percentages that correspond to the original amount and to the new amount.

1. The gas tank in dad's car holds 12 gallons. The gas tank in mom's truck holds 50% more than that. How much gas does the truck's tank hold?

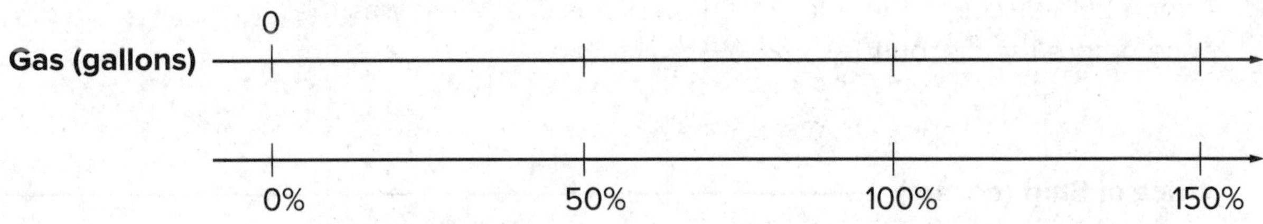

2. At a movie theater, the size of popcorn bags decreased 20%. If the old bags held 15 cups of popcorn, how much do the new bags hold?

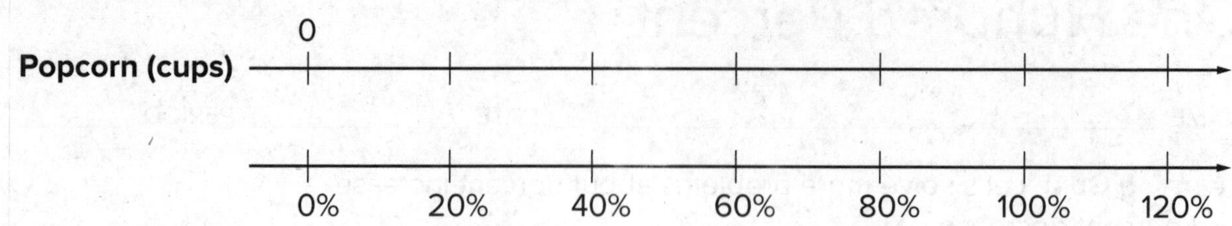

3. A school had 1,200 students last year and only 1,080 students this year. What was the percentage decrease in the number of students?

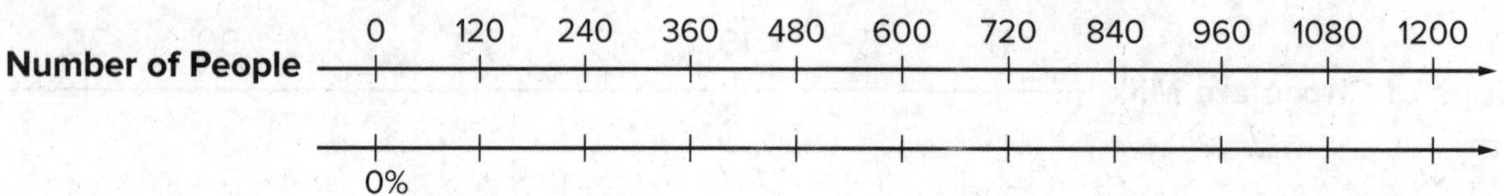

4. One week gas was $1.25 per gallon. The next week gas was $1.50 per gallon. By what percentage did the price increase?

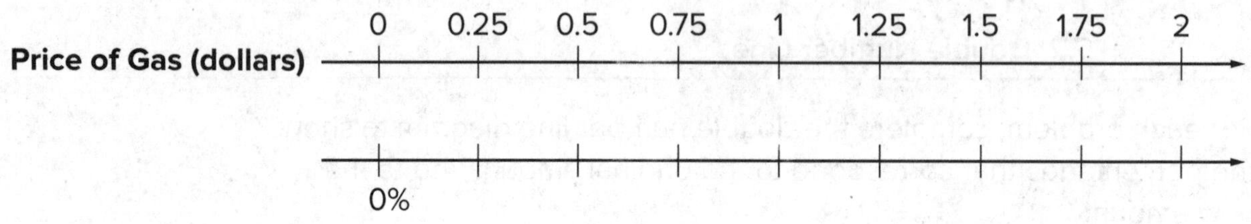

5. After a 25% discount, the price of a T-shirt was $12. What was the price before the discount?

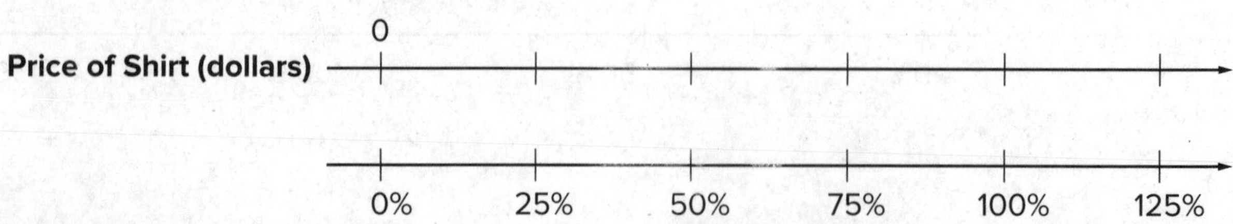

NAME ______________________ DATE ____________ PERIOD ____________

6. Compared to last year, the population of Boom Town has increased 25%. The population is now 6,600. What was the population last year?

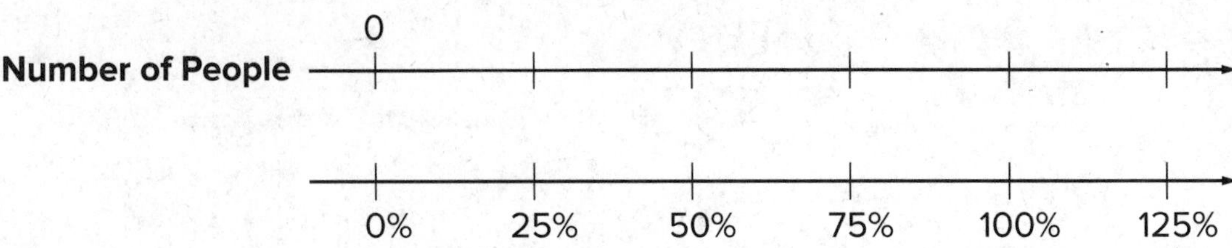

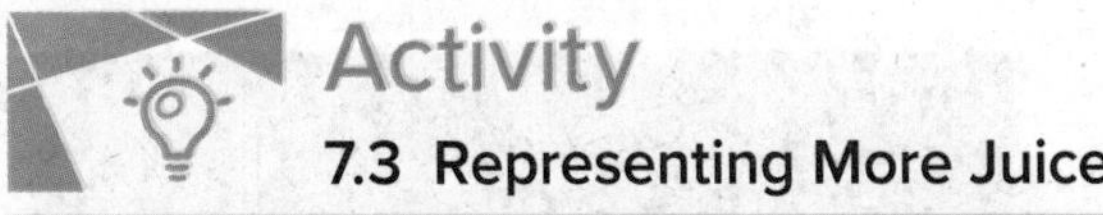

Activity

7.3 Representing More Juice

Two students are working on the same problem:

A juice box has 20% more juice in its new packaging. The original packaging held 12 fluid ounces. How much juice does the new packaging hold?

- Here is how Priya set up her double number line.

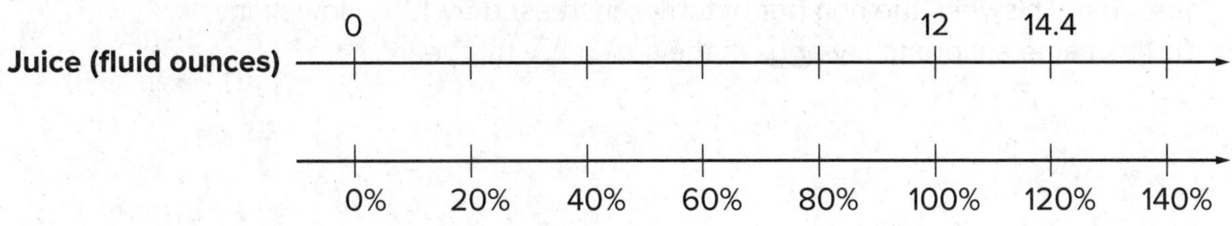

- Here is how Clare set up her double number line.

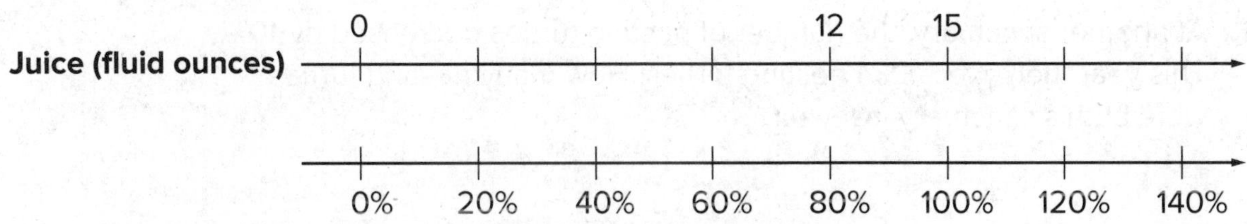

Do you agree with either of them? Explain or show your reasoning.

Are you ready for more?

Clare's diagram could represent a percent decrease. Describe a situation that could be represented with Clare's diagram.

Activity

7.4 Protecting the Green Sea Turtle

Green sea turtles live most of their lives in the ocean but come ashore to lay their eggs. Some beaches where turtles often come ashore have been made into protected sanctuaries so the eggs will not be disturbed.

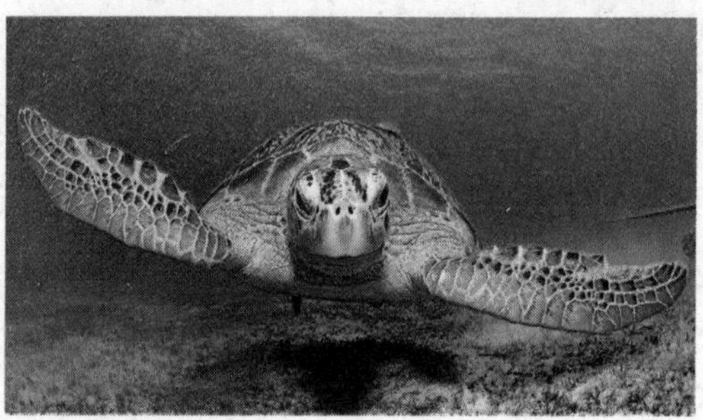

Gerald Nowak/Westend61/Getty Images

1. One sanctuary had 180 green sea turtles come ashore to lay eggs last year. This year, the number of turtles increased by 10%. How many turtles came ashore to lay eggs in the sanctuary this year?

2. At another sanctuary, the number of nesting turtles decreased by 10%. This year there were 234 nesting turtles. How many nesting turtles were at this sanctuary last year?

NAME ______________________ DATE ______________ PERIOD ______________

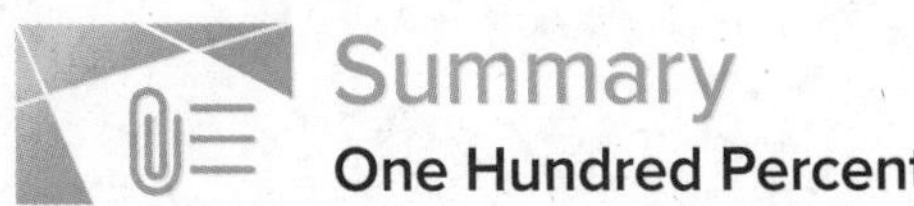

Summary
One Hundred Percent

We can use a double number line diagram to show information about percent increase and percent decrease:

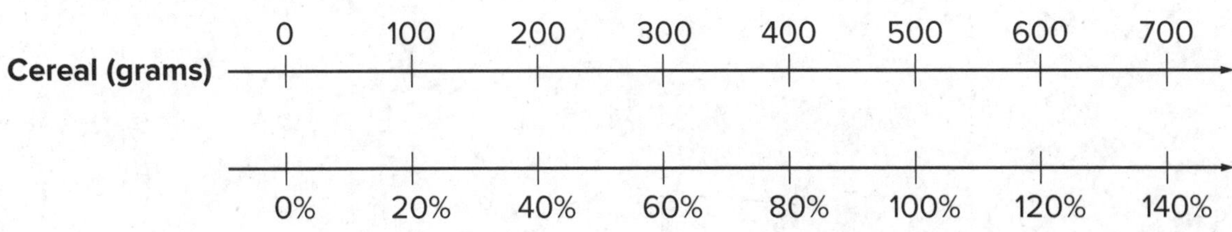

The initial amount of cereal is 500 grams, which is lined up with 100% in the diagram. We can find a 20% *increase* to 500 by adding 20% of 500:

$$500 + (0.2) \cdot 500 = (1.20) \cdot 500$$
$$= 600$$

In the diagram, we can see that 600 corresponds to 120%.

If the initial amount of 500 grams is *decreased* by 40%, we can find how much cereal there is by subtracting 40% of the 500 grams:

$$500 - (0.4) \cdot 500 = (0.6) \cdot 500$$
$$= 300$$

So, a 40% decrease is the same as 60% of the initial amount. In the diagram, we can see that 300 is lined up with 60%.

To solve percentage problems, we need to be clear about what corresponds to 100%. For example, suppose there are 20 students in a class, and we know this is an increase of 25% from last year. In this case, the number of students in the class *last* year corresponds to 100%. So, the initial amount (100%) is unknown and the final amount (125%) is 20 students.

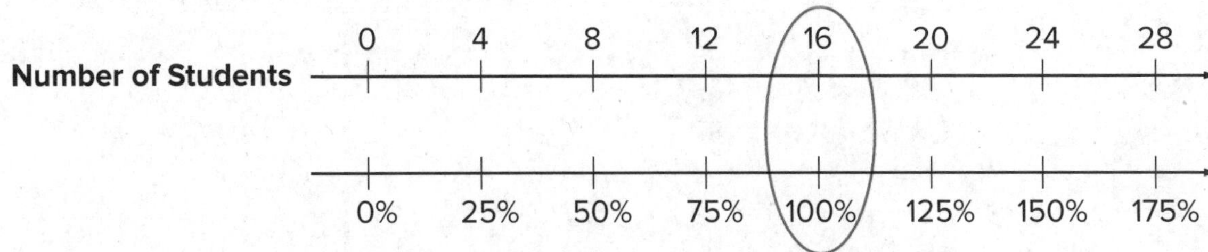

Looking at the double number line, if 20 students is a 25% increase from the previous year, then there were 16 students in the class last year.

Practice

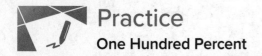

One Hundred Percent

1. A bakery used 25% more butter this month than last month. If the bakery used 240 kilograms of butter last month, how much did it use this month?

2. Last week, the price of oranges at the farmer's market was $1.75 per pound. This week, the price has decreased by 20%. What is the price of oranges this week?

3. Noah thinks the answers to these two questions will be the same. Do you agree with him? Explain your reasoning.

   - This year, a herd of bison had a 10% increase in population. If there were 550 bison in the herd last year, how many are in the herd this year?
   - This year, another herd of bison had a 10% decrease in population. If there are 550 bison in the herd this year, how many bison were there last year?

NAME ______________________ DATE ____________ PERIOD ________

4. Elena walked 12 miles. Then she walked 0.25 that distance. How far did she walk all together? Select **all** that apply. **(Lesson 4-5)**

A. $12 + 0.25 \cdot 12$

B. $12(1 + 0.25)$

C. $12 \cdot 1.25$

D. $12 \cdot 0.25$

E. $12 + 0.25$

5. A circle's circumference is 600 m. What is a good approximation of the circle's area? **(Lesson 3-8)**

A. 300 m^2

B. 3,000 m^2

C. 30,000 m^2

D. 300,000 m^2

6. The equation $d = 3t$ represents the relationship between the distance (d) in inches that a snail is from a certain rock and the time (t) in minutes. **(Lesson 1-6)**

   a. What does the number 3 represent?

   b. How many minutes does it take the snail to get 9 inches from the rock?

   c. How far will the snail be from the rock after 9 minutes?

Lesson 4-8

Percent Increase and Decrease with Equations

NAME ____________________ DATE ____________ PERIOD ________

Learning Goal Let's use equations to represent increases and decreases.

Warm Up

8.1 Number Talk: From 100 to 106

How do you get from one number to the next using multiplication or division?

1. From 100 to 106
2. From 100 to 90
3. From 90 to 100
4. From 106 to 100

Activity

8.2 Interest and Depreciation

1. Money in a particular savings account increases by about 6% after a year. How much money will be in the account after one year if the initial amount is \$100? \$50? \$200? \$125? x dollars? If you get stuck, consider using diagrams or a table to organize your work.

2. The value of a new car decreases by about 15% in the first year. How much will a car be worth after one year if its initial value was \$1,000? \$5,000? \$5,020? x dollars? If you get stuck, consider using diagrams or a table to organize your work.

Activity
8.3 Matching Equations

Match an equation to each of these situations. Be prepared to share your reasoning.

Situations	Equations
1. The water level in a reservoir is now 52 meters. If this was a 23% increase, what was the initial depth?	$0.23x = 52$ $1.23x = 52$
2. The snow is now 52 inches deep. If this was a 77% decrease, what was the initial depth?	$0.77x = 52$ $1.77x = 52$

Are you ready for more?

An astronaut was exploring the moon of a distant planet and found some glowing goo at the bottom of a very deep crater. She brought a 10-gram sample of the goo to her laboratory. She found that when the goo was exposed to light, the total amount of goo increased by 100% every hour.

1. How much goo will she have after 1 hour? After 2 hours? After 3 hours? After n hours?

2. When she put the goo in the dark, it shrank by 75% every hour. How many hours will it take for the goo that was exposed to light for n hours to return to the original size?

NAME ______________________ DATE ____________ PERIOD ____________

Activity

8.4 Representing Percent Increase and Decrease: Equations

1. The gas tank in dad's car holds 12 gallons. The gas tank in mom's truck holds 50% more than that. How much gas does the truck's tank hold?

   Explain why this situation can be represented by the equation $(1.5) \cdot 12 = t$. Make sure that you explain what t represents.

2. Write an equation to represent each of the following situations.

   a. A movie theater decreased the size of its popcorn bags by 20%. If the old bags held 15 cups of popcorn, how much do the new bags hold?

   b. After a 25% discount, the price of a T-shirt was $12. What was the price before the discount?

   c. Compared to last year, the population of Boom Town has increased by 25%. The population is now 6,600. What was the population last year?

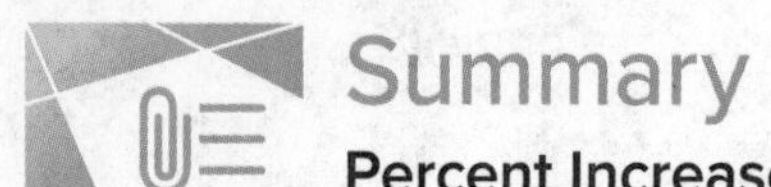

Summary

Percent Increase and Decrease with Equations

We can use equations to express percent increase and percent decrease. For example, if y is 15% more than x,

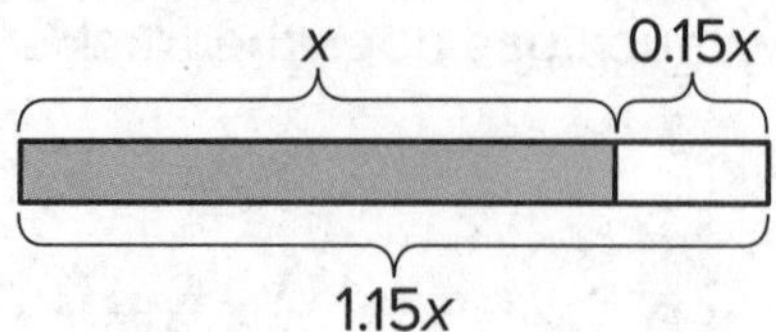

we can represent this using any of these equations:

$y = x + 0.15x$ $\quad$ $y = (1 + 0.15)x$ $\quad$ $y = 1.15x$

So, if someone makes an investment of x dollars, and its value increases by 15% to \$1,250, then we can write and solve the equation $1.15x = 1{,}250$ to find the value of the initial investment.

Here is another example: if a is 7% less than b,

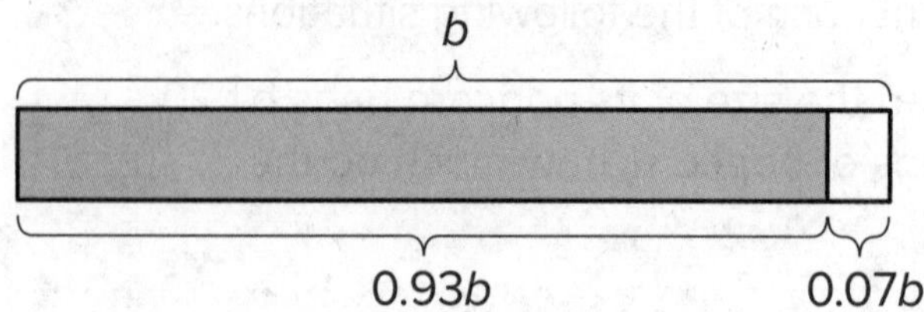

we can represent this using any of these equations:

$a = b - 0.07b$ $\quad$ $a = (1 - 0.07)b$ $\quad$ $a = 0.93b$

So if the amount of water in a tank decreased 7% from its starting value of b to its ending value of 348 gallons, then you can write $0.93b = 348$.

Often, an equation is the most efficient way to solve a problem involving percent increase or percent decrease.

NAME ______________________ DATE ____________ PERIOD ________

Practice

Percent Increase and Decrease with Equations

1. A pair of designer sneakers was purchased for $120. Since they were purchased, their price has increased by 15%. What is the new price?

2. Elena's aunt bought her a $150 savings bond when she was born. When Elena is 20 years old, the bond will have earned 105% in interest. How much will the bond be worth when Elena is 20 years old?

3. In a video game, Clare scored 50% more points than Tyler. If c is the number of points that Clare scored and t is the number of points that Tyler scored, which equations are correct? Select **all** that apply.

   A. $c = 1.5t$

   B. $c = t + 0.5$

   C. $c = t + 0.5t$

   D. $c = t + 50$

   E. $c = (1 + 0.5)t$

4. Draw a diagram to represent each situation: **(Lesson 4-6)**

a. The number of miles driven this month was a 30% decrease of the number of miles driven last month.

b. The amount of paper that the copy shop used this month was a 25% increase of the amount of paper they used last month.

5. Which decimal is the best estimate of the fraction $\frac{29}{40}$? **(Lesson 4-5)**

(A.) 0.5

(B.) 0.6

(C.) 0.7

(D.) 0.8

6. Could 7.2 inches and 28 inches be the diameter and circumference of the same circle? Explain why or why not. **(Lesson 3-3)**

Lesson 4-9

More and Less Than 1%

NAME ______________________ DATE ______________ PERIOD ____________

Learning Goal Let's explore percentages smaller than 1%.

Warm Up

9.1 Number Talk: What Percentage?

Determine the percentage mentally.

1. 10 is what percentage of 50?

2. 5 is what percentage of 50?

3. 1 is what percentage of 50?

4. 17 is what percentage of 50?

Activity

9.2 Waiting Tables

During one waiter's shift, he delivered 13 appetizers, 17 entrées, and 10 desserts.

1. What percentage of the dishes he delivered were:
   a. desserts?

   b. appetizers?

   c. entrées?

2. What do your percentages add up to?

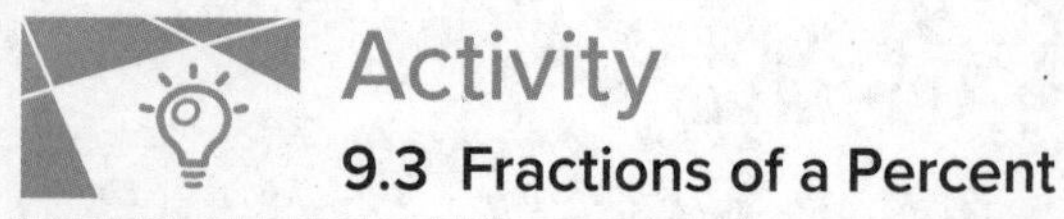

9.3 Fractions of a Percent

1. Find each percentage of 60. What do you notice about your answers?

   30% of 60 | 3% of 60 | 0.3% of 60 | 0.03% of 60

2. 20% of 5,000 is 1,000 and 21% of 5,000 is 1,050. Find each percentage of 5,000 and be prepared to explain your reasoning. If you get stuck, consider using the double number line diagram.

   a. 1% of 5,000

   b. 0.1% of 5,000

   c. 20.1% of 5,000

   d. 20.4% of 5,000

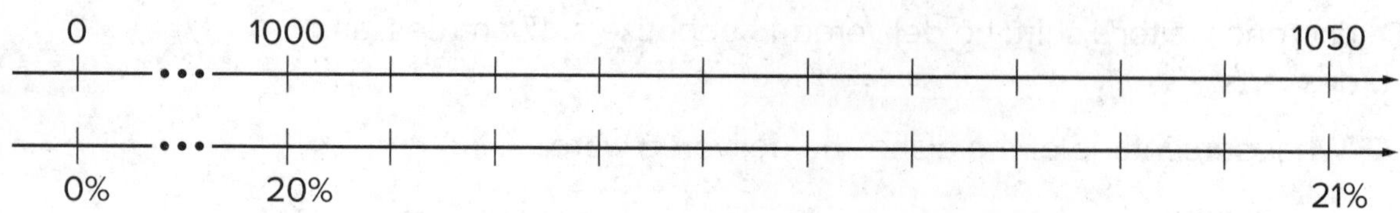

3. 15% of 80 is 12 and 16% of 80 is 12.8. Find each percentage of 80 and be prepared to explain your reasoning.

   a. 15.1% of 80

   b. 15.7% of 80

NAME ______________________ DATE ______________ PERIOD ____________

Are you ready for more?

To make Sierpinski's triangle,

- Start with an equilateral triangle. This is Step 1.
- Connect the midpoints of every side, and remove the middle triangle, leaving three smaller triangles. This is Step 2.
- Do the same to each of the remaining triangles. This is Step 3.
- Keep repeating this process.

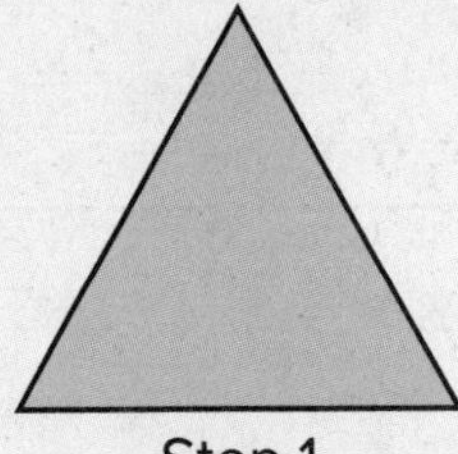
Step 1

Step 2

Step 3

1. What percentage of the area of the original triangle is left after Step 2? Step 3? Step 10?

2. At which step does the percentage first fall below 1%?

Activity

9.4 Population Growth

1. The population of City A was approximately 243,000 people, and it increased by 8% in one year. What was the new population?

2. The population of City B was approximately 7,150,000, and it increased by 0.8% in one year. What was the new population?

Summary

More and Less Than 1%

A percentage, such as 30%, is a rate per 100. To find 30% of a quantity, we multiply it by $30 \div 100$, or 0.3.

The same method works for percentages that are not whole numbers, like 7.8% or 2.5%.

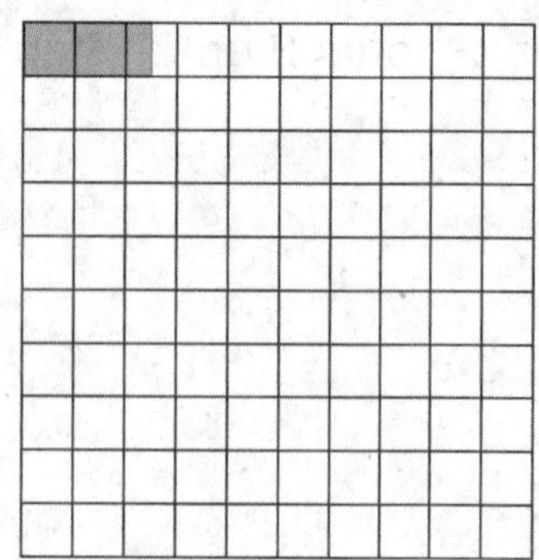

In the square, 2.5% of the area is shaded. To find 2.5% of a quantity, we multiply it by $2.5 \div 100$, or 0.025.

- For example, to calculate 2.5% interest on a bank balance of \$80, we multiply $(0.025) \cdot 80 = 2$, so the interest is \$2.

We can sometimes find percentages like 2.5% mentally by using convenient whole number percentages. For example, 25% of 80 is one-fourth of 80, which is 20. Since 2.5 is one tenth of 25, we know that 2.5% of 80 is one tenth of 20, which is 2.

NAME ______________________ DATE ______________ PERIOD ________

Practice

More and Less Than 1%

1. The student government snack shop sold 32 items this week.

   For each snack type, what percentage of all snacks sold were of that type?

Snack Type	Number of Items Sold
Fruit Cup	8
Veggie Sticks	6
Chips	14
Water	4

2. Select **all** the options that have the same value as $3\frac{1}{2}\%$ of 20.

   A. 3.5% of 20

   B. $3\frac{1}{2} \cdot 20$

   C. $(0.35) \cdot 20$

   D. $(0.035) \cdot 20$

   E. 7% of 10

3. 22% of 65 is 14.3. What is 22.6% of 65? Explain your reasoning.

4. A bakery used 30% more sugar this month than last month. If the bakery used 560 kilograms of sugar last month, how much did it use this month? **(Lesson 4-7)**

5. Match each situation to a diagram. The diagrams can be used more than once. **(Lesson 4-6)**

Situations

a. The amount of apples this year decreased by 15% compared with last year's amount.

b. The amount of pears this year is 85% of last year's amount.

c. The amount of cherries this year increased by 15% compared with last year's amount.

d. The amount of oranges this year is 115% of last year's amount.

Diagrams

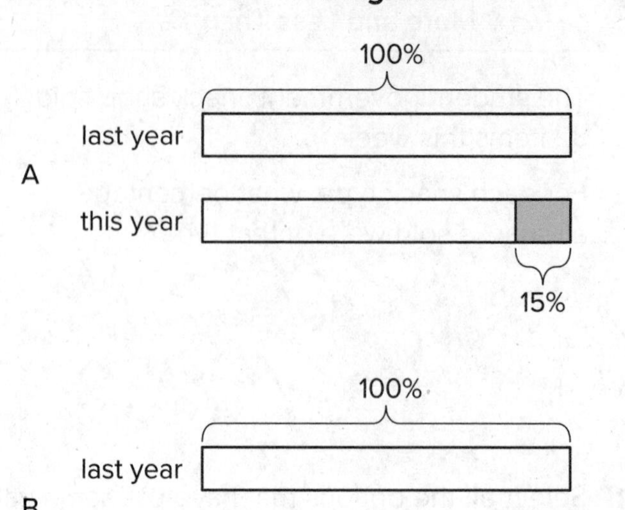

6. A certain type of car has room for 4 passengers. **(Lesson 2-6)**

a. Write an equation relating the number of cars (n) to the number of passengers (p).

b. How many passengers could fit in 78 cars?

c. How many cars would be needed to fit 78 passengers?

Lesson 4-10

Tax and Tip

NAME ______________________ DATE ____________ PERIOD ____________

Learning Goal Let's learn about sales tax and tips.

Warm Up

10.1 Notice and Wonder: The Price of Sunglasses

You are on vacation and want to buy a pair of sunglasses for $10 or less. You find a pair with a price tag of $10. The cashier says the total cost will be $10.45.

What do you notice? What do you wonder?

Activity

10.2 Shopping in Two Different Cities

Different cities have different sales tax rates. Here are the sales tax charges on the same items in two different cities. Complete the tables.

City 1:

Item	Price (dollars)	Sales Tax (dollars)	Total Cost (dollars)
Paper Towels	8.00	0.48	8.48
Lamp	25.00	1.50	
Pack of Gum	1.00		
Laundry Soap	12.00		
	x		

City 2:

Item	Price (dollars)	Sales Tax (dollars)	Total Cost (dollars)
Paper Towels	8.00	0.64	8.64
Lamp	25.00	2.00	
Pack of Gum	1.00		
Laundry Soap	12.00		
	x		

NAME ______________________________ DATE ______________ PERIOD ____________

Activity
10.3 Shopping in a Third City

Here is the sales tax on the same items in City 3.

Item	Price (dollars)	Sales Tax (dollars)
Paper Towels	8.00	0.58
Lamp	25.00	1.83
Pack of Gum	1.00	0.07
Laundry Soap	12.00	

1. What is the tax rate in this city?

2. For the sales tax on the laundry soap, Kiran says it should be $0.84. Lin says it should be $0.87. Do you agree with either of them? Explain your reasoning.

Activity

10.4 Dining at a Restaurant

1. Jada has a meal in a restaurant. She adds up the prices listed on the menu for everything they ordered and gets a subtotal of $42.00.

Date: Sep. 12th
Time: 6:55 PM
Server: #27

Bread Stix	9.50
Chicken Parm	15.50
Chef Salad	12.00
Lemon Soda	2.00
Tea	3.00
Subtotal	42.00
Sales Tax	3.99
Total	45.99

a. When the check comes, it says they also need to pay $3.99 in sales tax. What percentage of the subtotal is the sales tax?

b. After tax, the total is $45.99. What percentage of the subtotal is the total?

c. They actually pay $52.99. The additional $7 is a tip for the server. What percentage of the subtotal is the tip?

NAME ______________________ DATE ____________ PERIOD ____________

2. The tax rate at this restaurant is 9.5%.

Date: Sep. 12th	
Time: 6:04 PM	
Server: #27	
Bread Stix	9.50
Ravioli Bites	10.50
Cheesecake	4.95
Subtotal	24.95
Sales Tax	____
Total	____

Date: Sep. 12th	
Time: 7:12 PM	
Server: #27	
Garden Salad	____
Broccoli Bites	____
Subtotal	____
Sales Tax	1.61
Total	____

a. Another person's subtotal is $24.95. How much will their sales tax be?

b. Some other person's sales tax is $1.61. How much was their subtotal?

Are you ready for more?

Elena's cousins went to a restaurant. The part of the entire cost of the meal that was tax and tip together was 25% of the cost of the food alone. What could the tax rate and tip rate be?

Summary

Tax and Tip

Many places have *sales tax*. A sales tax is an amount of money that a government agency collects on the sale of certain items.

For example, a state might charge a tax on all cars purchased in the state. Often the tax rate is given as a percentage of the cost.

- For example, a state's tax rate on car sales might be 2%, which means that for every car sold in that state, the buyer has to pay a tax that is 2% of the sales price of the car.

Fractional percentages often arise when a state or city charges a sales tax on a purchase.

For example, the sales tax in Arizona is 7.5%. This means that when someone buys something, they have to add 0.075 times the amount on the price tag to determine the total cost of the item.

- For example, if the price tag on a T-shirt in Arizona says \$11.50, then the sales tax is $(0.075) \cdot 11.5 = 0.8625$, which rounds to 86 cents. The customer pays $11.50 + 0.86$, or \$12.36 for the shirt.

The total cost to the customer is the item price plus the sales tax. We can think of this as a percent increase. For example, in Arizona, the total cost to a customer is 107.5% of the price listed on the tag.

A *tip* is an amount of money that a person gives someone who provides a service. It is customary in many restaurants to give a tip to the server that is between 10% and 20% of the cost of the meal. If a person plans to leave a 15% tip on a meal, then the total cost will be 115% of the cost of the meal.

NAME ________________________ DATE ____________ PERIOD ________

Practice

Tax and Tip

1. In a city in Ohio, the sales tax rate is 7.25%. Complete the table to show the sales tax and the total price including tax for each item.

Item	Price Before Tax ($)	Sales Tax ($)	Price Including Tax ($)
Pillow	8.00		
Blanket	22.00		
Trash Can	14.50		

2. The sales tax rate in New Mexico is 5.125%. Select **all** the equations that represent the sales tax, t, you would pay in New Mexico for an item of cost c?

   A. $t = 5.125c$

   B. $t = 0.5125c$

   C. $t = 0.05125c$

   D. $t = c \div 0.05125$

   E. $t = \frac{5.125}{100}c$

3. Here are the prices of some items and the amount of sales tax charged on each in Nevada.

Cost of Item ($)	Sales Tax ($)
10	0.46
50	2.30
5	0.23

   a. What is the sales tax rate in Nevada?

   b. Write an expression for the amount of sales tax charged, in dollars, on an item that costs c dollars.

4. Find each amount. **(Lesson 4-9)**

a. 3.8% of 25

b. 0.2% of 50

c. 180.5% of 99

5. On Monday, the high was 60 degrees Fahrenheit. On Tuesday, the high was 18% more. How much did the high increase from Monday to Tuesday? **(Lesson 4-8)**

6. Complete the table. Explain or show your reasoning. **(Lesson 3-4)**

Object	Radius	Circumference
Ceiling Fan	2.8 ft	
Water Bottle Cap	13 mm	
Bowl		56.5 cm
Drum		75.4 in

Lesson 4-11

Percentage Contexts

NAME ______________________ DATE ______________ PERIOD ______________

Learning Goal Let's learn about more situations that involve percentages.

Warm Up

11.1 Leaving a Tip

Which of these expressions represent a 15% tip on a $20 meal?
Which represent the total bill?

$15 \cdot 20$

$20 + 0.15 \cdot 20$

$1.15 \cdot 20$

$\frac{15}{100} \cdot 20$

Activity

11.2 A Car Dealership

A car dealership pays a wholesale price of $12,000 to purchase a vehicle.

1. The car dealership wants to make a 32% profit.

   a. By how much will they mark up the price of the vehicle?

   b. After the markup, what is the retail price of the vehicle?

2. During a special sales event, the dealership offers a 10% discount off of the retail price. After the discount, how much will a customer pay for this vehicle?

Are you ready for more?

This car dealership pays the salesperson a bonus for selling the car equal to 6.5% of the sale price. How much commission did the salesperson lose when they decided to offer a 10% discount on the price of the car?

NAME ______________________ DATE ____________ PERIOD ____________

Activity
11.3 Commission at a Gym

1. For each gym membership sold, the gym keeps $42 and the employee who sold it gets $8. What is the commission the employee earns as a percentage of the total cost of the gym membership?

2. If an employee sells a family pass for $135, what is the amount of the commission they get to keep?

Activity
11.4 Card Sort: Percentage Situations

Your teacher will give you a set of cards. Take turns with your partner matching a situation with a descriptor. For each match, explain your reasoning to your partner. If you disagree, work to reach an agreement.

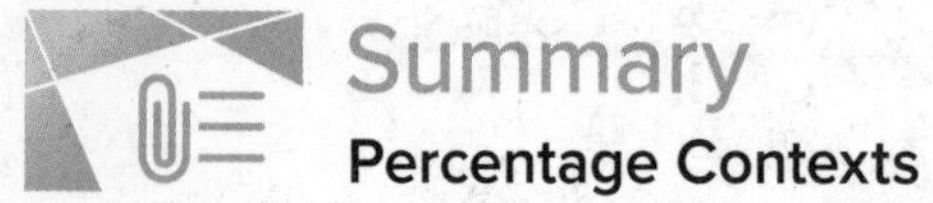

Summary

Percentage Contexts

There are many everyday situations where a percentage of an amount of money is added to or subtracted from that amount, in order to be paid to some other person or organization.

	Goes To	How It Works
Sales Tax	the government	added to the price of the item
Gratuity (tip)	the server	added to the cost of the meal
Interest	the lender (or account holder)	added to the balance of the loan, credit card, or bank account
Markup	the seller	added to the price of an item so the seller can make a profit
Markdown (discount)	the customer	subtracted from the price of an item to encourage the customer to buy it
Commission	the salesperson	subtracted from the payment that is collected

Some examples...

- If a restaurant bill is \$34 and the customer pays \$40, they left \$6 as a tip for the server.
    - That is 18% of \$34, so they left an 18% tip.
    - From the customer's perspective, we can think of this as an 18% increase of the restaurant bill.
- If a realtor helps a family sell their home for \$200,000 and earns a 3% commission, then...
    - the realtor makes \$6,000, because $(0.03) \cdot 200{,}000 = 6{,}000$,
    - and the family gets \$194,000, because $200{,}000 - 6{,}000 = 194{,}000$.
    - From the family's perspective, we can think of this as a 3% decrease on the sale price of the home.

NAME _______________ DATE _______________ PERIOD _______________

Practice

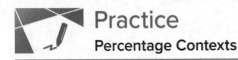

Percentage Contexts

1. A car dealership pays $8,350 for a car. They mark up the price by 17.4% to get the retail price. What is the retail price of the car at this dealership?

2. A store has a 20% off sale on pants. With this discount, the price of one pair of pants before tax is $15.20. What was the original price of the pants?

   A. $3.04

   B. $12.16

   C. $18.24

   D. $19.00

3. Lin is shopping for a couch with her dad and hears him ask the salesperson, "How much is your commission?" The salesperson says that her commission is $5\frac{1}{2}$% of the selling price.

   a. How much commission will the salesperson earn by selling a couch for $495?

   b. How much money will the store get from the sale of the couch?

4. A college student takes out a $7,500 loan from a bank. What will the balance of the loan be after one year (assuming the student has not made any payments yet)... **(Lesson 4-9)**

   a. if the bank charges 3.8% interest each year?

   b. if the bank charges 5.3% interest each year?

5. Match the situations with the equations. **(Lesson 4-5)**

Situations

a. Mai slept for x hours, and Kiran slept for $\frac{1}{10}$ less than that.

b. Kiran practiced the piano for x hours, and Mai practiced for $\frac{2}{5}$ less than that.

c. Mai drank x oz of juice and Kiran drank $\frac{4}{3}$ more than that.

d. Kiran spent x dollars and Mai spent $\frac{1}{4}$ less than that.

e. Mai ate x grams of almonds and Kiran ate 1.5 times more than that.

f. Kiran collected x pounds of recycling and Mai collected $\frac{3}{10}$ less than that.

g. Mai walked x kilometers and Kiran walked $\frac{3}{8}$ more than that.

h. Kiran completed x puzzles and Mai completed $\frac{3}{5}$ more than that.

Equations

$y = 2.33x$

$y = 1.375x$

$y = 0.6x$

$y = 0.9x$

$y = 0.75x$

$y = 1.6x$

$y = 0.7x$

$y = 2.5x$

Lesson 4-12

Finding the Percentage

NAME ______________________ DATE ______________ PERIOD ______________

Learning Goal Let's find unknown percentages.

Warm Up

12.1 Tax, Tip, and Discount

1. What percentage of the car price is the tax?

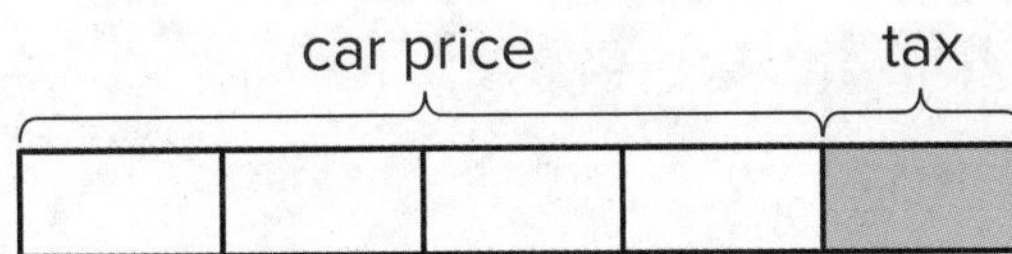

2. What percentage of the food cost is the tip?

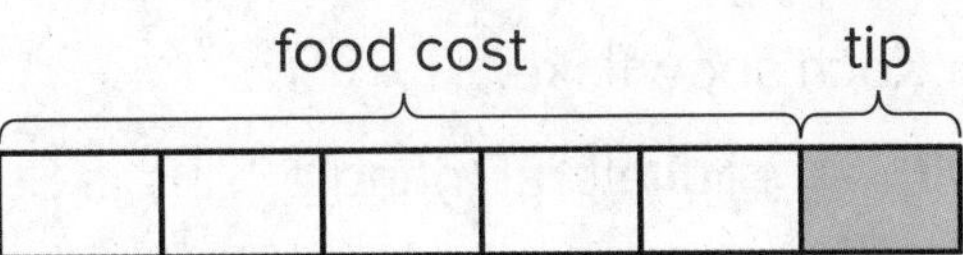

3. What percentage of the shirt cost is the discount?

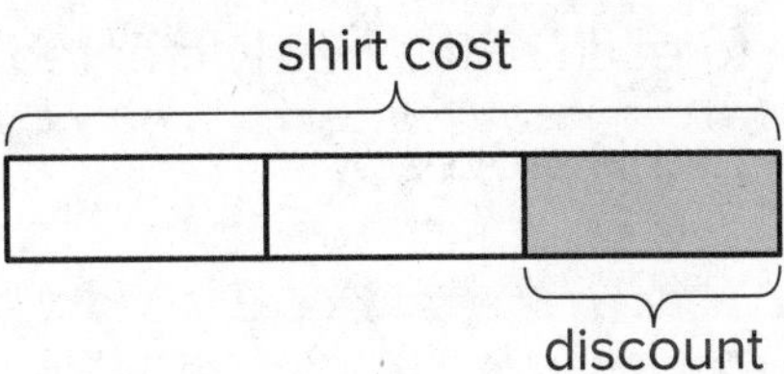

Activity

12.2 What Is the Percentage?

1. A salesperson sold a car for $18,250 and the commission is $693.50. What percentage of the sale price is their commission?

2. The bill for a meal was $33.75. The customer left $40. What percentage of the bill was the tip?

3. The original price of a bicycle was $375. Now it is on sale for $295. What percentage of the original price was the markdown?

Are you ready for more?

To make a Koch snowflake,

- Start with an equilateral triangle. This is Step 1.
- Divide each side into 3 equal pieces. Construct a smaller equilateral triangle on the middle third. This is Step 2.
- Do the same to each of the newly created sides. This is Step 3.
- Keep repeating this process.

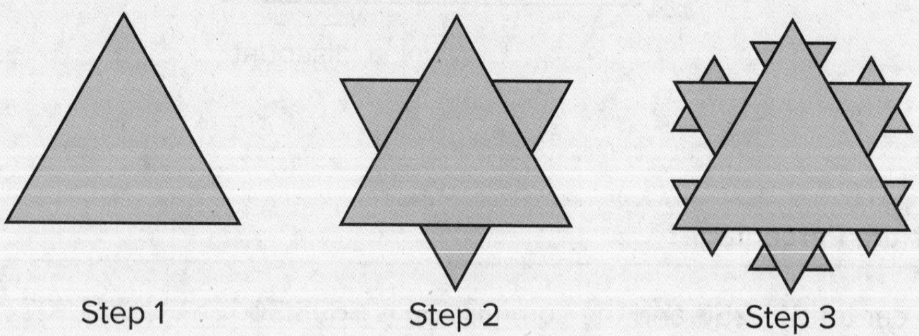

By what percentage does the perimeter increase at Step 2? Step 3? Step 10?

NAME ______________________ DATE ____________ PERIOD ________

Activity
12.3 Info Gap: Sporting Goods

Your teacher will give you either a *problem card* or a *data card*.
Do not show or read your card to your partner.

If your teacher gives you the ***problem card***:	If your teacher gives you the ***data card***:
1. Silently read your card and think about what information you need to be able to answer the question. **2.** Ask your partner for the specific information that you need. **3.** Explain how you are using the information to solve the problem. Continue to ask questions until you have enough information to solve the problem. **4.** Share the *problem card* and solve the problem independently. **5.** Read the *data card* and discuss your reasoning.	**1.** Silently read your card. **2.** Ask your partner *"What specific information do you need?"* and wait for them to *ask* for information. If your partner asks for information that is not on the card, do not do the calculations for them. Tell them you don't have that information. **3.** Before sharing the information, ask "*Why do you need that information?*" Listen to your partner's reasoning and ask clarifying questions. **4.** Read the *problem card* and solve the problem independently. **5.** Share the *data card* and discuss your reasoning.

Pause here so your teacher can review your work. Ask your teacher for a new set of cards and repeat the activity, trading roles with your partner.

Summary

Finding the Percentage

To find a 30% increase over 50, we can find 130% of 50. $1.3 \cdot 50 = 65$

To find a 30% decrease from 50, we can find 70% of 50. $0.7 \cdot 50 = 35$

If we know the initial amount and the final amount, we can also find the percent increase or percent decrease.

- For example, a plant was 12 inches tall and grew to be 15 inches tall. What percent increase is this? Here are two ways to solve this problem:
    - The plant grew 3 inches, because $15 - 12 = 3$. We can divide this growth by the original height, $3 \div 12 = 0.25$. So, the height of the plant increased by 25%.
    - The plant's new height is 125% of the original height, because $15 \div 12 = 1.25$. This means the height increased by 25%, because $125 - 100 = 25$.
- A rope was 2.4 meters long. Someone cut it down to 1.9 meters. What percent decrease is this? Here are two ways to solve the problem:
    - The rope is now $2.4 - 1.9$, or 0.5 meters shorter. We can divide this decrease by the original length, $0.5 \div 2.4 = 0.208\overline{3}$. So the length of the rope decreased by approximately 20.8%.
    - The rope's new length is about 79.2% of the original length, because $1.9 \div 2.4 = 0.791\overline{6}$. The length decreased by approximately 20.8%, because $100 - 79.2 = 20.8$.

NAME________________________ DATE____________ PERIOD__________

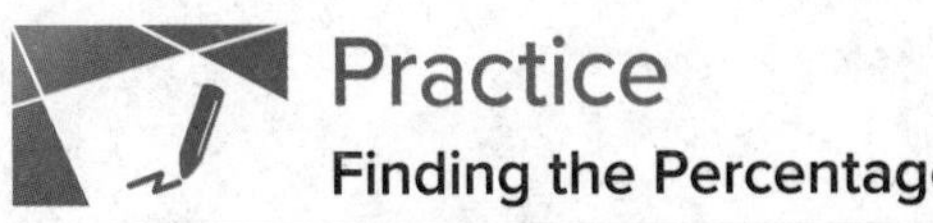

Practice
Finding the Percentage

1. A music store marks up the instruments it sells by 30%.

   a. If the store bought a guitar for $45, what will be its store price?

   b. If the price tag on a trumpet says $104, how much did the store pay for it?

   c. If the store paid $75 for a clarinet and sold it for $100, did the store mark up the price by 30%?

2. A family eats at a restaurant. The bill is $42 The family leaves a tip and spends $49.77.

   a. How much was the tip in dollars?

   b. How much was the tip as a percentage of the bill?

3. The price of gold is often reported per ounce. At the end of 2005, this price was \$513. At the end of 2015, it was \$1,060. By what percentage did the price per ounce of gold increase?

4. A phone keeps track of the number of steps taken and the distance traveled. Based on the information in the table, is there a proportional relationship between the two quantities? Explain your reasoning. **(Lesson 2-7)**

Number of Steps	Distance in Kilometers
950	1
2,852	3
4,845	5.1

5. Noah picked 3 kg of cherries. Mai picked half as many cherries as Noah. How many total kg of cherries did Mai and Noah pick? **(Lesson 4-4)**

A. $3 + \frac{1}{2}$

B. $3 - \frac{1}{2}$

C. $\left(1 + \frac{1}{2}\right) \cdot 3$

D. $1 + \frac{1}{2} \cdot 3$

Lesson 4-13

Measurement Error

NAME ______________________ DATE ____________ PERIOD ____________

Learning Goal Let's use percentages to describe how accurately we can measure.

WarmUp

13.1 Measuring to the Nearest

Your teacher will give you two rulers and three line segments labeled A, B, and C.

1. Use the centimeter ruler to measure each line segment to the nearest centimeter. Record these lengths in the first column of the table.

2. Use the millimeter ruler to measure each line segment to the nearest tenth of a centimeter. Record these lengths in the second column of the table.

Line Segment	Length (cm) as Measured With the First Ruler	Length (cm) as Measured With the Second Ruler
A		
B		
C		

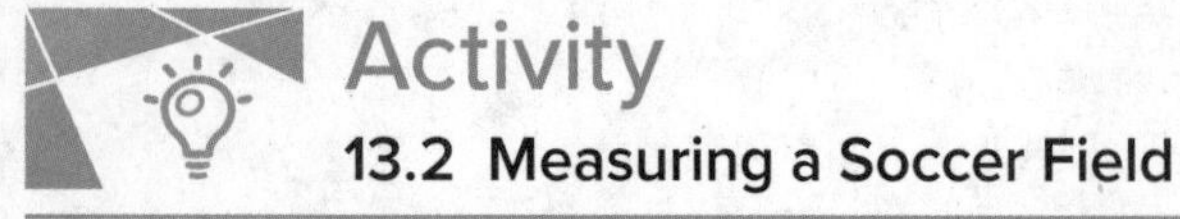

Activity
13.2 Measuring a Soccer Field

A soccer field is 120 yards long. Han measures the length of the field using a 30-foot-long tape measure and gets a measurement of 358 feet, 10 inches.

1. What is the amount of the error?

2. Express the error as a percentage of the actual length of the field.

Activity
13.3 Measuring Your Classroom

Your teacher will tell you which three items to measure. Keep using the paper rulers from the earlier activity.

1. Between you and your partner, decide who will use which ruler.

2. Measure the three items assigned by your teacher and record your measurements in the first column of the appropriate table.

   Using the cm ruler:

Item	Measured Length (cm)	Actual Length (cm)	Difference	Percentage

   Using the mm ruler:

Item	Measured Length (cm)	Actual Length (cm)	Difference	Percentage

NAME ______________________ DATE ____________ PERIOD __________

3. After you finish measuring the items, share your data with your partner. Next, ask your teacher for the actual lengths.

4. Calculate the difference between your measurements and the actual lengths in both tables.

5. For each difference, what percentage of the actual length is this amount? Record your answers in the last column of the tables.

Are you ready for more?

Before there were standard units of measurement, people often measured things using their hands or feet.

1. Measure the length of your foot to the nearest centimeter with your shoe on.

2. How many foot-lengths long is your classroom? Try to determine this as precisely as possible by carefully placing your heel next to your toe as you pace off the room.

3. Use this information to estimate the length of your classroom in centimeters.

4. Use a tape measure to measure the length of your classroom. What is the difference between the two measurements? Which one do you think is more accurate?

Summary

Measurement Error

When we are measuring a length using a ruler or measuring tape, we can get a measurement that is different from the actual length. This could be because we positioned the ruler incorrectly, or it could be because the ruler is not very precise. There is always at least a small difference between the actual length and a measured length, even if it is a microscopic difference!

Here are two rulers with different markings.

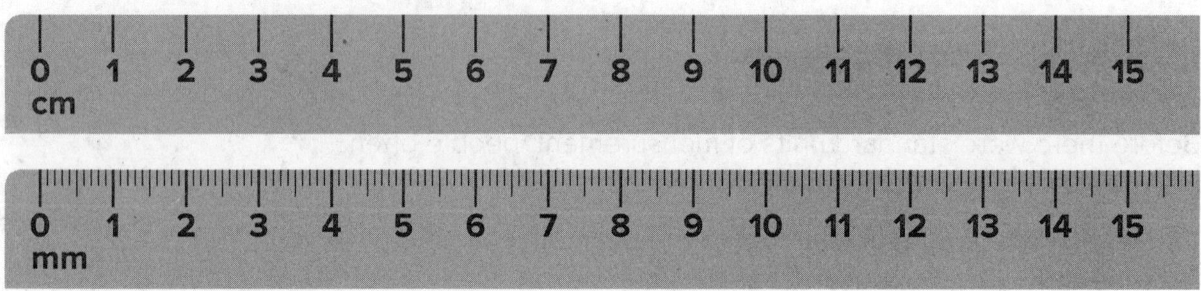

The second ruler is marked in millimeters, so it is easier to get a measurement to the nearest tenth of a centimeter with this ruler than with the first. For example, a line that is actually 6.2 cm long might be measured to be 6 cm long by the first ruler, because we measure to the nearest centimeter.

The **measurement error** is the positive difference between the measurement and the actual value.

- Measurement error is often expressed as a percentage of the actual value.
- We always use a positive number to express measurement error and, when appropriate, use words to describe whether the measurement is greater than or less than the actual value.

For example, if we get 6 cm when we measure a line that is actually 6.2 cm long, then the measurement error is 0.2 cm, or about 3.2%, because $0.2 \div 6.2 \approx 0.032$.

Glossary

measurement error

NAME ______________________ DATE ____________ PERIOD ________

Practice

Measurement Error

1. The depth of a lake is 15.8 m.

   a. Jada accurately measured the depth of the lake to the nearest meter. What measurement did Jada get?

   b. By how many meters does the measured depth differ from the actual depth?

   c. Express the measurement error as a percentage of the actual depth.

2. A watermelon weighs 8,475 grams. A scale measured the weight with an error of 12% under the actual weight. What was the measured weight?

3. Noah's oven thermometer gives a reading that is 2% greater than the actual temperature.

   a. If the actual temperature is 325°F, what will the thermometer reading be?

   b. If the thermometer reading is 76°F, what is the actual temperature?

4. At the beginning of the month, there were 80 ounces of peanut butter in the pantry. Now, there is $\frac{1}{3}$ less than that. How many ounces of peanut butter are in the pantry now? **(Lesson 4-4)**

A. $\frac{2}{3} \cdot 80$

B. $\frac{1}{3} \cdot 80$

C. $80 - \frac{1}{3}$

D. $1 + \frac{1}{3} \cdot 80$

5. a. Fill in the table for side length and area of different squares. **(Lesson 3-7)**

Side Length (cm)	Area (cm^2)
3	
100	
25	
s	

b. Is the relationship between the side length of a square and the area of a square proportional?

Lesson 4-14

Percent Error

NAME ______________________ DATE ____________ PERIOD ____________

Learning Goal Let's use percentages to describe other situations that involve error.

Warm Up

14.1 Number Talk: Estimating a Percentage of a Number

Estimate.

1. 25% of 15.8
2. 9% of 38
3. 1.2% of 127
4. 0.53% of 6
5. 0.06% of 202

Activity

14.2 Plants, Bicycles, and Crowds

1. Instructions to care for a plant say to water it with $\frac{3}{4}$ cup of water every day. The plant has been getting 25% too much water. How much water has the plant been getting?

2. The pressure on a bicycle tire is 63 psi. This is 5% higher than what the manual says is the correct pressure. What is the correct pressure?

3. The crowd at a sporting event is estimated to be 3,000 people. The exact attendance is 2,486 people. What is the **percent error**?

Are you ready for more?

A micrometer is an instrument that can measure lengths to the nearest micron (a micron is a millionth of a meter). Would this instrument be useful for measuring any of the following things? If so, what would the largest percent error be?

1. The thickness of an eyelash, which is typically about 0.1 millimeters.

2. The diameter of a red blood cell, which is typically about 8 microns.

3. The diameter of a hydrogen atom, which is about 100 picometers (a picometer is a trillionth of a meter).

NAME ______________________ DATE ______________ PERIOD ______________

Activity
14.3 Measuring in the Heat

A metal measuring tape expands when the temperature goes above 50° F. For every degree Fahrenheit above 50, its length increases by 0.00064%.

1. The temperature is 100 degrees Fahrenheit. How much longer is a 30-foot measuring tape than its correct length?

2. What is the percent error?

Summary
Percent Error

Percent error can be used to describe any situation where there is a correct value and an incorrect value, and we want to describe the relative difference between them.

For example, if a milk carton is supposed to contain 16 fluid ounces and it only contains 15 fluid ounces:

- the measurement error is 1 oz, and
- the percent error is 6.25% because $1 \div 16 = 0.0625$.

We can also use percent error when talking about estimates. For example, a teacher estimates there are about 600 students at their school. If there are actually 625 students, then the percent error for this estimate was 4%, because $625 - 600 = 25$ and $25 \div 625 = 0.04$.

Glossary

percent error

Practice

Percent Error

1. A student estimated that it would take 3 hours to write a book report, but it actually took her 5 hours. What is the percent error for her estimate?

2. A radar gun measured the speed of a baseball at 103 miles per hour. If the baseball was actually going 102.8 miles per hour, what was the percent error in this measurement?

3. It took 48 minutes to drive downtown. An app estimated it would be less than that. If the error was 20%, what was the app's estimate?

4. A farmer estimated that there were 25 gallons of water left in a tank. If this is an underestimate by 16%, how much water was actually in the tank?

NAME ________________ DATE ________ PERIOD ________

5. For each story, write an equation that describes the relationship between the two quantities. **(Lesson 4-4)**

a. Diego collected x kg of recycling. Lin collected $\frac{2}{5}$ more than that.

b. Lin biked x km. Diego biked $\frac{3}{10}$ less than that.

c. Diego read for x minutes. Lin read $\frac{4}{7}$ of that.

6. For each diagram, decide if y is an increase or a decrease of x. Then determine the percentage. **(Lesson 4-12)**

a.

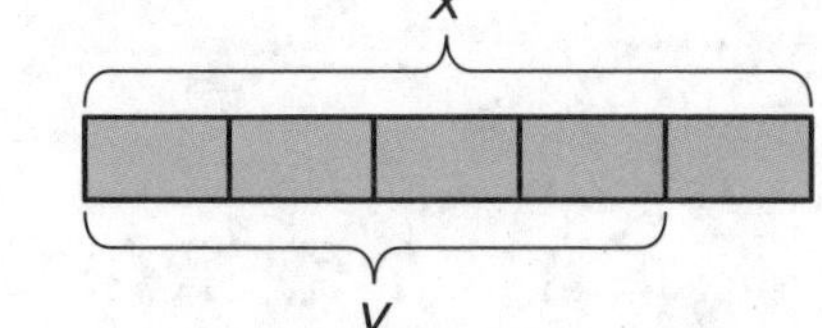

b.

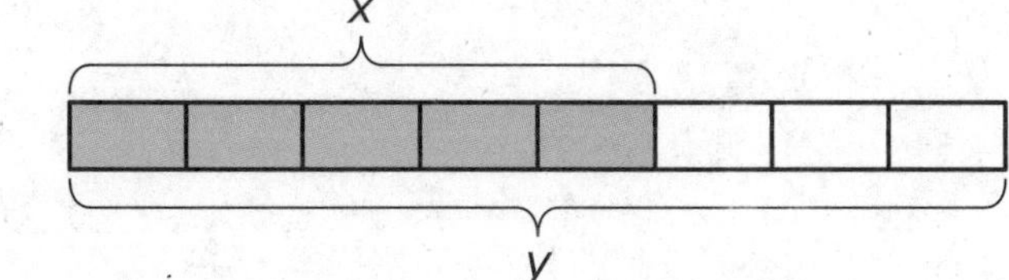

7. Lin is making a window covering for a window that has the shape of a half circle on top of a square of side length 3 feet. How much fabric does she need? **(Lesson 3-10)**

Lesson 4-15

Error Intervals

NAME ________________________ DATE ____________ PERIOD ________

Learning Goal Let's solve more problems about percent error.

Warm Up

15.1 A Lot of Iron Ore

An industrial scale is guaranteed by the manufacturer to have a percent error of no more than 1%. What is a possible reading on the scale if you put 500 kilograms of iron ore on it?

Activity

15.2 Saw Mill

1. A saw mill cuts boards that are 16 ft long. After they are cut, the boards are inspected and rejected if the length has a percent error of 1.5% or more.
   a. List some board lengths that should be accepted.
   b. List some board lengths that should be rejected.
2. The saw mill also cuts boards that are 10, 12, and 14 feet long. An inspector rejects a board that was 2.3 inches too long. What was the intended length of the board?

Activity

15.3 Info Gap: Quality Control

Your teacher will give you either a *problem card* or a *data card*.
Do not show or read your card to your partner.

If your teacher gives you the *problem card*:	If your teacher gives you the *data card*:
1. Silently read your card and think about what information you need to be able to answer the question. 2. Ask your partner for the specific information that you need. 3. Explain how you are using the information to solve the problem. Continue to ask questions until you have enough information to solve the problem. 4. Share the *problem card* and solve the problem independently. 5. Read the *data card* and discuss your reasoning.	1. Silently read your card. 2. Ask your partner *"What specific information do you need?"* and wait for them to *ask* for information. If your partner asks for information that is not on the card, do not do the calculations for them. Tell them you don't have that information. 3. Before sharing the information, ask "*Why do you need that information?*" Listen to your partner's reasoning and ask clarifying questions. 4. Read the *problem card* and solve the problem independently. 5. Share the *data card* and discuss your reasoning.

Pause here so your teacher can review your work. Ask your teacher for a new set of cards and repeat the activity, trading roles with your partner.

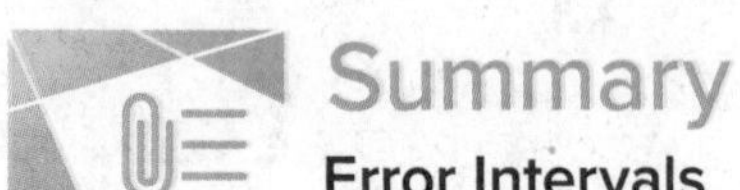

Summary

Error Intervals

Percent error is often used to express a range of possible values.
For example, if a box of cereal is guaranteed to have 750 grams of cereal, with a margin of error of less than 5%, what are possible values for the actual number of grams of cereal in the box? The error could be as large as $(0.05) \cdot 750 = 37.5$ and could be either above or below than the correct amount.

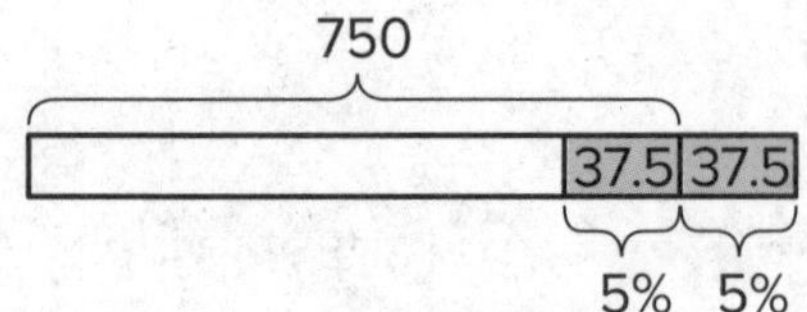

Therefore, the box can have anywhere between 712.5 and 787.5 grams of cereal in it, but it should not have 700 grams or 800 grams, because both of those are more than 37.5 grams away from 750 grams.

NAME ______________________ DATE ____________ PERIOD ________

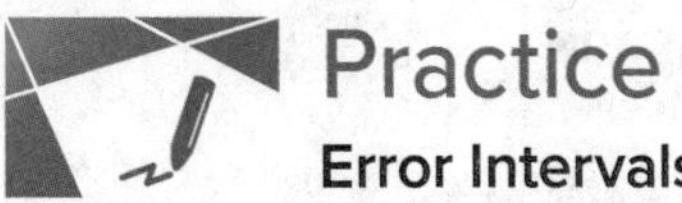

Practice
Error Intervals

1. Jada measured the height of a plant in a science experiment and finds that, to the nearest $\frac{1}{4}$ of an inch, it is $4\frac{3}{4}$ inches.

    a. What is the largest the actual height of the plant could be?

    b. What is the smallest the actual height of the plant could be?

    c. How large could the percent error in Jada's measurement be?

2. The reading on a car's speedometer has 1.6% maximum error. The speed limit on a road is 65 miles per hour.

    a. The speedometer reads 64 miles per hour. Is it possible that the car is going over the speed limit?

    b. The speedometer reads 66 miles per hour. Is the car definitely going over the speed limit?

3. Water is running into a bathtub at a constant rate. After 2 minutes, the tub is filled with 2.5 gallons of water. Write two equations for this proportional relationship. Use w for the amount of water (gallons) and t for time (minutes). In each case, what does the constant of proportionality tell you about the situation? **(Lesson 2-5)**

4. Noah picked 3 kg of cherries. Jada picked half as many cherries as Noah. How many total kg of cherries did Jada and Noah pick? **(Lesson 4-5)**

(A.) $3 + 0.5$

(B.) $3 - 0.5$

(C.) $(1 + 0.5) \cdot 3$

(D.) $1 + 0.5 \cdot 3$

5. Here is a shape with some measurements in cm. **(Lesson 3-7)**

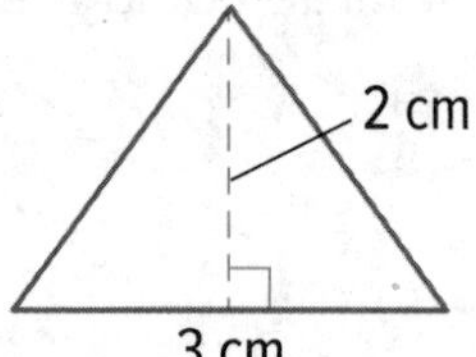

a. Complete the table showing the area of different scaled copies of the triangle.

Scale Factor	Area (cm^2)
1	
2	
5	
s	

b. Is the relationship between the scale factor and the area of the scaled copy proportional?

Lesson 4-16

Posing Percentage Problems

NAME _______________ DATE _______________ PERIOD _______________

Learning Goal Let's explore how percentages are used in the news.

Warm Up

16.1 Sorting the News

Your teacher will give you a variety of news clippings that include percentages.

1. Sort the clippings into two piles: those that are about increases and those that are about decreases.

2. Were there any clippings that you had trouble deciding which pile they should go in?

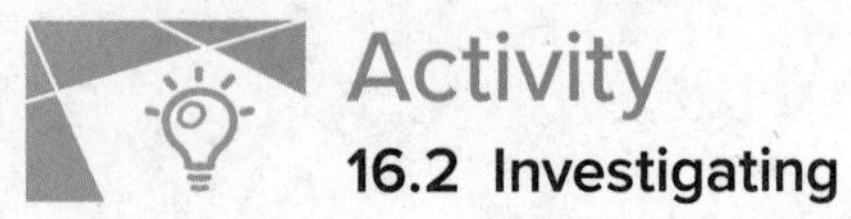

Activity

16.2 Investigating

In the previous activity, you sorted news clippings into two piles.

1. For each pile, choose one example. Draw a diagram that shows how percentages are being used to describe the situation.

   a. Increase Example:

   b. Decrease Example:

2. For each example, write *two* questions that you can answer with the given information. Next, find the answers. Explain or show your reasoning.

NAME ______________________ DATE ____________ PERIOD ________

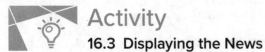

Activity
16.3 Displaying the News

1. Choose the example that you find the most interesting. Create a visual display that includes:
   - a title that describes the situation
   - the news clipping
   - your diagram of the situation
   - the two questions you asked about the situation
   - the answers to each of your questions
   - an explanation of how you calculated each answer

   Pause here so your teacher can review your work.

2. Examine each display. Write one comment and one question for the group.

3. Next, read the comments and questions your classmates wrote for your group. Revise your display using the feedback from your classmates.

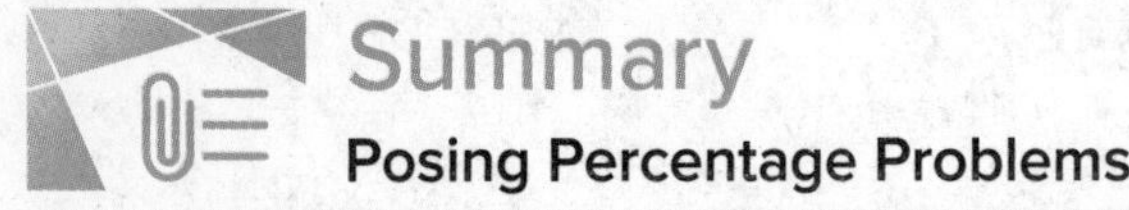

Summary

Posing Percentage Problems

Statements about percentage increase or decrease need to specify what the whole is to be mathematically meaningful.

Sometimes advertisements, media, etc. leave the whole ambiguous in order to make somewhat misleading claims. We should be careful to think critically about what mathematical claim is being made.

For example, if a disinfectant claims to “kill 99% of all bacteria,” does it mean that

- it kills 99% of the number of bacteria on a surface?
- or is it 99% of the types of bacteria commonly found inside the house?
- or 99% of the total mass or volume of bacteria?
- does it even matter if the remaining 1% are the most harmful bacteria?

Resolving questions of this type is an important step in making informed decisions.

Learning Targets

Lesson		Learning Target(s)
4-1	Lots of Flags	• I can find dimensions on scaled copies of a rectangle. • I remember how to compute percentages.
4-2	Ratios and Rates With Fractions	• I can solve problems about ratios of fractions and decimals.
4-3	Revisiting Proportional Relationships	• I can use a table with 2 rows and 2 columns to find an unknown value in a proportional relationship. • When there is a constant rate, I can identify the two quantities that are in a proportional relationship.

(continued on the next page)

(continued from the previous page)

Lesson		Learning Target(s)
4-4	Half as Much Again	• I can use the distributive property to rewrite an expression like $x + \frac{1}{2}x$ as $\left(1 + \frac{1}{2}\right)x$. • I understand that "half as much again" and "multiply by $\frac{3}{2}$" mean the same thing.
4-5	Say It with Decimals	• I can use the distributive property to rewrite an equation like $x + 0.5x = 1.5x$. • I can write fractions as decimals. • I understand that "half as much again" and "multiply by 1.5" mean the same thing.
4-6	Increasing and Decreasing	• I can draw a tape diagram that represents a percent increase or decrease. • When I know a starting amount and the percent increase or decrease, I can find the new amount.
4-7	One Hundred Percent	• I can use a double number line diagram to help me solve percent increase and decrease problems. • I understand that if I know how much a quantity has grown, then the original amount represents 100%. • When I know the new amount and the percentage of increase or decrease, I can find the original amount.

Lesson		Learning Target(s)
4-8	Percent Increase and Decrease with Equations	• I can solve percent increase and decrease problems by writing an equation to represent the situation and solving it.
4-9	More and Less than 1%	• I can find percentages of quantities like 12.5% and 0.4%. • I understand that to find 0.1% of an amount I have to multiply by 0.001.
4-10	Tax and Tip	• I understand and can solve problems about sales tax and tip.
4-11	Percentage Contexts	• I understand and can solve problems about commission, interest, markups, and discounts.
4-12	Finding the Percentage	• I can find the percentage increase or decrease when I know the original amount and the new amount.

(continued on the next page)

(continued from the previous page)

Lesson	Learning Target(s)
4-13 Measurement Error	• I can represent measurement error as a percentage of the correct measurement. • I understand that all measurements include some error.
4-14 Percent Error	• I can solve problems that involve percent error.
4-15 Error Intervals	• I can find a range of possible values for a quantity if I know the maximum percent error and the correct value.
4-16 Posing Percentage Problems	• I can write and solve problems about real-world situations that involve percent increase and decrease.

Notes:

Unit 5

Rational Number Arithmetic

Creative Travel Projects/Shutterstock

How much colder? In one of the upcoming lessons, you'll use rational numbers to compare winter temperatures from various locations.

Topics

- Interpreting Negative Numbers
- Adding and Subtracting Rational Numbers
- Multiplying and Dividing Rational Numbers
- Four Operations with Rational Numbers
- Solving Equations When There are Negative Numbers
- Let's Put It to Work

Unit 5

Rational Number Arithmetic

Lesson 5-1

Interpreting Negative Numbers

NAME _______________ DATE _______________ PERIOD _______________

Learning Goal Let's review what we know about signed numbers.

Here is a weather thermometer. Three of the numbers have been left off.

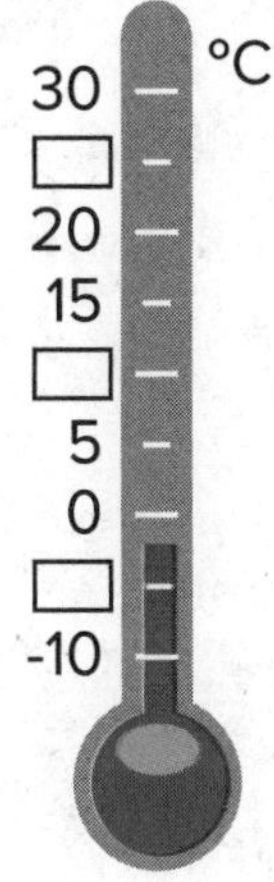

1. What numbers go in the boxes?

2. What temperature does the thermometer show?

Activity

1.2 Fractions of a Degree

1. What temperature is shown on each thermometer?

2. Which thermometer shows the highest temperature?

3. Which thermometer shows the lowest temperature?

4. Suppose the temperature outside is -4°C. Is that colder or warmer than the coldest temperature shown? How do you know?

NAME ______________________ DATE ____________ PERIOD ________

Activity
1.3 Seagulls Soar, Sharks Swim

Here is a picture of some sea animals. The number line on the left shows the vertical position of each animal above or below sea level, in meters.

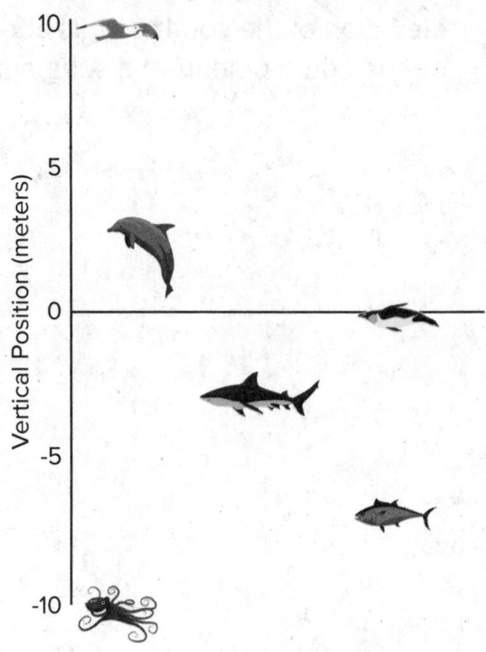

1. How far above or below sea level is each animal? Measure to their eye level.

2. A mobula ray is 3 meters above the surface of the ocean. How does its vertical position compare to the height or depth of:

   a. the jumping dolphin? b. the flying seagull? c. the octopus?

3. An albatross is 5 meters above the surface of the ocean. How does its vertical position compare to the height or depth of:

   a. the jumping dolphin? b. the flying seagull? c. the octopus?

4. A clownfish is 2 meters below the surface of the ocean. How does its vertical position compare to the height or depth of:

   a. the jumping dolphin? b. the flying seagull? c. the octopus?

5. The vertical distance of a new dolphin from the dolphin in the picture is 3 meters. What is its distance from the surface of the ocean?

Are you ready for more?

The north pole is in the middle of the ocean. A person at sea level at the north pole would be 3,949 miles from the center of the earth. The sea floor below the north pole is at an elevation of approximately -2.7 miles. The elevation of the south pole is about 1.7 miles. How far is a person standing on the south pole from a submarine at the sea floor below the north pole?

Activity

1.4 Card Sort: Rational Numbers

1. Your teacher will give your group a set of cards. Order the cards from least to greatest.
2. Pause here so your teacher can review your work. Then, your teacher will give you a second set of cards.
3. Add the new set of cards to the first set so that all of the cards are ordered from least to greatest.

NAME ______________________ DATE ______________ PERIOD ____________

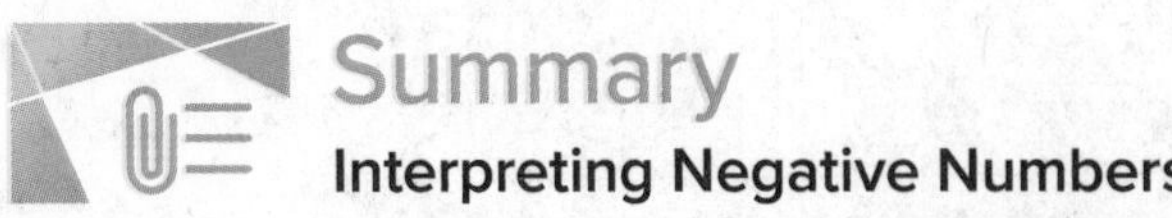

Summary

Interpreting Negative Numbers

We can use **positive numbers** and **negative numbers** to represent temperature and elevation.

When numbers represent temperatures, positive numbers indicate temperatures that are warmer than zero and negative numbers indicate temperatures that are colder than zero.

This thermometer shows a temperature of -1 degree Celsius, which we write -1° C.

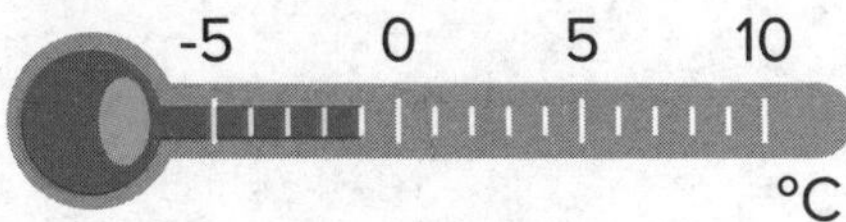

When numbers represent elevations, positive numbers indicate positions above sea level and negative numbers indicate positions below sea level.

We can see the order of signed numbers on a number line.

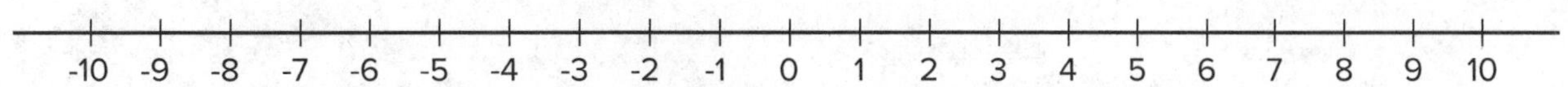

A number is always less than numbers to its right. So $-7 < -3$.

We use **absolute value** to describe how far a number is from 0.

- The numbers 15 and -15 are both 15 units from 0, so $|15| = 15$ and $|-15| = 15$.
- We call 15 and -15 *opposites.* They are on opposite sides of 0 on the number line, but the same distance from 0.

Glossary

absolute value
negative number
positive number

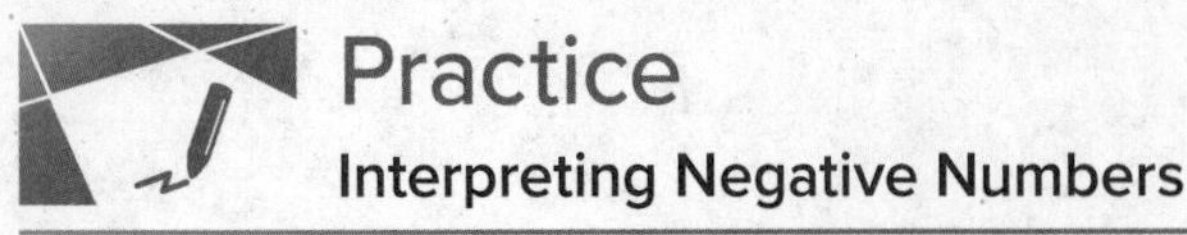

Practice

Interpreting Negative Numbers

1. It was -5° C in Copenhagen and -12° C in Oslo. Which city was colder?

2. Respond to each of the following.
   a. A fish is 12 meters below the surface of the ocean. What is its elevation?

   b. A sea bird is 28 meters above the surface of the ocean. What is its elevation?

   c. If the bird is directly above the fish, how far apart are they?

3. Compare using >, =, or <.

   a. 3 ______ -3 b. 12 ______ 24 c. -12 ______ -24 d. 5 ______ -(-5)

   e. 7.2 ______ 7 f. -7.2 ______ -7 g. -1.5 ______ $\frac{-3}{2}$ h. $\frac{-4}{5}$ ______ $\frac{-5}{4}$

   i. $\frac{-3}{5}$ ______ $\frac{-6}{10}$ j. $\frac{-2}{3}$ ______ $\frac{1}{3}$

4. Han wants to buy a $30 ticket to a game, but the pre-order tickets are sold out. He knows there will be more tickets sold the day of the game, with a markup of 200%. How much should Han expect to pay for the ticket if he buys it the day of the game? **(Lesson 4-7)**

5. A type of green paint is made by mixing 2 cups of yellow with 3.5 cups of blue. **(Lesson 2-1)**
   a. Find a mixture that will make the same shade of green but a smaller amount.

   b. Find a mixture that will make the same shade of green but a larger amount.

   c. Find a mixture that will make a different shade of green that is bluer.

   d. Find a mixture that will make a different shade of green that is more yellow.

Lesson 5-2

Changing Temperatures

NAME ______________________ DATE ____________ PERIOD ____________

Learning Goal Let's add signed numbers.

Warm Up

2.1 Which One Doesn't Belong: Arrows

Which pair of arrows doesn't belong?

A.

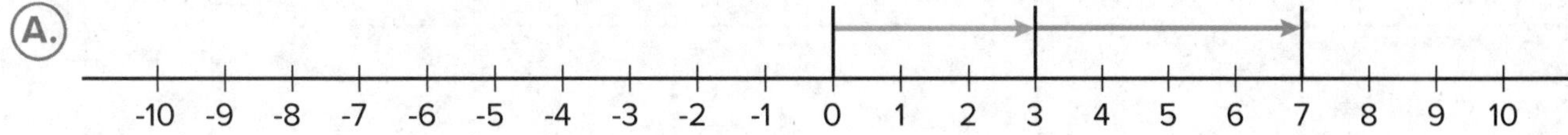

B.

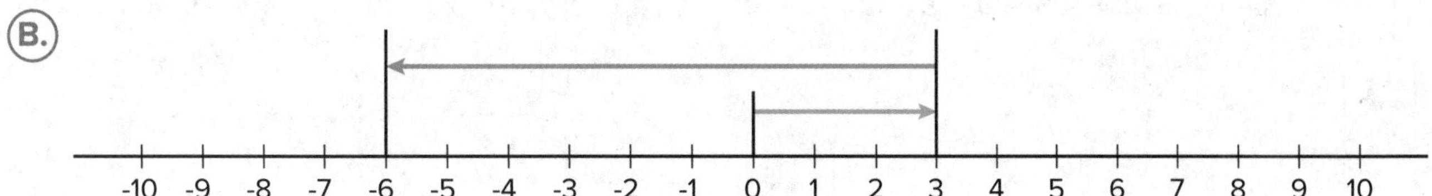

C.

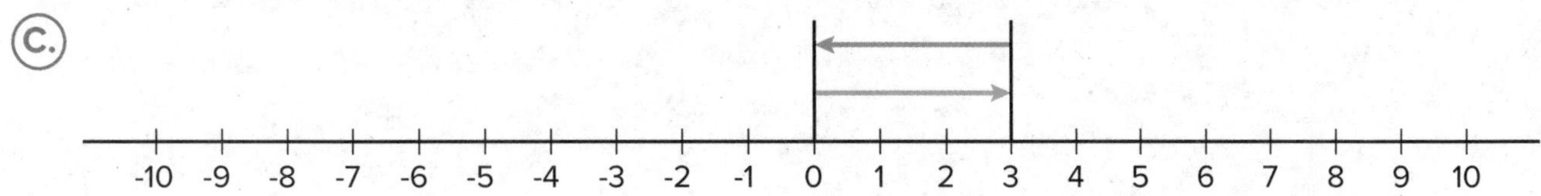

D.

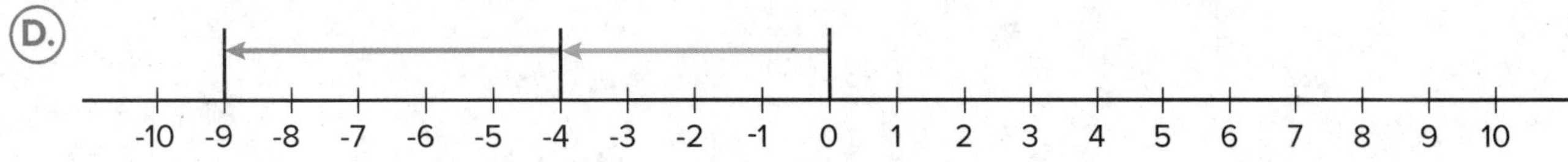

Activity

2.2 Warmer and Colder

1. Complete the table and draw a number line diagram for each situation.

	Start (°C)	Change (°C)	Final (°C)	Addition Equation
a.	+40	10 degrees warmer	+50	40 + 10 = 50
b.	+40	5 degrees colder		
c.	+40	30 degrees colder		
d.	+40	40 degrees colder		
e.	+40	50 degrees colder		

a. 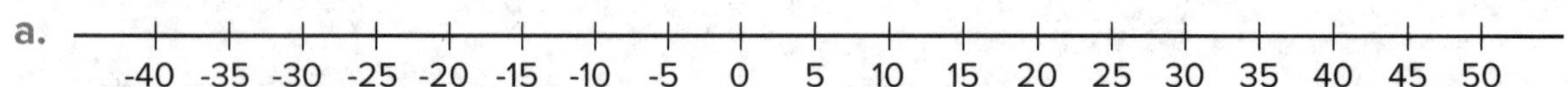

b. 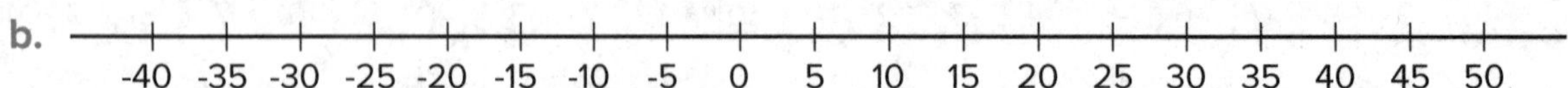

c. 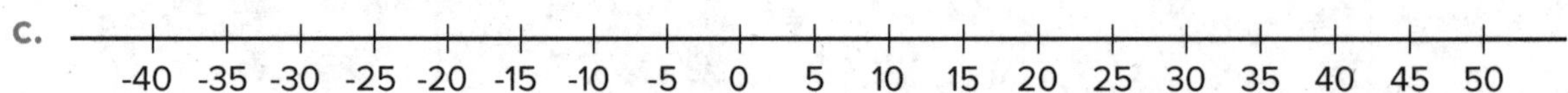

d. 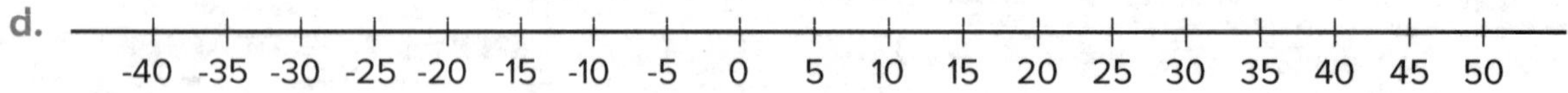

e. 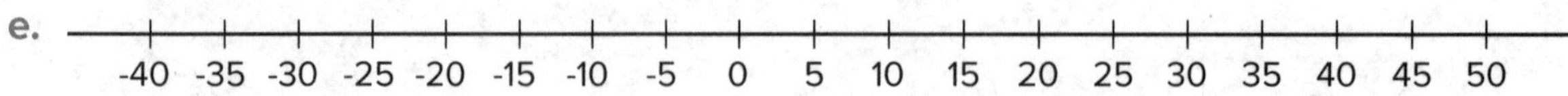

NAME ______________________ DATE ____________ PERIOD ____________

2. Complete the table and draw a number line diagram for each situation.

	Start (°C)	Change (°C)	Final (°C)	Addition Equation
a.	-20	30 degrees warmer		
b.	-20	35 degrees warmer		
c.	-20	15 degrees warmer		
d.	-20	15 degrees colder		

a. 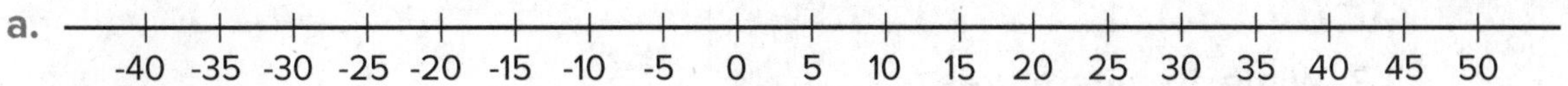

b. 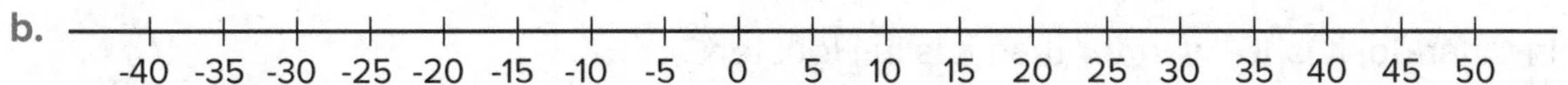

c. 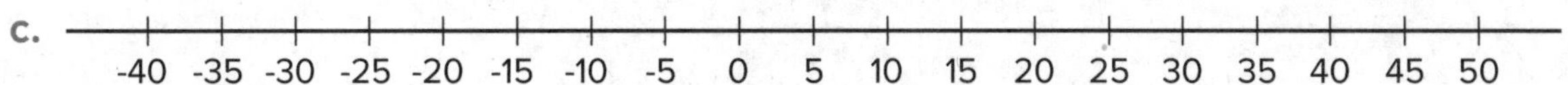

d. 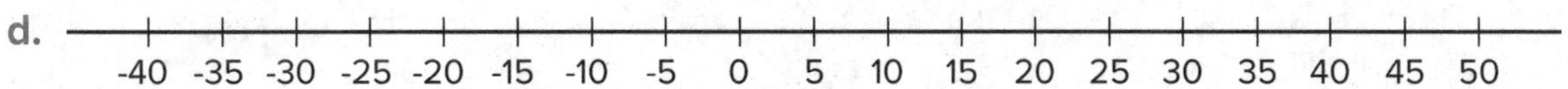

Are you ready for more?

For the numbers a and b represented in the figure, which expression is equal to $|a + b|$?

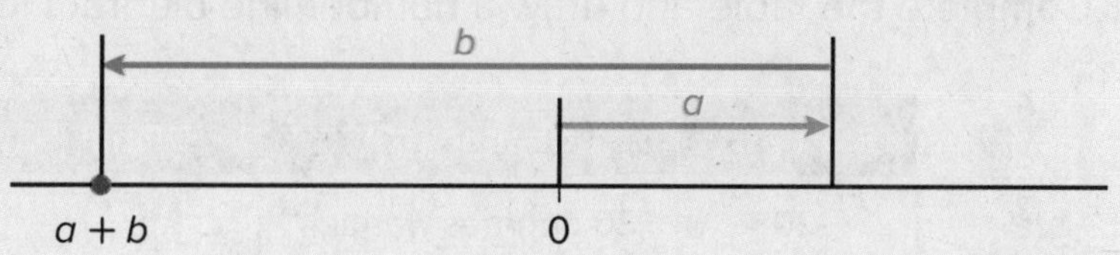

$|a| + |b|$ $|a| - |b|$ $|b| - |a|$

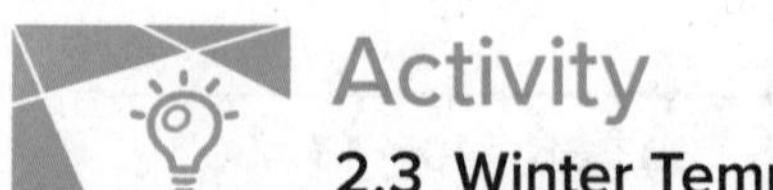

Activity

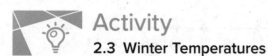

2.3 Winter Temperatures

One winter day, the temperature in Houston is 8° Celsius. Find the temperatures in these other cities. Explain or show your reasoning.

1. In Orlando, it is 10° warmer than it is in Houston.

2. In Salt Lake City, it is 8° colder than it is in Houston.

3. In Minneapolis, it is 20° colder than it is in Houston.

4. In Fairbanks, it is 10° colder than it is in *Minneapolis*.

5. Write an addition equation that represents the relationship between the temperature in Houston and the temperature in Fairbanks.

NAME ______________________ DATE ____________ PERIOD ________

Summary

Changing Temperatures

If it is 42° outside and the temperature increases by 7°, then we can add the initial temperature and the change in temperature to find the final temperature.

$42 + 7 = 49$

If the temperature decreases by 7°, we can either subtract $42 - 7$ to find the final temperature, or we can think of the change as -7°. Again, we can add to find the final temperature.

$42 + (-7) = 35$

In general, we can represent a change in temperature with a positive number if it increases and a negative number if it decreases. Then we can find the final temperature by adding the initial temperature and the change. If it is 3° and the temperature decreases by 7°, then we can add to find the final temperature.

$3 + (-7) = -4$

We can represent signed numbers with arrows on a number line. We can represent positive numbers with arrows that start at 0 and points to the right.

- For example, this arrow represents +10 because it is 10 units long and it points to the right.

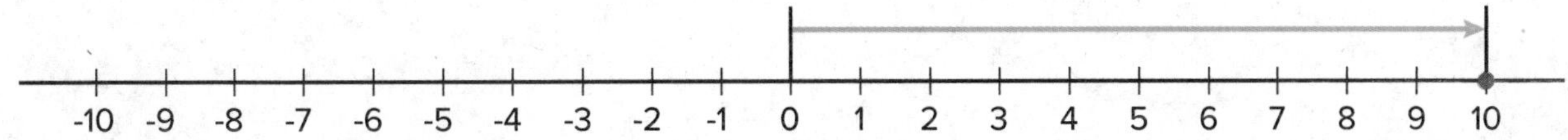

We can represent negative numbers with arrows that start at 0 and point to the left.

- For example, this arrow represents -4 because it is 4 units long and it points to the left.

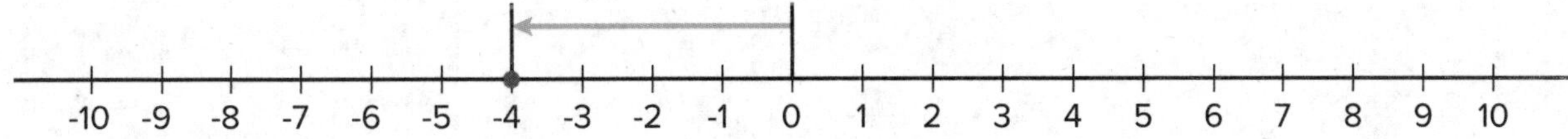

To represent addition, we put the arrows "tip to tail." So this diagram represents 3 + 5.

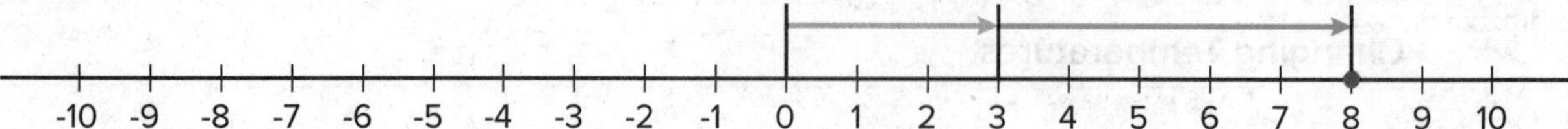

And this represents 3 + (-5).

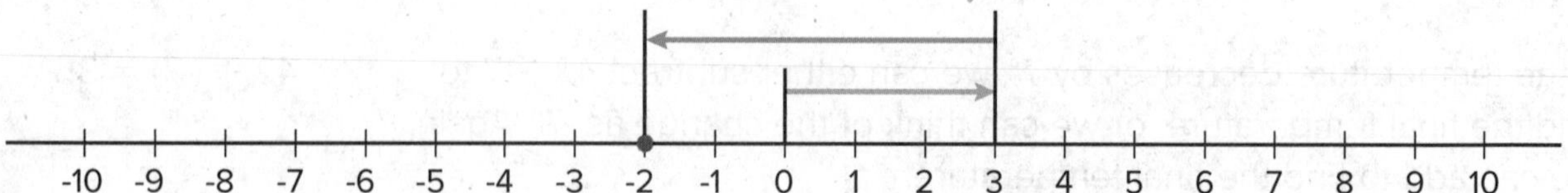

NAME ______________________ DATE ____________ PERIOD ____________

Practice

Changing Temperatures

1. Respond to each question.

   a. The temperature is -2° C. If the temperature rises by 15° C, what is the new temperature?

   b. At midnight the temperature is -6° C. At midday the temperature is 9° C. By how much did the temperature rise?

2. Draw a diagram to represent each of these situations. Then write an addition expression that represents the final temperature.

   a. The temperature was 80° F and then fell 20° F.

   b. The temperature was -13° F and then rose 9° F.

   c. The temperature was -5° F and then fell 8° F.

3. Complete each statement with a number that makes the statement true. **(Lesson 5-1)**

   a. ______ < 7° C

   b. ______ < -3° C

   c. -0.8° C < ______ < -0.1° C

   d. ______ > -2° C

4. Decide whether each table could represent a proportional relationship. If the relationship could be proportional, what would be the constant of proportionality? **(Lesson 2-7)**

   a. The number of wheels on a group of buses.

Number of Buses	Number of Wheels	Wheels per Bus
5	30	
8	48	
10	60	
15	90	

   b. The number of wheels on a train.

Number of Train Cars	Number of Wheels	Wheels per Train Car
20	184	
30	264	
40	344	
50	424	

5. Noah was assigned to make 64 cookies for the bake sale. He made 125% of that number. 90% of the cookies he made were sold. How many of Noah's cookies were left after the bake sale? **(Lesson 4-7)**

Lesson 5-3

Changing Elevation

NAME ______________________ DATE ____________ PERIOD ________

Learning Goal Let's solve problems about adding signed numbers.

Warm Up

3.1 That's the Opposite

1. Draw arrows on a number line to represent these situations:

   a. The temperature was -5 degrees. Then the temperature rose 5 degrees.

   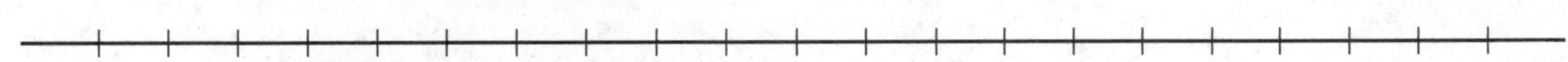

   b. A climber was 30 feet above sea level. Then she descended 30 feet.

   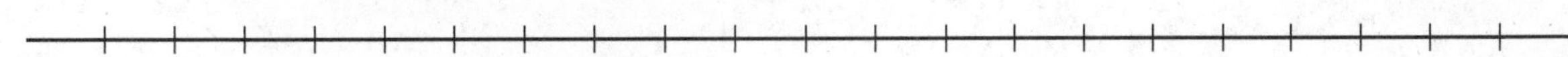

2. What's the opposite?

   a. Running 150 feet east.

   b. Jumping down 10 steps.

   c. Pouring 8 gallons into a fish tank.

Activity

3.2 Cliffs and Caves

1. A mountaineer is climbing on a cliff. She is 400 feet above the ground. If she climbs up, this will be a positive change. If she climbs down, this will be a negative change.

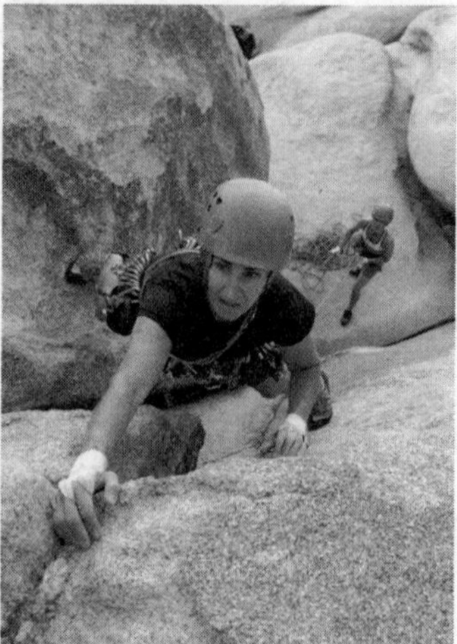

a. Complete the table.

	Starting Elevation (feet)	Change (feet)	Final Elevation (feet)
A	+400	300 up	
B	+400	150 down	
C	+400	400 down	
D	+400		+50

b. Write an addition equation and draw a number line diagram for B. Include the starting elevation, change, and final elevation in your diagram.

NAME ______________________ DATE ____________ PERIOD ____________

2. A spelunker is down in a cave next to the cliff. If she climbs down deeper into the cave, this will be a negative change. If she climbs up, whether inside the cave or out of the cave and up the cliff, this will be a positive change.

a. Complete the table.

	Starting Elevation (feet)	Change (feet)	Final Elevation (feet)
A	-200	150 down	
B	-200	100 up	
C	-200	200 up	
D	-200	250 up	
E	-200		-500

b. Write an addition equation and draw a number line diagram for C and D. Include the starting elevation, change, and final elevation in your diagram.

c. What does the expression -75 + 100 tell us about the spelunker? What does the value of the expression tell us?

Activity

3.3 Adding Rational Numbers

Find the sums.

1. -35 + (30 + 5)

2. -0.15 + (-0.85) + 12.5

3. $\frac{1}{2} + \left(-\frac{3}{4}\right)$

Are you ready for more?

Find the sum without a calculator.

10 + 21 + 32 + 43 + 54 + (-54) + (-43) + (-32) + (-21) + (-10)

NAME ______________________ DATE ______________ PERIOD ____________

Activity

3.4 School Supply Number Line

Your teacher will give you a long strip of paper.

Follow these instructions to create a number line.

1. Fold the paper in half along its length and along its width.
2. Unfold the paper and draw a line along each crease.
3. Label the line in the middle of the paper 0. Label the right end of the paper + and the left end of the paper −.
4. Select two objects of different lengths, for example a pen and a glue stick. The length of the longer object is a and the length of the shorter object is b.
5. Use the objects to measure and label each of the following points on your number line.

a	$2b$	$-b$
b	$a + b$	$a + -b$
$2a$	$-a$	$b + -a$

6. Complete each statement using <, >, or =. Use your number line to explain your reasoning.

   a. a ______ b

   b. $-a$ ______ $-b$

   c. $a + -a$ ______ $b + -b$

   d. $a + -b$ ______ $b + -a$

   e. $a + -b$ ______ $-a + b$

Summary

Changing Elevation

The opposite of a number is the same distance from 0 but on the other side of 0. The opposite of -9 is 9.

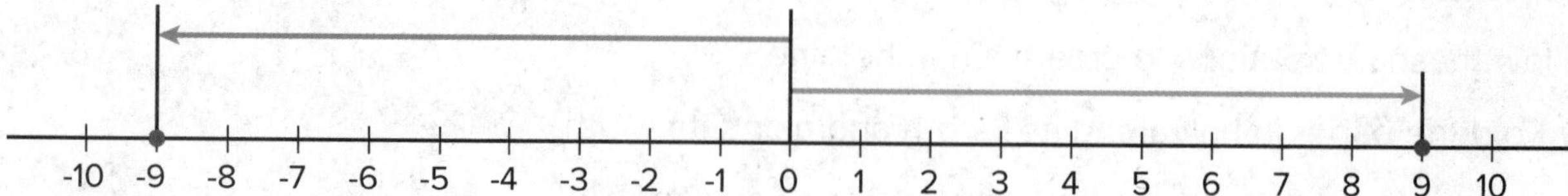

When we add opposites, we always get 0. This diagram shows that $9 + -9 = 0$.

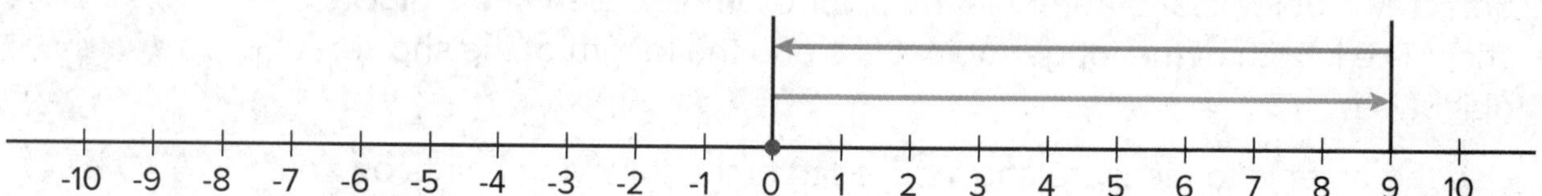

When we add two numbers with the same sign, the arrows that represent them point in the same direction. When we put the arrows tip to tail, we see the sum has the same sign.

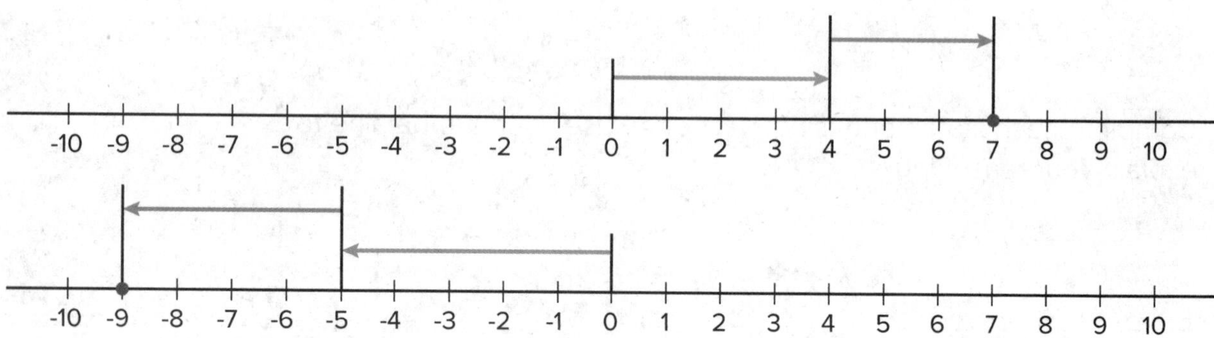

To find the sum, we add the magnitudes and give it the correct sign. For example, $(-5) + (-4) = -(5 + 4)$.

On the other hand, when we add two numbers with different signs, we subtract their magnitudes (because the arrows point in the opposite direction) and give it the sign of the number with the larger magnitude.
For example, $(-5) + 12 = +(12 - 5)$.

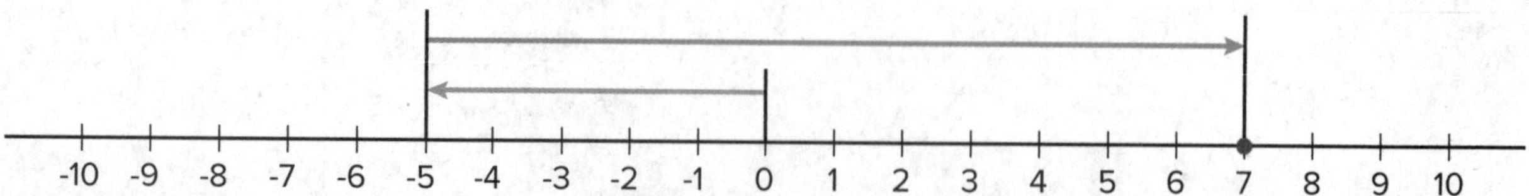

NAME ______________________ DATE ____________ PERIOD ____________

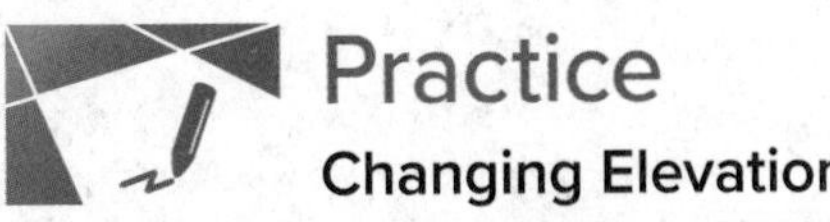

Practice
Changing Elevation

1. What is the final elevation if:

   a. a bird starts at 20 m and changes 16 m?

   b. a butterfly starts at 20 m and changes -16 m?

   c. a diver starts at 5 m and changes -16 m?

   d. a whale starts at -9 m and changes 11 m?

   e. a fish starts at -9 meters and changes -11 m?

2. One of the particles in an atom is called an electron. It has a charge of -1. Another particle in an atom is a proton. It has charge of +1. The charge of an atom is the sum of the charges of the electrons and the protons. A carbon atom has an overall charge of 0, because it has 6 electrons and 6 protons and -6 + 6 = 0. Find the overall charge for the rest of the elements on the list.

	Charge from Electrons	Charge from Protons	Overall Charge
Carbon	-6	+6	0
Neon	-10	+10	
Oxide	-10	+8	
Copper	-27	+29	
Tin	-50	+50	

3. Add.

a. $14.7 + 28.9$

b. $-9.2 + 4.4$

c. $-81.4 + (-12)$

d. $51.8 + (-0.8)$

4. Last week, the price, in dollars, of a gallon of gasoline was g. This week, the price of gasoline per gallon increased by 5%. Which expressions represent this week's price, in dollars, of a gallon of gasoline? Select **all** that apply. (Lesson 4-8)

(A.) $g + 0.05$

(B.) $g + 0.05g$

(C.) $1.05g$

(D.) $0.05g$

(E.) $(1 + 0.05)g$

5. Decide whether each table could represent a proportional relationship. If the relationship could be proportional, what would be the constant of proportionality? (Lesson 2-7)

a. Annie's Attic is giving away $5 off coupons.

Original Price	Sale Price
$15	$10
$25	$20
$35	$30

b. Bettie's Boutique is having a 20% off sale.

Original Price	Sale Price
$15	$12
$25	$20
$35	$28

Lesson 5-4

Money and Debts

NAME ______________________ DATE ____________ PERIOD ____________

Learning Goal Let's apply what we know about signed numbers to money.

Warm Up

4.1 Concert Tickets

Priya wants to buy three tickets for a concert. She has earned $135 and each ticket costs $50. She borrows the rest of the money she needs from a bank and buys the tickets.

1. How can you represent the amount of money that Priya has after buying the tickets?

2. How much more money will Priya need to earn to pay back the money she borrowed from the bank?

3. How much money will she have after she pays back the money she borrowed from the bank?

Activity

4.2 Cafeteria Food Debt

At the beginning of the month, Kiran had $24 in his school cafeteria account. Use a variable to represent the unknown quantity in each transaction below and write an equation to represent it. Then, represent each transaction on a number line. What is the unknown quantity in each case?

1. In the first week he spent $16 on lunches. How much was in his account then?

2. Then he deposited some more money and his account balance was $28. How much did he deposit?

3. Then he spent $34 on lunches the next week. How much was in his account then?

4. Then he deposited enough money to pay off his debt to the cafeteria. How much did he deposit?

5. Explain why it makes sense to use a negative number to represent Kiran's account balance when he owes money.

NAME _______________ DATE _______________ PERIOD _______________

Activity
4.3 Bank Statement

Here is a bank statement.

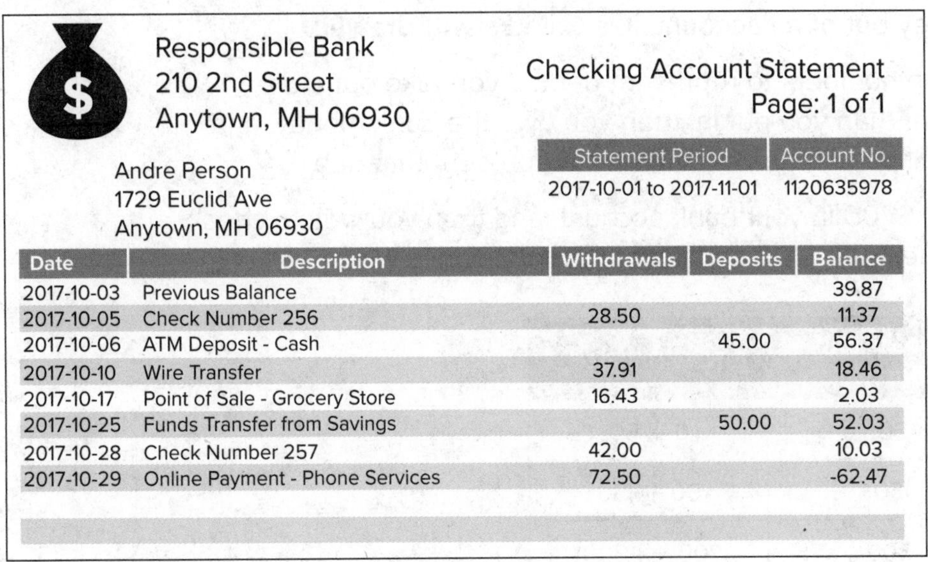

Responsible Bank
210 2nd Street
Anytown, MH 06930

Checking Account Statement
Page: 1 of 1

Andre Person
1729 Euclid Ave
Anytown, MH 06930

Statement Period	Account No.
2017-10-01 to 2017-11-01	1120635978

Date	Description	Withdrawals	Deposits	Balance
2017-10-03	Previous Balance			39.87
2017-10-05	Check Number 256	28.50		11.37
2017-10-06	ATM Deposit - Cash		45.00	56.37
2017-10-10	Wire Transfer	37.91		18.46
2017-10-17	Point of Sale - Grocery Store	16.43		2.03
2017-10-25	Funds Transfer from Savings		50.00	52.03
2017-10-28	Check Number 257	42.00		10.03
2017-10-29	Online Payment - Phone Services	72.50		-62.47

1. If we put withdrawals and deposits in the same column, how can they be represented?

2. Andre withdraws $40 to buy a music player. What is his new balance?

3. If Andre deposits $100 in this account, will he still be in debt? How do you know?

Are you ready for more?

The *national debt* of a country is the total amount of money the government of that country owes. Imagine everyone in the United States was asked to help pay off the national debt. How much would each person have to pay?

Summary

Money and Debts

Banks use positive numbers to represent money that gets put into an account and negative numbers to represent money that gets taken out of an account.

- When you put money into an account, it is called a **deposit**.
- When you take money out of an account, it is called a **withdrawal**.

People also use negative numbers to represent debt. If you take out more money from your account than you put in, then you owe the bank money, and your account balance will be a negative number to represent that debt.

For example, if you have $200 in your bank account, and then you write a check for $300, you will owe the bank $100 and your account balance will be -$100.

Starting Balance	Deposits and Withdrawals	New Balance
0	50	0 + 50
50	150	50 + 150
200	-300	200 + (-300)
-100		

In general, you can find a new account balance by adding the value of the deposit or withdrawal to it. You can also tell quickly how much money is needed to repay a debt using the fact that to get to zero from a negative value you need to add its opposite.

Glossary

deposit
withdrawal

NAME ______________________ DATE ____________ PERIOD ________

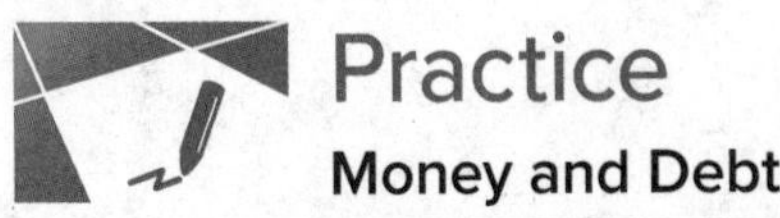

Practice

Money and Debts

1. The table shows five transactions and the resulting account balance in a bank account, except some numbers are missing. Fill in the missing numbers.

	Transaction Amount	Account Balance
Transaction 1	200	200
Transaction 2	-147	53
Transaction 3	90	
Transaction 4	-229	
Transaction 5		0

2. Respond to each of the following.

   a. Clare has $54 in her bank account. A store credits her account with a $10 refund. How much does she now have in the bank?

   b. Mai's bank account is overdrawn by $60, which means her balance is -$60. She gets $85 for her birthday and deposits it into her account. How much does she now have in the bank?

c. Tyler is overdrawn at the bank by $180. He gets $70 for his birthday and deposits it. What is his account balance now?

d. Andre has $37 in his bank account and writes a check for $87. After the check has been cashed, what will the bank balance show?

3. Last week, it rained g inches. This week, the amount of rain decreased by 5%. Which expressions represent the amount of rain that fell this week? Select **all** that apply. **(Lesson 4-8)**

A. $g - 0.05$

B. $g - 0.05g$

C. $0.95g$

D. $0.05g$

E. $(1 - 0.05)g$

NAME ______________________ DATE ______________ PERIOD ______________

4. Decide whether or not each equation represents a proportional relationship. **(Lesson 2-8)**

a. Volume measured in cups (c) vs. the same volume measured in ounces (z): $c = \frac{1}{8}z$

b. Area of a square (A) vs. the side length of the square (s): $A = s^2$

c. Perimeter of an equilateral triangle (P) vs. the side length of the triangle (s): $3s = P$

d. Length (L) vs. width (w) for a rectangle whose area is 60 square units: $L = \frac{60}{w}$

5. Add. **(Lesson 5-3)**

a. $5\frac{3}{4} + \left(-\frac{1}{4}\right)$

b. $-\frac{2}{3} + \frac{1}{6}$

c. $-\frac{8}{5} + \left(-\frac{3}{4}\right)$

6. In each diagram, x represents a different value. For each diagram,
(Lesson 5-1)

a. What is something that is *definitely* true about the value of x?

b. What is something that *could be* true about the value of x?

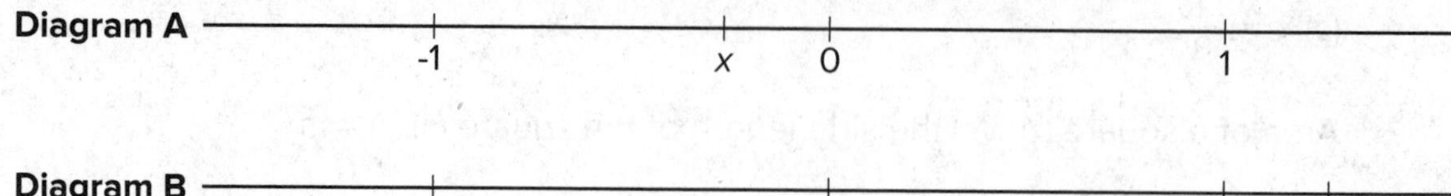

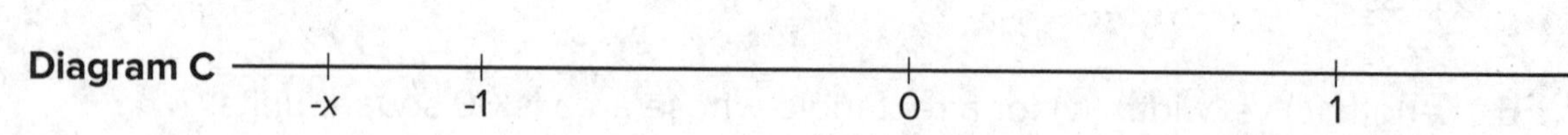

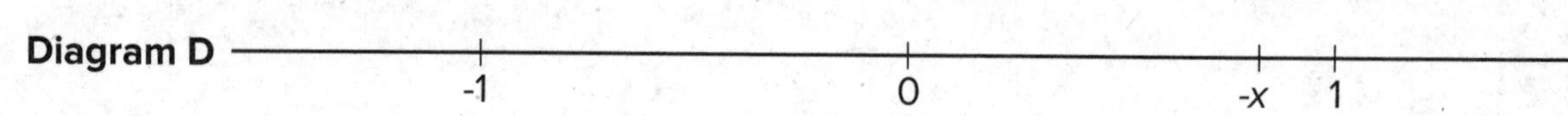

Lesson 5-5

Representing Subtraction

NAME ________________ DATE ________ PERIOD ________

Learning Goal Let's subtract signed numbers.

Warm Up

5.1 Equivalent Equations

Consider the equation $2 + 3 = 5$. Here are some more equations, using the same numbers, that express the same relationship in a different way:

$3 + 2 = 5$ $\quad$ $5 - 3 = 2$ $\quad$ $5 - 2 = 3$

For each equation, write two more equations, using the same numbers, that express the same relationship in a different way.

1. $9 + (-1) = 8$

2. $-11 + x = 7$

Activity

5.2 Subtraction with Number Lines

1. Here is an unfinished number line diagram that represents a sum of 8.

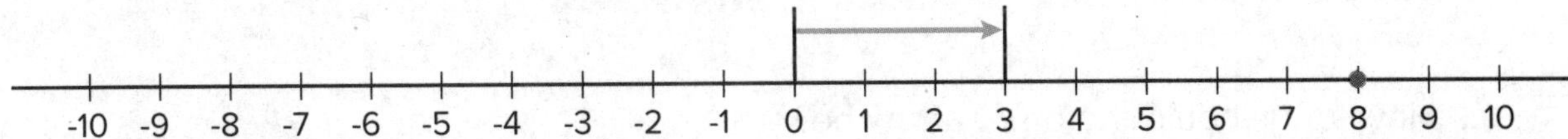

a. How long should the other arrow be?

b. For an equation that goes with this diagram, Mai writes $3 + ? = 8$. Tyler writes $8 - 3 = ?$. Do you agree with either of them?

c. What is the unknown number? How do you know?

2. Here are two more unfinished diagrams that represent sums.

First Diagram:

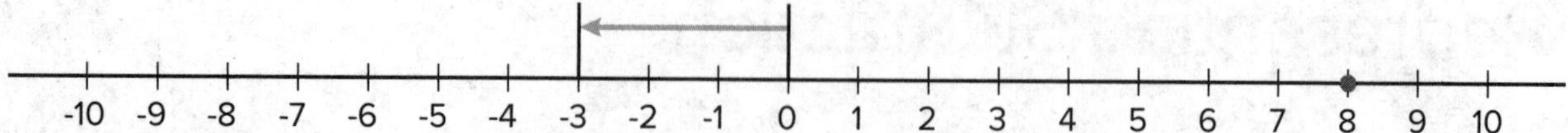

a. What equation would Mai write if she used the same reasoning as before?

b. What equation would Tyler write if he used the same reasoning as before?

c. How long should the other arrow be?

d. What number would complete each equation? Be prepared to explain your reasoning.

Second Diagram:

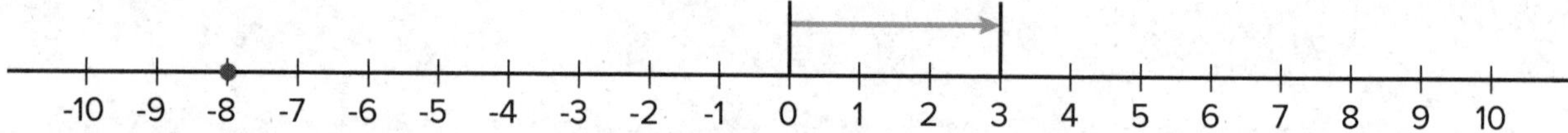

a. What equation would Mai write if she used the same reasoning as before?

b. What equation would Tyler write if he used the same reasoning as before?

c. How long should the other arrow be?

d. What number would complete each equation? Be prepared to explain your reasoning.

3. Draw a number line diagram for $(-8) - (-3) = ?$ What is the unknown number? How do you know?

NAME ______________________ DATE ____________ PERIOD ________

Activity
5.3 We Can Add Instead

1. Match each diagram to one of these expressions:

   $3 + 7$ $3 + (-7)$ $3 - 7$ $3 - (-7)$

a.
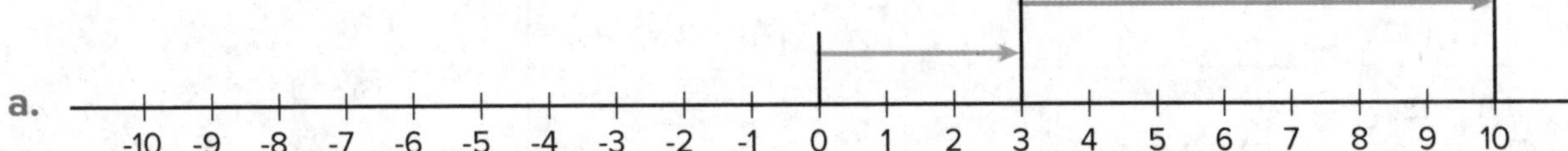

b.
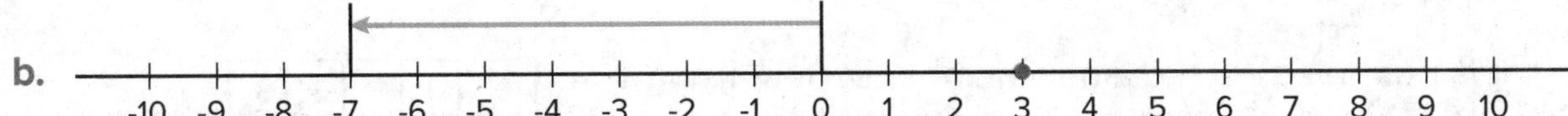

c.
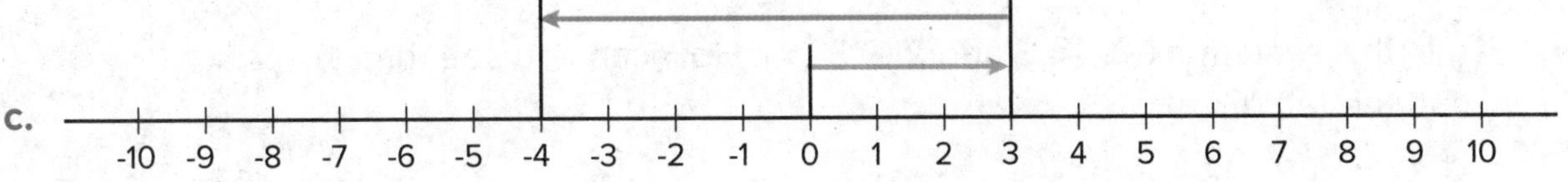

d.
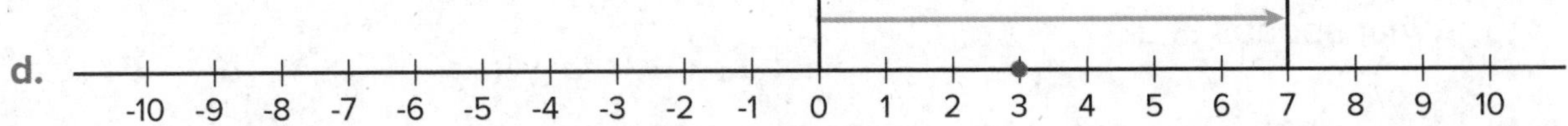

2. Which expressions in the first question have the same value? What do you notice?

3. Complete each of these tables. What do you notice?

Expression	Value
8 + (-8)	
8 – 8	
8 + (-5)	
8 – 5	
8 + (-12)	
8 – 12	

Expression	Value
-5 + 5	
-5 – (-5)	
-5 + 9	
-5 – (-9)	
-5 + 2	
-5 – (-2)	

Are you ready for more?

It is possible to make a new number system using *only* the numbers 0, 1, 2, and 3. We will write the symbols for adding and subtracting in this system like this: $2 \oplus 1 = 3$ and $2 \ominus 1 = 1$. The table shows some of the sums.

⊕	0	1	2	3
0	0	1	2	3
1	1	2	3	0
2	2	3	0	1
3				

1. In this system, $1 \oplus 2 = 3$ and $2 \oplus 3 = 1$. How can you see that in the table?
2. What do you think $3 \oplus 1$ should be?
3. What about $3 \oplus 3$?
4. What do you think $3 \ominus 1$ should be?
5. What about $2 \ominus 3$?
6. Can you think of any uses for this number system?

NAME ____________________ DATE ____________ PERIOD ____________

Summary

Representing Subtraction

The equation $7 - 5 = ?$ is equivalent to $? + 5 = 7$. The diagram illustrates the second equation.

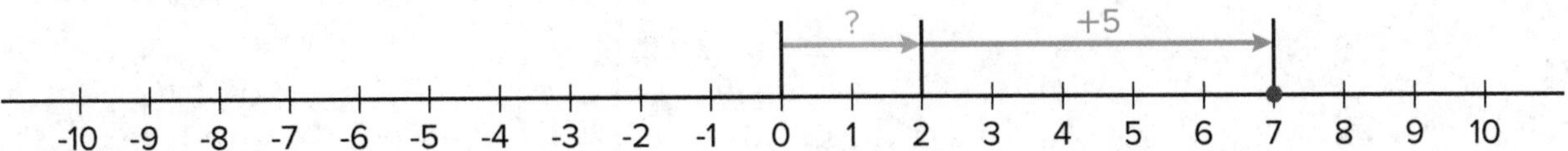

Notice that the value of $7 + (-5)$ is 2.

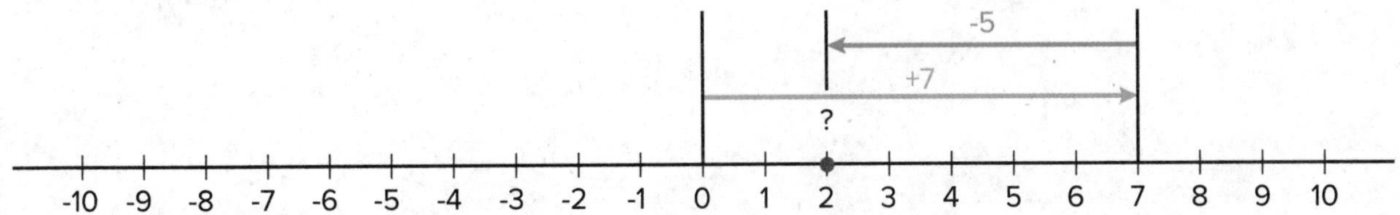

We can solve the equation $? + 5 = 7$ by adding -5 to both sides. This shows that $7 - 5 = 7 + (-5)$.

Likewise, $3 - 5 = ?$ is equivalent to $? + 5 = 3$.

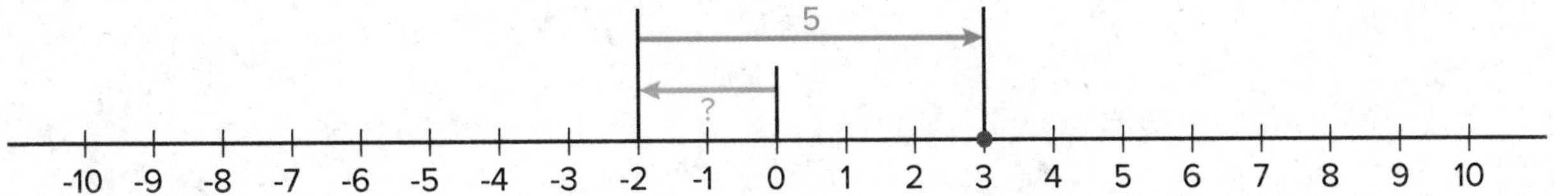

Notice that the value of $3 + (-5)$ is -2.

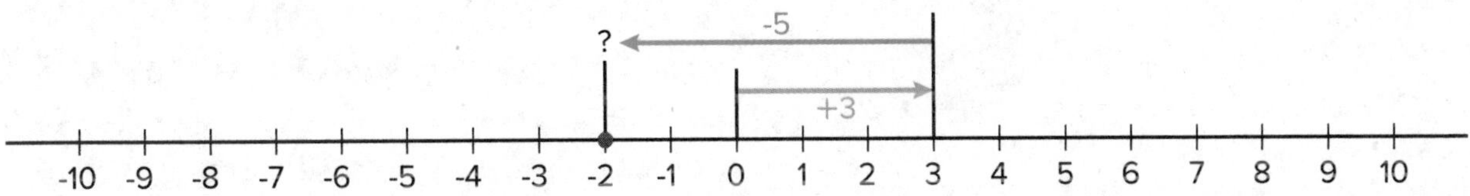

We can solve the equation $? + 5 = 3$ by adding -5 to both sides. This shows that $3 - 5 = 3 + (-5)$.

In general, $a - b = a + (-b)$.

If $a - b = x$, then $x + b = a$. We can add $-b$ to both sides of this second equation to get that $x = a + (-b)$.

Practice

Representing Subtraction

1. Write each subtraction equation as an addition equation.

   a. $a - 9 = 6$

   b. $p - 20 = -30$

   c. $z - (-12) = 15$

   d. $x - (-7) = -10$

2. Find each difference. If you get stuck, consider drawing a number line diagram.

   a. $9 - 4$

   b. $4 - 9$

   c. $9 - (-4)$

   d. $-9 - (-4)$

   e. $-9 - 4$

   f. $4 - (-9)$

   g. $-4 - (-9)$

   h. $-4 - 9$

NAME ______________________ DATE ____________ PERIOD ________

3. A restaurant bill is $59 and you pay $72. What percentage gratuity did you pay? **(Lesson 4-10)**

4. Find the solution to each equation mentally.

   a. $30 + a = 40$

   b. $500 + b = 200$

   c. $-1 + c = -2$

   d. $d + 3{,}567 = 0$

5. One kilogram is 2.2 pounds. Complete the tables. What is the interpretation of the constant of proportionality in each case? **(Lesson 2-3)**

Pounds	Kilograms
2.2	1
11	
5.5	
1	

__________ kilogram per pound

Kilograms	Pounds
1	2.2
7	
30	
0.5	

__________ pounds per kilogram

Lesson 5-6

Subtracting Rational Numbers

NAME ______________________ DATE ______________ PERIOD ______________

Learning Goal Let's bring addition and subtraction together.

Warm Up

6.1 Number Talk: Missing Addend

1. Solve each equation mentally.

   a. $247 + c = 458$ b. $c + 43.87 = 58.92$ c. $\frac{15}{8} + c = \frac{51}{8}$

2. Rewrite each addition equation as a subtraction equation.

Activity

6.2 Expressions with Altitude

A mountaineer is changing elevations. Write an expression that represents the difference between the final elevation and beginning elevation. Then write the value of the change. The first one is done for you.

Beginning Elevation (feet)	Final Elevation (feet)	Difference between Final and Beginning	Change
+400	+900	900 – 400	+500
+400	+50		
+400	-120		
-200	+610		
-200	-50		
-200	-500		
-200	0		

Are you ready for more?

Fill in the table so that every row and every column sums to 0.
Can you find another way to solve this puzzle?

	-12	0		5
0			-18	25
25		-18	5	-12
-12				-18
	-18	25	-12	

	-12	0		5
0			-18	25
25		-18	5	-12
-12				-18
	-18	25	-12	

Activity

6.3 Does the Order Matter?

1. Find the value of each subtraction expression.

A
3 − 2
5 − (-9)
(-11) − 2
(-6) − (-3)
(-1.2) − (-3.6)
$\left(-2\frac{1}{2}\right) - \left(-3\frac{1}{2}\right)$

B
2 − 3
(-9) − 5
2 − (-11)
(-3) − (-6)
(-3.6) − (-1.2)
$\left(-3\frac{1}{2}\right) - \left(-2\frac{1}{2}\right)$

2. What do you notice about the expressions in Column A compared to Column B?

3. What do you notice about their values?

NAME ______________________ DATE ____________ PERIOD ____________

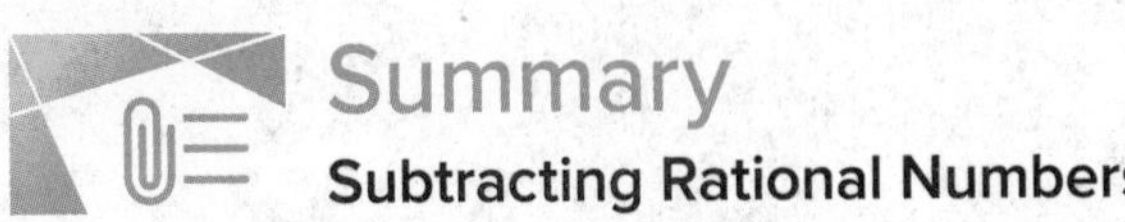

Summary

Subtracting Rational Numbers

When we talk about the difference of two numbers, we mean, "subtract them." Usually, we subtract them in the order they are named. For example, the difference of +8 and -6 is 8 − (-6).

The difference of two numbers tells you how far apart they are on the number line. 8 and -6 are 14 units apart, because 8 − (-6) = 14:

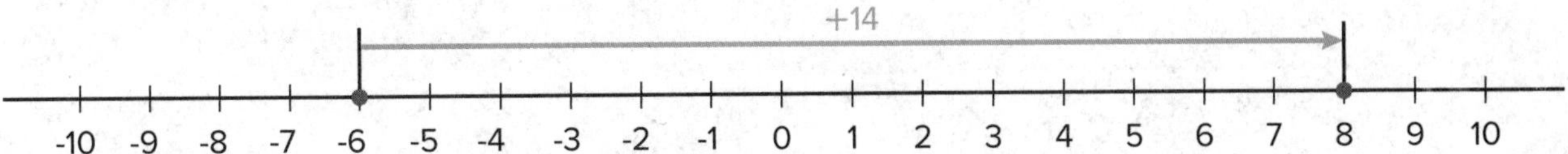

Notice that if you subtract them in the opposite order, you get the opposite number, or (-6) − 8 = -14.

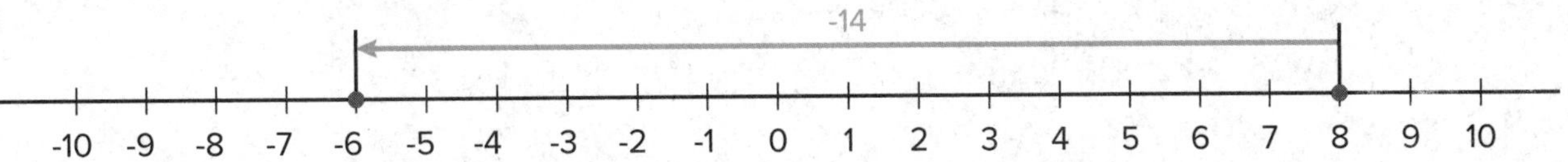

In general, the distance between two numbers a and b on the number line is $|a - b|$.

- Note that the *distance* between two numbers is always positive, no matter the order.
- But the *difference* can be positive or negative, depending on the order.

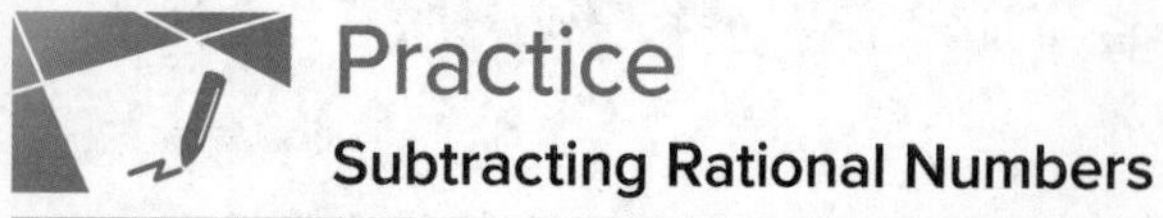

Practice

Subtracting Rational Numbers

1. Write a sentence to answer each question:

   a. How much warmer is 82 than 40?

   b. How much warmer is 82 than -40?

2. Respond to the following questions.

   a. What is the difference in height between 30 m up a cliff and 87 m up a cliff? What is the distance between these positions?

   b. What is the difference in height between an albatross flying at 100 m above the surface of the ocean and a shark swimming 30 m below the surface? What is the distance between them if the shark is right below the albatross?

3. A company produces screens of different sizes. Based on the table, could there be a relationship between the number of pixels and the area of the screen? If so, write an equation representing the relationship. If not, explain your reasoning. **(Lesson 2-8)**

Square Inches of Screen	Number of Pixels
6	31,104
72	373,248
105	544,320
300	1,555,200

NAME ______________________ DATE ____________ PERIOD ________

4. Find each difference.

a. (-5) – 6

b. 35 – (-8)

c. $\frac{2}{5} - \frac{3}{5}$

d. $-4\frac{3}{8} - \left(-1\frac{1}{4}\right)$

5. A family goes to a restaurant. When the bill comes, this is printed at the bottom of it:

Gratuity Guide for Your Convenience:
15% would be \$4.89
18% would be \$5.87
20% would be \$6.52

How much was the price of the meal? Explain your reasoning. **(Lesson 4-10)**

6. Which is a scaled copy of Polygon A? Identify a pair of corresponding sides and a pair of corresponding angles. Compare the areas of the scaled copies. **(Lesson 1-2)**

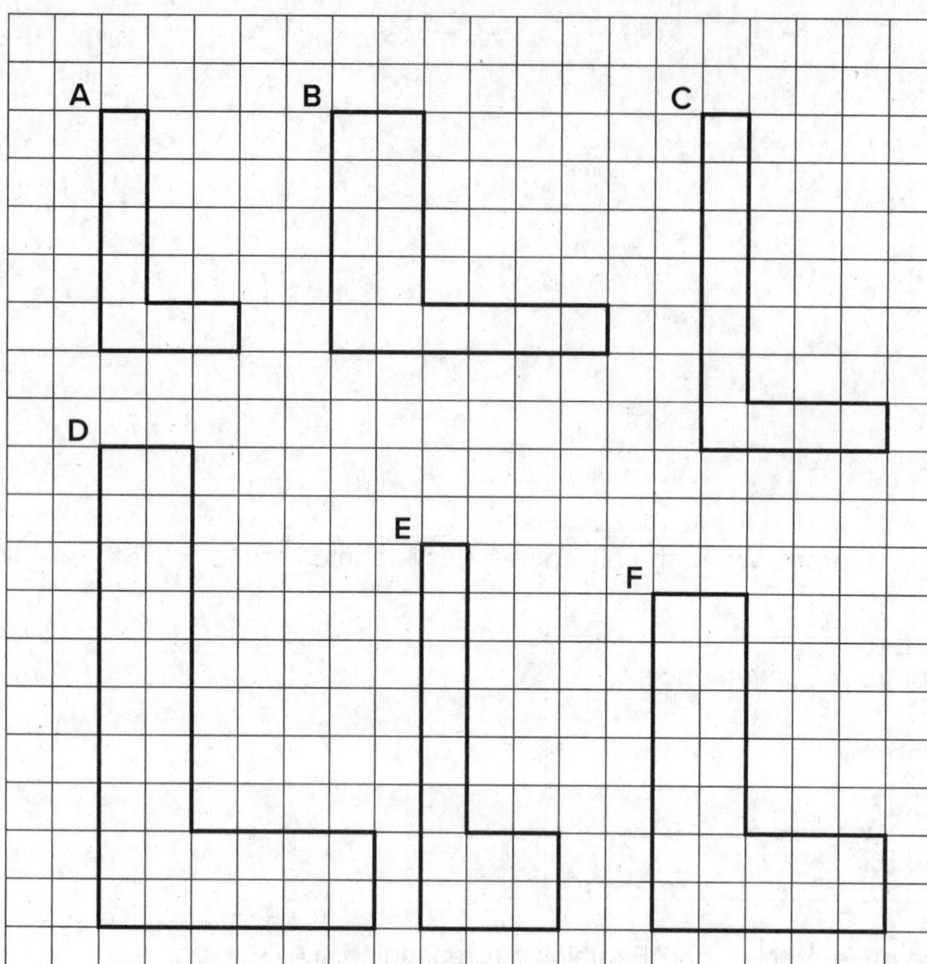

Lesson 5-7

Adding and Subtracting to Solve Problems

NAME ______________________ DATE ______________ PERIOD ______________

Learning Goal Let's apply what we know about signed numbers to different situations.

Warm Up

7.1 Positive or Negative?

Without computing:

1. Is the solution to $-2.7 + x = -3.5$ positive or negative?

2. Select all the expressions that are solutions to $-2.7 + x = -3.5$.

   A. $-3.5 + 2.7$

   B. $3.5 - 2.7$

   C. $-3.5 - (-2.7)$

   D. $-3.5 - 2.7$

Activity

7.2 Phone Inventory

A store tracks the number of cell phones it has in stock and how many phones it sells. The table shows the inventory for one phone model at the beginning of each day last week. The inventory changes when they sell phones or get shipments of phones into the store.

	Inventory	Change
Monday	18	-2
Tuesday	16	-5
Wednesday	11	-7
Thursday	4	-6
Friday	-2	20

1. What do you think it means when the change is positive? Negative?

2. What do you think it means when the inventory is positive? Negative?

3. Based on the information in the table, what do you think the inventory will be at on Saturday morning? Explain your reasoning.

4. What is the difference between the greatest inventory and the least inventory?

Activity
7.3 Solar Power

Han's family got a solar panel. Each month they get a credit to their account for the electricity that is generated by the solar panel. The credit they receive varies based on how sunny it is.

In January they used $83.56 worth of electricity and generated $6.75 worth of electricity. Here is their electricity bill from January.

Current Charges: $83.56
Solar Credit: -$6.75
Amount Due: $76.81

1. In July they were traveling away from home and only used $19.24 of electricity. Their solar panel generated $22.75 worth of electricity. What was their amount due in July?

2. The table shows the value of the electricity they used and the value of the electricity they generated each week for a month. What amount is due for this month?

	Used ($)	Generated ($)
Week 1	13.45	-6.33
Week 2	21.78	-8.94
Week 3	18.12	-7.70
Week 4	24.05	-5.36

3. What is the difference between the value of the electricity generated in week 1 and week 2? Between week 2 and week 3?

NAME ______________________ DATE ____________ PERIOD ________

Are you ready for more?

While most rooms in any building are all at the same level of air pressure, hospitals make use of “positive pressure rooms” and “negative pressure rooms.” What do you think it means to have negative pressure in this setting? What could be some uses of these rooms?

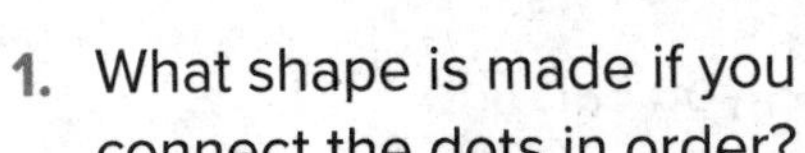

Activity
7.4 Differences and Distances

Plot these points on the coordinate grid: $A = (5, 4)$, $B = (5, -2)$, $C = (-3, -2)$, $D = (-3, 4)$

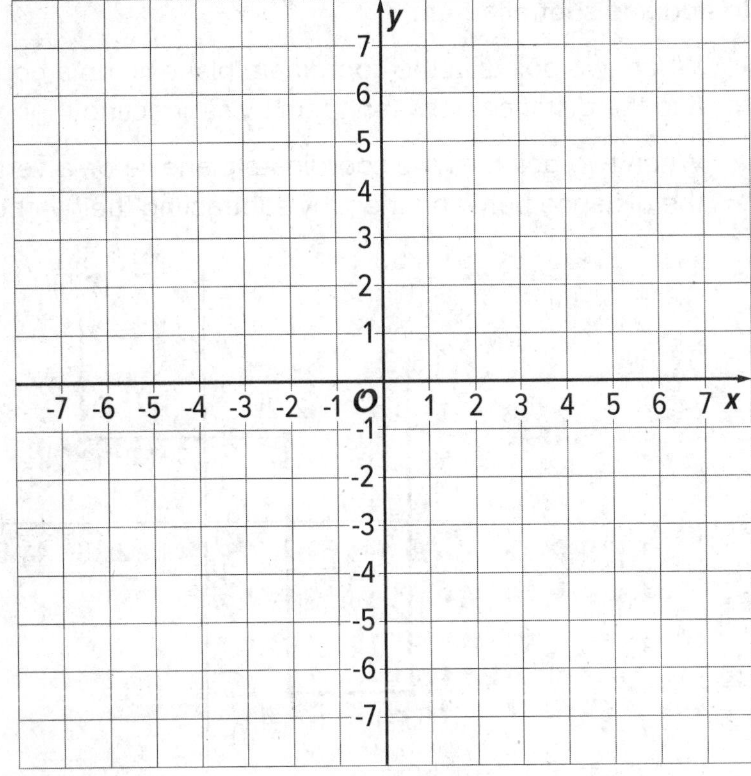

1. What shape is made if you connect the dots in order?

2. What are the side lengths of figure *ABCD*?

3. What is the difference between the *x*-coordinates of *B* and *C*?

4. What is the difference between the *x*-coordinates of *C* and *B*?

5. How do the differences of the coordinates relate to the distances between the points?

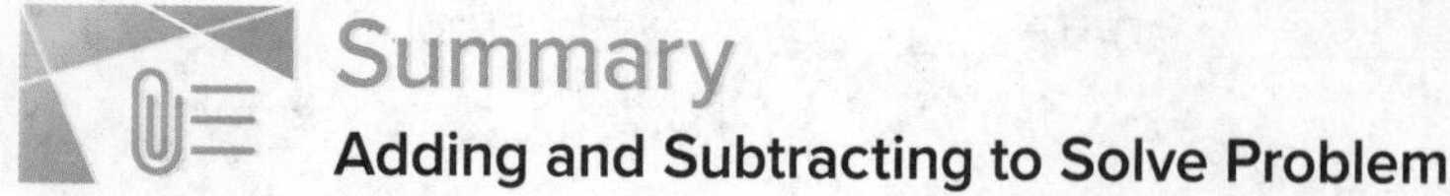

Summary

Adding and Subtracting to Solve Problems

Sometimes we use positive and negative numbers to represent quantities in context. Here are some contexts we have studied that can be represented with positive and negative numbers:

- temperature
- elevation
- inventory
- an account balance
- electricity flowing in and flowing out

In these situations, using positive and negative numbers, and operations on positive and negative numbers, helps us understand and analyze them.

To solve problems in these situations, we just have to understand what it means when the quantity is positive, when it is negative, and what it means to add and subtract them.

- When two points in the coordinate plane lie on a horizontal line, you can find the distance between them by subtracting their *x*-coordinates.
- When two points in the coordinate plane lie on a vertical line, you can find the distance between them by subtracting their *y*-coordinates.

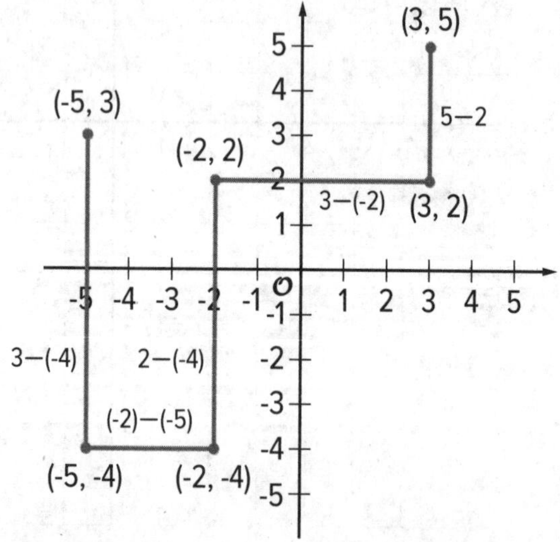

Remember: the *distance* between two numbers is independent of the order, but the *difference* depends on the order.

NAME ______________________ DATE ____________ PERIOD ____________

Practice

Adding and Subtracting to Solve Problems

1. The table shows four transactions and the resulting account balance in a bank account, except some numbers are missing. Fill in the missing numbers.

	Transaction Amount	Account Balance
Transaction 1	360	360
Transaction 2	-22.50	337.50
Transaction 3		182.35
Transaction 4		-41.40

2. The *departure from the average* is the difference between the actual amount of rain and the average amount of rain for a given month.

The historical average for rainfall in Albuquerque, NM for June, July, and August is shown in the table.

June	July	August
0.67	1.5	1.57

a. Last June only 0.17 inches of rain fell all month. What is the difference between the average rainfall and the actual rainfall for last June?

b. The departure from the average rainfall last July was -0.36 inches. How much rain fell last July?

c. How much rain would have to fall in August so that the total amount of rain equals the average rainfall for these three months? What would the departure from the average be in August in that situation?

3. Respond to each question. **(Lesson 5-6)**

a. How much higher is 500 than 400 m?

b. How much higher is 500 than -400 m?

c. What is the change in elevation from 8,500 m to 3,400 m?

d. What is the change in elevation between 8,500 m and -300 m?

e. How much higher is -200 m than 450 m?

4. Tyler orders a meal that costs $15. **(Lesson 4-10)**

a. If the tax rate is 6.6%, how much will the sales tax be on Tyler's meal?

b. Tyler also wants to leave a tip for the server. How much do you think he should pay in all? Explain your reasoning.

5. In a video game, a character is healed at a constant rate as long as they are standing in a certain circle. Complete the table. **(Lesson 2-3)**

Time in Circle (seconds)	Health Gained (points)
4	100
10	
3	
	1,000

Lesson 5-8

Position, Speed, and Direction

NAME ______________________ DATE ____________ PERIOD ____________

Learning Goal Let's use signed numbers to represent movement.

Warm Up

8.1 Distance, Rate, Time

1. An airplane moves at a constant speed of 120 miles per hour for 3 hours. How far does it go?

2. A train moves at constant speed and travels 6 miles in 4 minutes. What is its speed in miles per minute?

3. A car moves at a constant speed of 50 miles per hour. How long does it take the car to go 200 miles?

Activity

8.2 Going Left, Going Right

1. After each move, record your location in the table. Then write an expression to represent the ending position that uses the starting position, the speed, and the time. The first row is done for you.

Starting Position	Direction	Speed (units per second)	Time (seconds)	Ending Position (units)	Expression
0	right	5	3	+15	$0 + 5 \cdot 3$
0	left	4	6		
0	right	2	8		
0	right	6	2		
0	left	1.1	5		

You may wish to use the number line to help you with the position changes in the table.

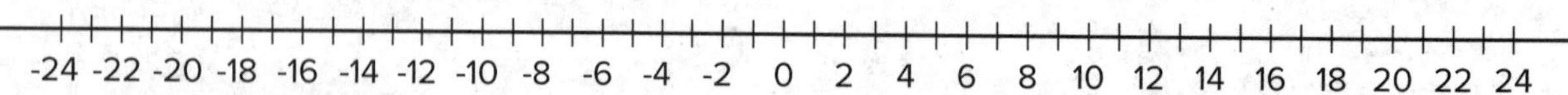

2. How can you see the *direction* of movement in the expression?

3. Using a starting position p, a speed s, and a time t, write two expressions for an ending position. One expression should show the result of moving right, and one expression should show the result of moving left.

NAME ____________________ DATE __________ PERIOD __________

Activity
8.3 Velocity

A traffic safety engineer was studying travel patterns along a highway. She set up a camera and recorded the speed and direction of cars and trucks that passed by the camera. Positions to the east of the camera are positive, and to the west are negative.

Vehicles that are traveling towards the east have a positive velocity, and vehicles that are traveling towards the west have a negative velocity.

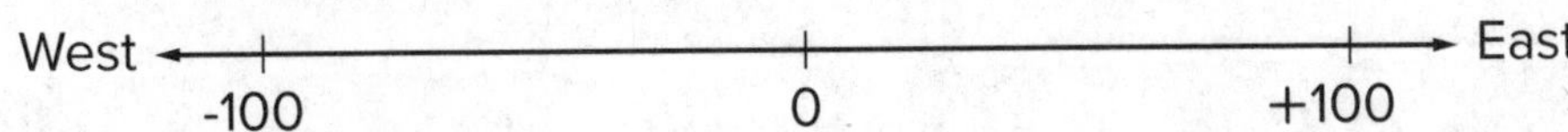

1. Complete the table with the position of each vehicle if the vehicle is traveling at a constant speed for the indicated time period. Then write an equation.

Velocity (meters per second)	Time after Passing the Camera (seconds)	Ending Position (meters)	Equation Describing the Position
+25	+10	+250	$25 \cdot 10 = 250$
-20	+30		
+32	+40		
-35	+20		
+28	0		

2. If a car is traveling east when it passes the camera, will its position be positive or negative 60 seconds after it passes the camera? If we multiply two positive numbers, is the result positive or negative?

3. If a car is traveling west when it passes the camera, will its position be positive or negative 60 seconds after it passes the camera? If we multiply a negative and a positive number, is the result positive or negative?

Are you ready for more?

In many contexts we can interpret negative rates as "rates in the opposite direction." For example, a car that is traveling -35 miles per hour is traveling in the opposite direction of a car that is traveling 40 miles per hour.

1. What could it mean if we say that water is flowing at a rate of -5 gallons per minute?

2. Make up another situation with a negative rate and explain what it could mean.

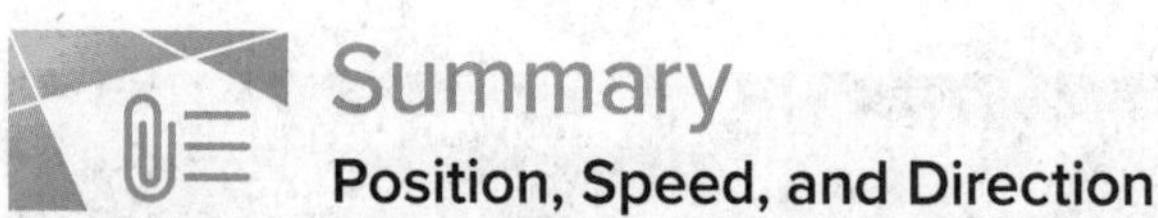

Summary

Position, Speed, and Direction

We can use signed numbers to represent the position of an object along a line. We pick a point to be the reference point and call it zero. Positions to the right of zero are positive. Positions to the left of zero are negative.

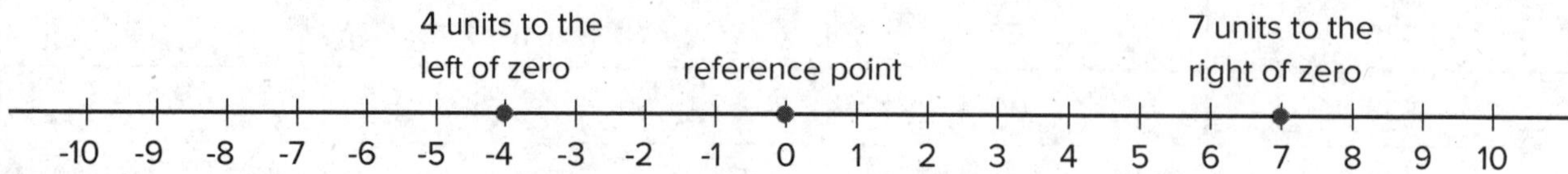

When we combine speed with direction indicated by the sign of the number, it is called *velocity*. For example, if you are moving 5 meters per second to the right, then your velocity is +5 meters per second. If you are moving 5 meters per second to the left, then your velocity is -5 meters per second.

- If you start at zero and move 5 meters per second for 10 seconds, you will be $5 \cdot 10 = 50$ meters to the right of zero. In other words, $5 \cdot 10 = 50$.
- If you start at zero and move -5 meters per second for 10 seconds, you will be $5 \cdot 10 = 50$ meters to the *left* of zero. In other words, $-5 \cdot 10 = -50$.

In general, a negative number times a positive number is a negative number.

NAME ______________________ DATE ____________ PERIOD ____________

Practice

Position, Speed, and Direction

1. A number line can represent positions that are north and south of a truck stop on a highway. Decide whether you want positive positions to be north or south of the truck stop. Then plot the following positions on a number line.

   a. the truck stop

   b. 5 miles north of the truck stop

   c. 3.5 miles south of the truck stop

2. Respond to each of the following.

   a. How could you distinguish between traveling west at 5 miles per hour and traveling east at 5 miles per hour without using the words "east" and "west"?

   b. Four people are cycling. They each start at the same point. (0 represents their starting point.) Plot their finish points after five seconds of cycling on a number line.

      - Lin cycles at 5 meters per second.
      - Elena cycles at 3 meters per second.
      - Diego cycles at -4 meters per second.
      - Noah cycles at -6 meters per second.

3. Find the value of each expression. **(Lesson 5-6)**

   a. $16.2 + -8.4$

   b. $\frac{2}{5} - \frac{3}{5}$

   c. $-9.2 + -7$

   d. $-4\frac{3}{8} - \left(-1\frac{1}{4}\right)$

4. For each equation, write two more equations using the same numbers that express the same relationship in a different way. **(Lesson 5-5)**

a. $3 + 2 = 5$

b. $7.1 + 3.4 = 10.5$

c. $15 - 8 = 7$

d. $\frac{3}{2} + \frac{9}{5} = \frac{33}{10}$

5. A shopper bought a watermelon, a pack of napkins, and some paper plates. In his state, there is no tax on food. The tax rate on non-food items is 5%. The total for the three items he bought was \$8.25 before tax, and he paid \$0.19 in tax. How much did the watermelon cost? **(Lesson 4-10)**

6. Which graphs could not represent a proportional relationship? Explain how you decided. **(Lesson 2-10)**

Graph A

Graph C

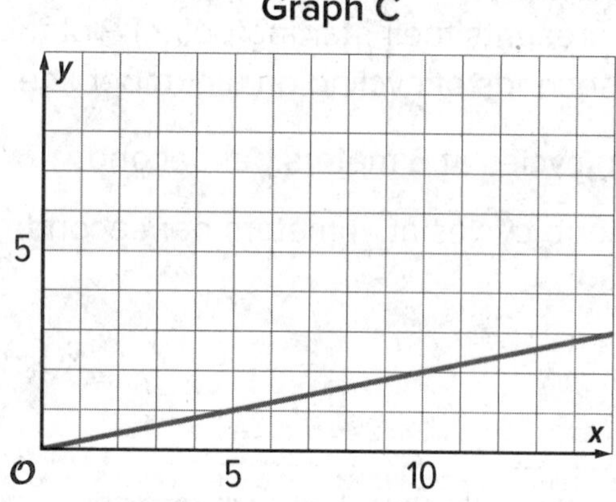

Graph B

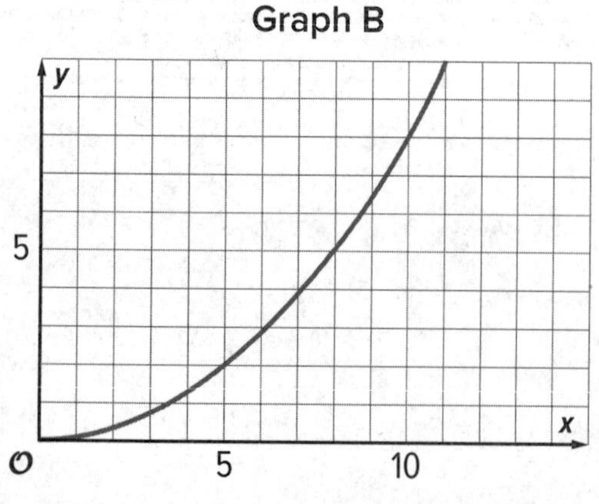

Graph D

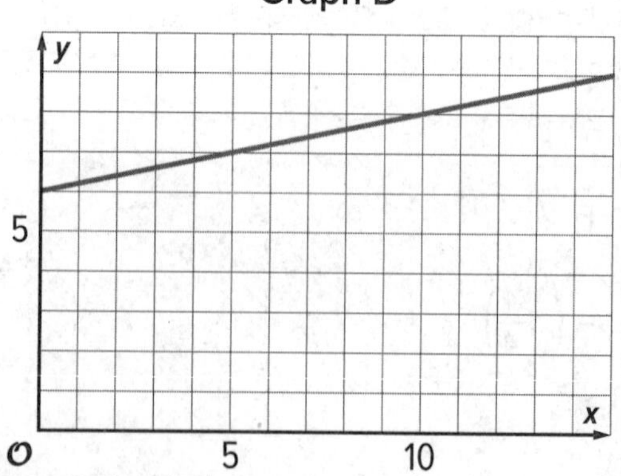

Lesson 5-9

Multiplying Rational Numbers

NAME ____________________ DATE ____________ PERIOD ____________

Learning Goal Let's multiply signed numbers.

Warm Up

9.1 Before and After

Where was the girl...

1. 5 seconds *after* this picture was taken? Mark her approximate location on the picture.
2. 5 seconds *before* this picture was taken? Mark her approximate location on the picture.

Activity

9.2 Backwards in Time

A traffic safety engineer was studying travel patterns along a highway. She set up a camera and recorded the speed and direction of cars and trucks that passed by the camera. Positions to the east of the camera are positive, and to the west are negative.

1. Here are some positions and times for one car:

Position (feet)	-180	-120	-60	0	60	120
Time (seconds)	-3	-2	-1	0	1	2

   a. In what direction is this car traveling?

   b. What is its velocity?

2. Respond to the following questions.

a. What does it mean when the time is zero?

b. What could it mean to have a negative time?

3. Here are the positions and times for a different car whose velocity is -50 feet per second:

Position (feet)				0	-50	-100
Time (seconds)	-3	-2	-1	0	1	2

a. Complete the table with the rest of the positions.

b. In what direction is this car traveling? Explain how you know.

4. Complete the table for several different cars passing the camera.

	Velocity (meters per second)	Time after Passing Camera (seconds)	Ending Position (meters)	Equation
Car C	+25	+10	+250	$25 \cdot 10 = 250$
Car D	-20	+30		
Car E	+32	-40		
Car F	-35	-20		
Car G	-15	-8		

5. Respond to the following questions.

a. If a car is traveling east when it passes the camera, will its position be positive or negative 60 seconds *before* it passes the camera?

b. If we multiply a positive number and a negative number, is the result positive or negative?

6. Respond to the following questions.

a. If a car is traveling west when it passes the camera, will its position be positive or negative 60 seconds *before* it passes the camera?

b. If we multiply two negative numbers, is the result positive or negative?

NAME ______________________ DATE ______________ PERIOD ______________

Activity
9.3 Cruising

Around noon, a car was traveling -32 meters per second down a highway. At exactly noon (when time was 0), the position of the car was 0 meters.

1. Complete the table.

Time (s)	-10	-7	-4	-1	2	5	8	11
Position (m)								

2. Graph the relationship between the time and the car's position.

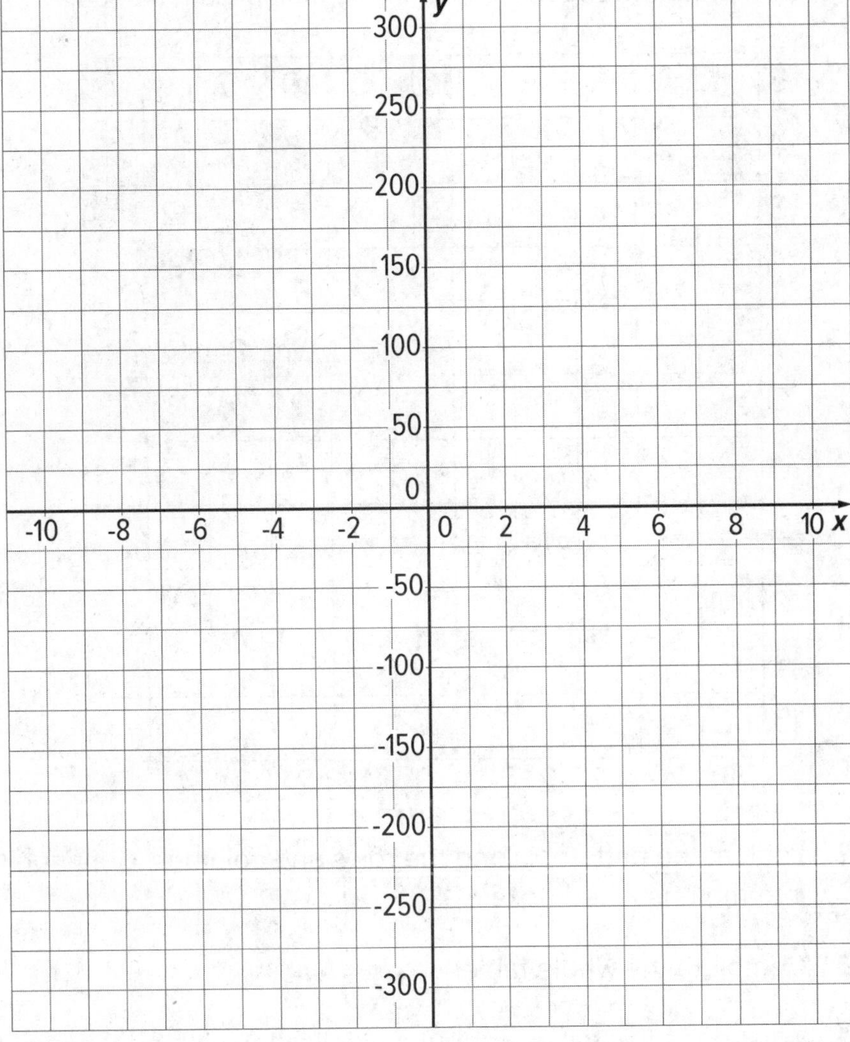

3. What was the position of the car at -3 seconds?

4. What was the position of the car at 6.5 seconds?

Are you ready for more?

Find the value of these expressions without using a calculator.

$(-1)^2$ $(-1)^3$ $(-1)^4$ $(-1)^{99}$

Activity

9.4 Rational Numbers Multiplication Grid

1. Complete the *shaded* boxes in the multiplication square.

5						0	5	10	15	20	
4						0	4	8	12	16	
3						0	3	6	9	12	
2					-2	0	2	4	6	8	
1						0	1	2	3	4	
0						0	0	0	0	0	
-1											
-2											
-3											
-4											
-5											
	-5	-4	-3	-2	-1	0	1	2	3	4	5

2. Look at the patterns along the rows and columns. Continue those patterns into the unshaded boxes.

3. Complete the whole table.

4. What does this tell you about multiplication with negative numbers?

NAME ____________________ DATE ____________ PERIOD ________

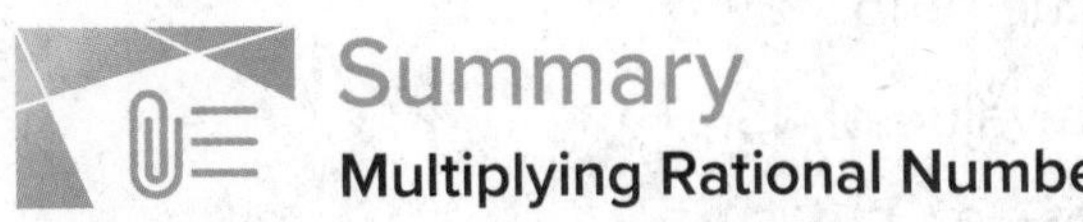

Summary
Multiplying Rational Numbers

We can use signed numbers to represent time relative to a chosen point in time. We can think of this as starting a stopwatch. The positive times are after the watch starts, and negative times are times before the watch starts.

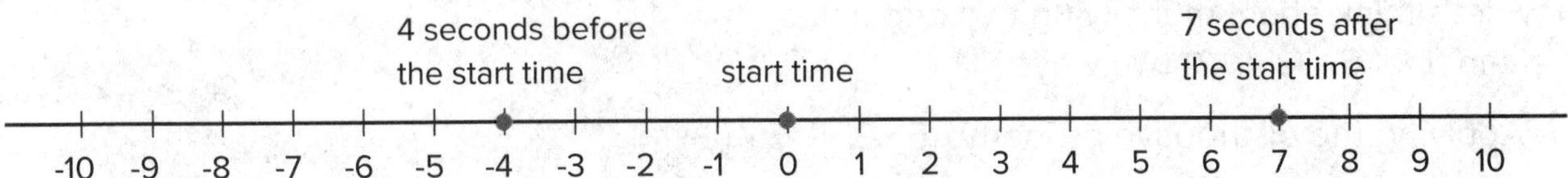

If a car is at position 0 and is moving in a positive direction, then for times after that (positive times), it will have a positive position. A positive times a positive is positive.

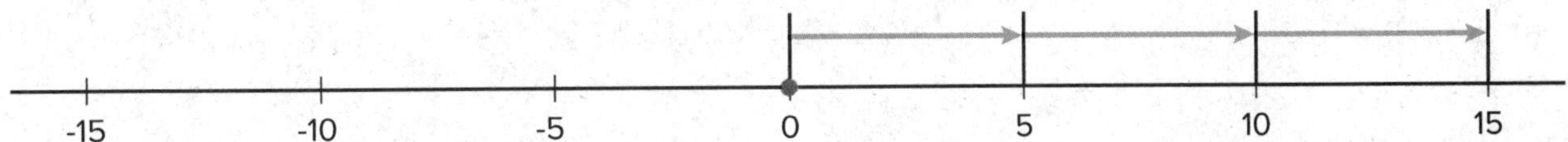

If a car is at position 0 and is moving in a negative direction, then for times after that (positive times), it will have a negative position. A negative times a positive is negative.

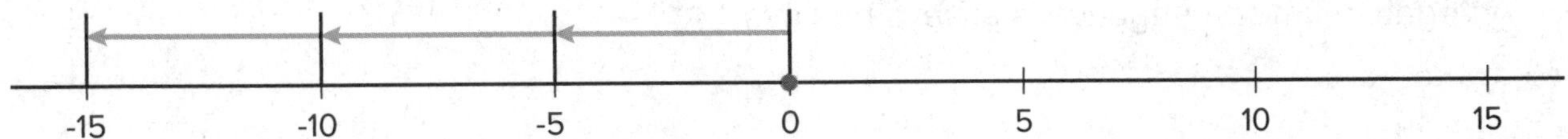

If a car is at position 0 and is moving in a positive direction, then for times *before* that (negative times), it must have had a negative position. A positive times a negative is negative.

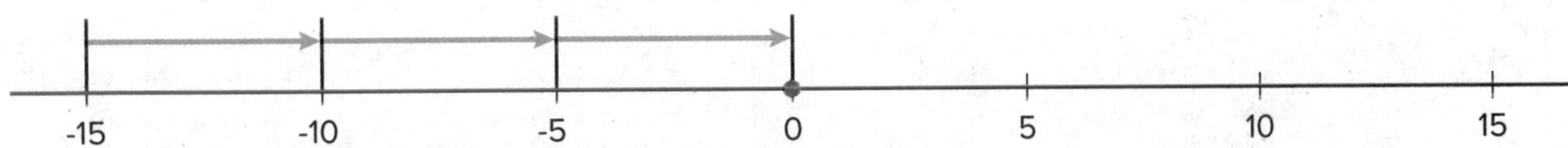

If a car is at position 0 and is moving in a negative direction, then for times *before* that (negative times), it must have had a positive position. A negative times a negative is positive.

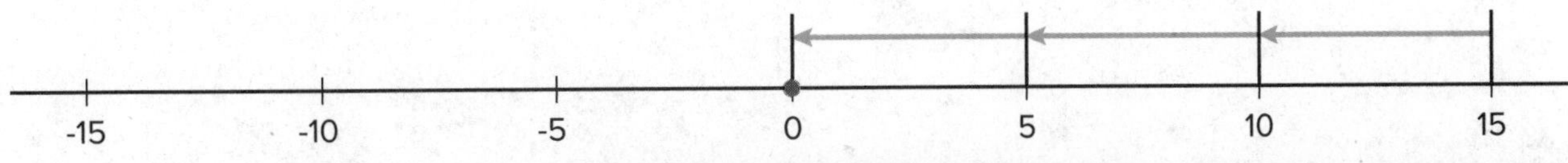

Here is another way of seeing this:

- We can think of $3 \cdot 5$ as $5 + 5 + 5$, which has a value of 15.
- We can think of $3 \cdot (-5)$ as $-5 + -5 + -5$, which has a value of -15.

We know we can multiply positive numbers in any order: $3 \cdot 5 = 5 \cdot 3$

If we can multiply signed numbers in any order, then $(-5) \cdot 3$ would also equal -15.

Now let's think about multiplying two negatives.
We can find $-5 \cdot (3 + -3)$ two ways:

- Applying the distributive property: $-5 \cdot 3 + -5 \cdot (-3)$
- Adding the numbers in parentheses: $-5 \cdot (0) = 0$

That means that these expressions must be equal.

$$-5 \cdot 3 + -5 \cdot (-3) = 0$$

Multiplying the first two numbers gives $-15 + -5 \cdot (-3) = 0$, which means that $-5 \cdot (-3) = 15$.

There was nothing special about these particular numbers. This always works!

- A positive times a positive is always positive.
- A negative times a positive or a positive times a negative is always negative.
- A negative times a negative is always positive.

NAME ______________________ DATE ____________ PERIOD ____________

Practice
Multiplying Rational Numbers

1. Fill in the missing numbers in these equations.

   a. $-2 \cdot (-4.5) = ?$

   b. $-8.7 \cdot -10 = ?$

   c. $-7 \cdot ? = 14$

   d. $? \cdot (-10) = 90$

2. A weather station on the top of a mountain reports that the temperature is currently 0° C and has been falling at a constant rate of 3° C per hour. Find each temperature. Explain or show your reasoning.

   If it continues to fall at this rate, what will the temperature be...

   a. in 2 hours?

   b. in 5 hours?

   c. in half an hour?

   What was the temperature...

   d. 1 hour ago?

   e. 3 hours ago?

   f. 4.5 hours ago?

3. Find the value of each expression.

   a. $\frac{1}{4} \cdot (-12)$

   b. $-\frac{1}{3} \cdot 39$

   c. $\left(-\frac{4}{5}\right) \cdot (-75)$

   d. $-\frac{2}{5} \cdot \left(-\frac{3}{4}\right)$

   e. $\frac{8}{3} \cdot -42$

4. To make a specific hair dye, a hair stylist uses a ratio of $1\frac{1}{8}$ oz of red tone, $\frac{3}{4}$ oz of gray tone, and $\frac{5}{8}$ oz of brown tone. **(Lesson 4-2)**

   a. If the stylist needs to make 20 oz of dye, how much of each dye color is needed?

   b. If the stylist needs to make 100 oz of dye, how much of each dye color is needed?

5. Respond to each of the following. **(Lesson 5-7)**

   a. Here are the vertices of rectangle *FROG*: (-2, 5), (-2, 1), (6, 5), (6, 1). Find the perimeter of this rectangle. If you get stuck, try plotting the points on a coordinate plane.

   b. Find the area of the rectangle *FROG*.

   c. Here are the coordinates of rectangle *PLAY*: (-11, 20), (-11, -3), (-1, 20), (-1, -3). Find the perimeter and area of this rectangle. See if you can figure out its side lengths without plotting the points.

Lesson 5-10

Multiply!

NAME ______________________________ DATE ______________ PERIOD ___________

Learning Goal Let's get more practice multiplying signed numbers.

Warm Up

10.1 Which One Doesn't Belong: Expressions

Which expression doesn't belong?

(A.) $7.9x$

(B.) $7.9 \cdot (-10)$

(C.) $7.9 + x$

(D.) -79

Activity

10.2 Card Sort: Matching Expressions

Your teacher will give you cards with multiplication expressions on them. Match expressions that are equal to each other. There will be 3 cards in each group.

Activity

10.3 Row Game: Multiplying Rational Numbers

Evaluate the expressions in one of the columns. Your partner will work on the other column. Check in with your partner after you finish each row. Your answers in each row should be the same. If your answers aren't the same, work together to find the error.

Column A	Column B
$790 \div 10$	$(7.9) \cdot 10$
$-\frac{6}{7} \cdot 7$	$(0.1) \cdot -60$
$(2.1) \cdot -2$	$(-8.4) \cdot \frac{1}{2}$
$(2.5) \cdot (-3.25)$	$-\frac{5}{2} \cdot \frac{13}{4}$
$-10 \cdot (3.2) \cdot (-7.3)$	$5 \cdot (-1.6) \cdot (-29.2)$

Are you ready for more?

A sequence of rational numbers is made by starting with 1, and from then on, each term is one more than the reciprocal of the previous term. Evaluate the first few expressions in the sequence. Can you find any patterns? Find the 10th term in this sequence.

$$1 \qquad 1+\frac{1}{1} \qquad 1+\frac{1}{1+1} \qquad 1+\frac{1}{1+\frac{1}{1+1}} \qquad 1+\frac{1}{1+\frac{1}{1+\frac{1}{1+1}}} \qquad \dots$$

Summary

Multiply!

A positive times a positive is always positive. For example, $\frac{3}{5} \cdot \frac{7}{8} = \frac{21}{40}$.

A negative times a negative is also positive. For example, $-\frac{3}{5} \cdot -\frac{7}{8} = \frac{21}{40}$.

A negative times a positive or a positive times a negative is always negative. For example, $\frac{3}{5} \cdot -\frac{7}{8} = -\frac{3}{5} \cdot \frac{7}{8} = -\frac{21}{40}$.

A negative times a negative times a negative is also negative. For example, $-3 \cdot -4 \cdot -5 = -60$.

NAME ______________________ DATE ____________ PERIOD ____________

Practice
Multiply!

1. Evaluate each expression:

   a. $-12 \cdot \frac{1}{3}$

   b. $-12 \cdot -\frac{1}{3}$

   c. $12 \cdot \left(-\frac{5}{4}\right)$

   d. $-12 \cdot \left(-\frac{5}{4}\right)$

2. Evaluate each expression:

   a. $-1 \cdot 2 \cdot 3$

   b. $-1 \cdot (-2) \cdot 3$

   c. $-1 \cdot (-2) \cdot (-3)$

3. Order each set of numbers from least to greatest. **(Lesson 5-1)**

   a. 4, 8, -2, -6, 0

   b. $-5, -5.2, 5.5, -5\frac{1}{2}, \frac{-5}{2}$

4. $30 + -30 = 0$. **(Lesson 5-3)**

   a. Write another sum of two numbers that equals 0.

   b. Write a sum of three numbers that equals 0.

   c. Write a sum of four numbers that equals 0, none of which are opposites.

5. A submarine is searching for underwater features. It is accompanied by a small aircraft and an underwater robotic vehicle. **(Lesson 5-6)**

At one time the aircraft is 200 m above the surface, the submarine is 55 m below the surface, and the underwater robotic vehicle is 227 m below the surface.

a. What is the difference in height between the submarine and the aircraft?

b. What is the distance between the underwater robotic vehicle and the submarine?

6. Respond to the following questions. **(Lesson 5-8)**

a. Clare is cycling at a speed of 12 miles per hour. If she starts at a position chosen as zero, what will her position be after 45 minutes?

b. Han is cycling at a speed of -8 miles per hour; if he starts at the same zero point, what will his position be after 45 minutes?

c. What will the distance between them be after 45 minutes?

7. Fill in the missing numbers in these equations. **(Lesson 5-9)**

a. $(-7) \cdot ? = -14$

b. $? \cdot 3 = -15$

c. $? \cdot 4 = 32$

d. $-49 \cdot 3 = ?$

Lesson 5-11

Dividing Rational Numbers

NAME ______________________ DATE ____________ PERIOD ____________

Learning Goal Let's divide signed numbers.

Warm Up

11.1 Tell Me Your Sign

Consider the equation $-27x = -35$.

Without computing...

1. is the **solution** to this equation positive or negative?

2. are either of these two numbers solutions to the equation?

$$\frac{35}{27} \qquad -\frac{35}{27}$$

Activity

11.2 Multiplication and Division

1. Find the missing values in the equations.
   a. $-3 \cdot 4 = ?$
   b. $-3 \cdot ? = 12$
   c. $3 \cdot ? = 12$
   d. $? \cdot -4 = 12$
   e. $? \cdot 4 = -12$

2. Rewrite the unknown factor problems as division problems.

3. Complete the sentences. Be prepared to explain your reasoning.

   a. The sign of a positive number divided by a positive number is always:

   b. The sign of a positive number divided by a negative number is always:

   c. The sign of a negative number divided by a positive number is always:

   d. The sign of a negative number divided by a negative number is always:

4. Han and Clare walk towards each other at a constant rate, meet up, and then continue past each other in opposite directions. We will call the position where they meet up 0 feet and the time when they meet up 0 seconds.

   - Han's velocity is 4 feet per second.
   - Clare's velocity is -5 feet per second.

   a. Where is each person 10 seconds before they meet up?

   b. When is each person at the position -10 feet from the meeting place?

NAME ______________________ DATE ______________ PERIOD ______________

Are you ready for more?

It is possible to make a new number system using *only* the numbers 0, 1, 2, and 3. We will write the symbols for multiplying in this system like this: $1 \otimes 2 = 2$. The table shows some of the products.

$\otimes$	0	1	2	3
0	0	0	0	0
1		1	2	3
2			0	2
3				

1. In this system, $1 \otimes 3 = 3$ and $2 \otimes 3 = 2$. How can you see that in the table?

2. What do you think $2 \otimes 1$ is?

3. What about $3 \otimes 3$?

4. What do you think the solution to $3 \otimes n = 2$ is?

5. What about $2 \otimes n = 3$?

Activity
11.3 Drilling Down

A water well drilling rig has dug to a height of -60 feet after one full day of continuous use.

1. Assuming the rig drilled at a constant rate, what was the height of the drill after 15 hours?

2. If the rig has been running constantly and is currently at a height of -147.5 feet, for how long has the rig been running?

3. Use the coordinate grid to show the drill's progress.

hours
-10 O 10 20 30 40 50 60 70 80 90 100
feet
-50
-100
-150
-200
-250

4. At this rate, how many hours will it take until the drill reaches -250 feet?

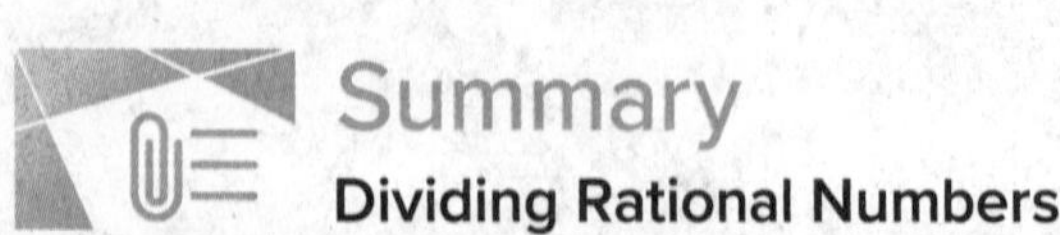

Summary

Dividing Rational Numbers

Any division problem is actually a multiplication problem:

- $6 \div 2 = 3$ because $2 \cdot 3 = 6$
- $6 \div -2 = -3$ because $-2 \cdot -3 = 6$
- $-6 \div 2 = -3$ because $2 \cdot -3 = -6$
- $-6 \div -2 = 3$ because $-2 \cdot 3 = -6$

Because we know how to multiply signed numbers, that means we know how to divide them.

- The sign of a positive number divided by a negative number is always negative.
- The sign of a negative number divided by a positive number is always negative.
- The sign of a negative number divided by a negative number is always positive.

A number that can be used in place of the variable that makes the equation true is called a **solution** to the equation. For example, for the equation $x \div -2 = 5$, the solution is -10, because it is true that $-10 \div -2 = 5$.

Glossary

solution to an equation

NAME ______________________ DATE ______________ PERIOD ____________

Practice
Dividing Rational Numbers

1. Find the quotients:

a. $24 \div -6$ **b.** $-15 \div 0.3$ **c.** $-4 \div -20$

2. Find the quotients.

a. $\frac{2}{5} \div \frac{3}{4}$ **b.** $\frac{9}{4} \div \frac{-3}{4}$

c. $\frac{-5}{7} \div \frac{-1}{3}$ **d.** $\frac{-5}{3} \div \frac{1}{6}$

3. Is the solution positive or negative?

a. $2 \cdot x = 6$ **b.** $-2 \cdot x = 6.1$

c. $2.9 \cdot x = -6.04$ **d.** $-2.473 \cdot x = -6.859$

4. Find the solution mentally.

a. $3 \cdot -4 = a$ **b.** $b \cdot (-3) = -12$

c. $-12 \cdot c = 12$ **d.** $d \cdot 24 = -12$

5. In order to make a specific shade of green paint, a painter mixes $1\frac{1}{2}$ quarts of blue paint, 2 cups of green paint, and $\frac{1}{2}$ gallon of white paint. How much of each color is needed to make 100 cups of this shade of green paint? **(Lesson 4-2)**

6. Here is a list of the highest and lowest elevation on each continent. **(Lesson 5-3)**

	Highest Point (m)	Lowest Point (m)
Europe	4,810	-28
Asia	8,848	-427
Africa	5,895	-155
Australia	4,884	-15
North America	6,198	-86
South America	6,960	-105
Antarctica	4,892	-50

a. Which continent has the largest difference in elevation? The smallest?

b. Make a display (dot plot, box plot, or histogram) of the data set and explain why you chose that type of display to represent this data set.

Lesson 5-12

Negative Rates

NAME ______________________ DATE __________ PERIOD __________

Learning Goal Let's apply what we know about signed numbers.

Warm Up

12.1 Grapes per Minute

1. If you eat 5 grapes per minute for 8 minutes, how many grapes will you eat?

2. If you hear 9 new songs per day for 3 days, how many new songs will you hear?

3. If you run 15 laps per practice, how many practices will it take you to run 30 laps?

Activity

12.2 Water Level in the Aquarium

1. A large aquarium should contain 10,000 liters of water when it is filled correctly. It will overflow if it gets up to 12,000 liters. The fish will get sick if it gets down to 4,000 liters. The aquarium has an automatic system to help keep the correct water level. If the water level is too low, a faucet fills it. If the water level is too high, a drain opens.

   One day, the system stops working correctly. The faucet starts to fill the aquarium at a rate of 30 liters per minute, and the drain opens at the same time, draining the water at a rate of 20 liters per minute.

   a. Is the water level rising or falling? How do you know?

   b. How long will it take until the tank starts overflowing or the fish get sick?

2. A different aquarium should contain 15,000 liters of water when filled correctly. It will overflow if it gets to 17,600 liters.

One day there is an accident and the tank cracks in 4 places. Water flows out of each crack at a rate of $\frac{1}{2}$ liter per hour. An emergency pump can re-fill the tank at a rate of 2 liters per minute. How many minutes must the pump run to replace the water lost each hour?

Activity

12.3 Up and Down with the Piccards

1. Challenger Deep is the deepest known point in the ocean, at 35,814 feet below sea level. In 1960, Jacques Piccard and Don Walsh rode down in the Trieste and became the first people to visit the Challenger Deep.

   a. If sea level is represented by 0 feet, explain how you can represent the depth of a submarine descending from sea level to the bottom of Challenger Deep.

   b. Trieste's descent was a change in depth of -3 feet per second. We can use the relationship $y = -3x$ to model this, where y is the depth (in feet) and x is the time (in seconds). Using this model, how much time would the Trieste take to reach the bottom?

NAME ______________________ DATE ____________ PERIOD ____________

c. It took the Trieste 3 hours to ascend back to sea level. This can be modeled by a different relationship $y = kx$. What is the value of k in this situation?

2. The design of the Trieste was based on the design of a hot air balloon built by Auguste Piccard, Jacques's father. In 1932, Auguste rode in his hot-air balloon up to a record-breaking height.

   a. Auguste's ascent took 7 hours and went up 51,683 feet. Write a relationship $y = kx$ to represent his ascent from his starting location.

   b. Auguste's descent took 3 hours and went down 52,940 feet. Write another relationship to represent his descent.

   c. Did Auguste Piccard end up at a greater or lesser altitude than his starting point? How much higher or lower?

Are you ready for more?

During which part of either trip was a Piccard changing vertical position the fastest? Explain your reasoning.

- Jacques's descent
- Jacques's ascent
- Auguste's ascent
- Auguste's descent

Summary

Negative Rates

We saw earlier that we can represent speed with direction using signed numbers.

Speed with direction is called *velocity*.

Positive velocities always represent movement in the opposite direction from negative velocities.

We can do this with vertical movement.
Moving up can be represented with positive numbers and moving down with negative numbers.

The magnitude tells you how fast, and the sign tells you which direction.
(We could actually do it the other way around if we wanted to, but usually we make up positive and down negative.)

NAME ____________________ DATE __________ PERIOD __________

Practice
Negative Rates

1. Describe a situation where each of the following quantities might be useful.
   a. -20 gallons per hour
   b. -10 feet per minute
   c. -0.1 kilograms per second

2. A submarine is only allowed to change its depth by rising toward the surface in 60-meter stages. It starts off at -340 meters.
   a. At what depth is it after:
      i. 1 stage
      ii. 2 stages
      iii. 4 stages
   b. How many stages will it take to return to the surface?

3. Some boats were traveling up and down a river. A satellite recorded the movements of several boats.
   a. A motor boat traveled -3.4 miles per hour for 0.75 hours. How far did it go?
   b. A tugboat traveled -1.5 miles in 0.3 hours. What was its velocity?
   c. What do you think that negative distances and velocities could mean in this situation?

4. Respond to each of the following. **(Lesson 4-3)**

a. A cookie recipe uses 3 cups of flour to make 15 cookies. How many cookies can you make with this recipe with 4 cups of flour? (Assume you have enough of the other ingredients.)

b. A teacher uses 36 centimeters of tape to hang up 9 student projects. At that rate, how much tape would the teacher need to hang up 10 student projects?

5. Evaluate each expression. When the answer is not a whole number, write your answer as a fraction. **(Lesson 5-11)**

a. $-4 \cdot -6$

b. $-24 \cdot \frac{-7}{6}$

c. $4 \div -6$

d. $\frac{4}{3} \div -24$

Lesson 5-13

Expressions with Rational Numbers

NAME ______________________ DATE ____________ PERIOD ____________

Learning Goal Let's develop our signed number sense.

Warm Up

13.1 True or False: Rational Numbers

Decide if each statement is true or false. Be prepared to explain your reasoning.

1. (-38.76)(-15.6) is negative

2. 10,000 − 99,999 < 0

3. $\left(\frac{3}{4}\right)\left(-\frac{4}{3}\right) = 0$

4. (30)(-80) − 50 = 50 − (30)(-80)

Activity

13.2 Card Sort: The Same but Different

Your teacher will give you a set of cards. Group them into pairs of expressions that have the same value.

Activity

13.3 Near and Far from Zero

1. For each set of values for a and b, evaluate the given expressions and record your answers in the table.

a	b	$-a$	$-4b$	$-a + b$	$a \div -b$	a^2	b^3
$-\frac{1}{2}$	6						
$\frac{1}{2}$	-6						
-6	$-\frac{1}{2}$						

2. When $a = -\frac{1}{2}$ and $b = 6$, which expression...
   a. has the largest value?
   b. has the smallest value?
   c. is the closest to zero?

3. When $a = \frac{1}{2}$ and $b = -6$, which expression...
   a. has the largest value?
   b. has the smallest value?
   c. is the closest to zero?

4. When $a = -6$ and $b = -\frac{1}{2}$, which expression...
   a. has the largest value?
   b. has the smallest value?
   c. is the closest to zero?

Are you ready for more?

Are there any values you could use for a and b that would make all of these expressions have the same value? Explain your reasoning.

NAME ______________________ DATE ______________ PERIOD ______________

Activity

13.4 Seagulls and Sharks Again

A seagull has a vertical position a, and a shark has a vertical position b. Draw and label a point on the vertical axis to show the vertical position of each new animal.

1. A dragonfly at d, where $d = -b$
2. A jellyfish at j, where $j = 2b$
3. An eagle at e, where $e = \frac{1}{4}a$.
4. A clownfish at c, where $c = \frac{-a}{2}$
5. A vulture at v, where $v = a + b$
6. A goose at g, where $g = a - b$

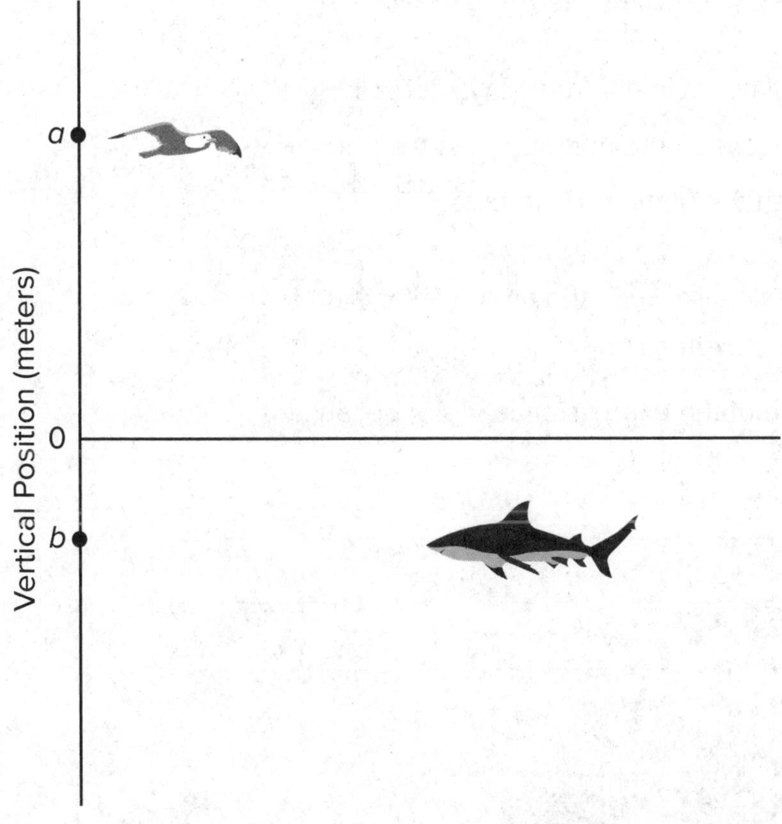

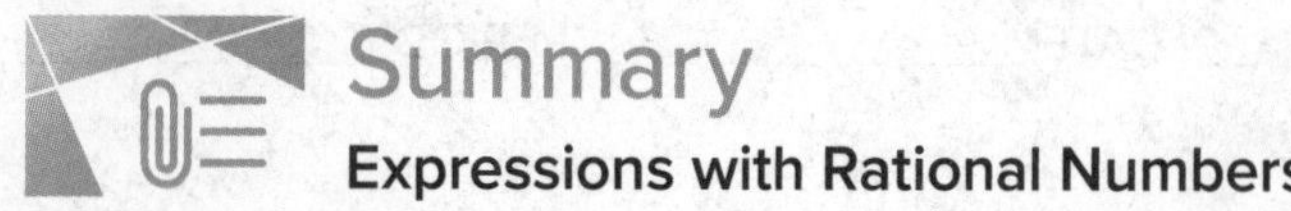

Summary

Expressions with Rational Numbers

We can represent sums, differences, products, and quotients of **rational numbers**, and combinations of these, with numerical and algebraic expressions.

Sums:

$\frac{1}{2} + -9$ $\quad\quad$ $-8.5 + x$

Differences:

$\frac{1}{2} - -9$ $\quad\quad$ $-8.5 - x$

Products:

$\left(\frac{1}{2}\right)(-9)$ $\quad\quad$ $-8.5x$

Quotients:

$\frac{1}{2} \div -9$ $\quad\quad$ $\frac{-8.5}{x}$

We can write the product of two numbers in different ways.

- By putting a little dot between the factors, like this: $-8.5 \cdot x$.
- By putting the factors next to each other without any symbol between them at all, like this: $-8.5x$.

We can write the quotient of two numbers in different ways as well.

- By writing the division symbol between the numbers, like this: $-8.5 \div x$.
- By writing a fraction bar between the numbers, like this: $\frac{-8.5}{x}$.

When we have an algebraic expression like $\frac{-8.5}{x}$ and are given a value for the variable, we can find the value of the expression.

For example, if x is 2, then the value of the expression is -4.25, because $-8.5 \div 2 = -4.25$.

Glossary

rational number

NAME ________________________ DATE ____________ PERIOD ____________

Practice

Expressions with Rational Numbers

1. The value of x is $\frac{-1}{4}$. Order these expressions from least to greatest:

x $\quad$ $1 - x$ $\quad$ $x - 1$ $\quad$ $-1 \div x$

2. Here are four expressions that have the value $\frac{-1}{2}$:

$\frac{-1}{4} + \left(\frac{-1}{4}\right)$ $\quad$ $\frac{1}{2} - 1$ $\quad$ $-2 \cdot \frac{1}{4}$ $\quad$ $-1 \div 2$

Write five expressions: a sum, a difference, a product, a quotient, and one that involves at least two operations that have the value $\frac{-3}{4}$.

3. Find the value of each expression.

a. $-22 + 5$

b. $-22 - (-5)$

c. $(-22)(-5)$

d. $-22 \div 5$

4. The price of an ice cream cone is $3.25, but it costs $3.51 with tax. What is the sales tax rate? **(Lesson 4-10)**

5. Two students are both working on the same problem: A box of laundry soap has 25% more soap in its new box. The new box holds 2 kg. How much soap did the old box hold? **(Lesson 4-7)**

Here is how Jada set up her double number line.

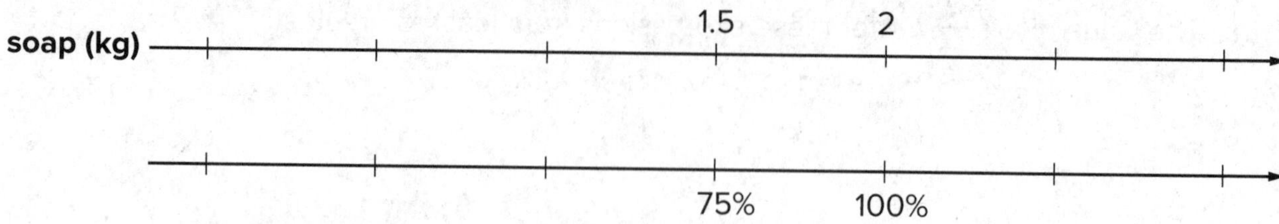

Here is how Lin set up her double number line.

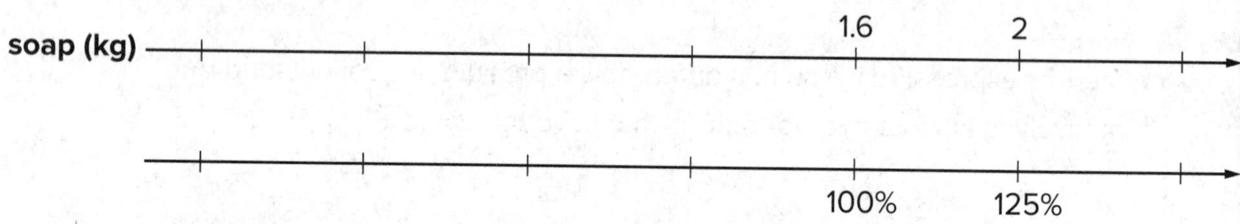

Do you agree with either of them? Explain or show your reasoning.

6. Respond to each of the following. **(Lesson 4-3)**

   a. A coffee maker's directions say to use 2 tablespoons of ground coffee for every 6 ounces of water. How much coffee should you use for 33 ounces of water?

   b. A runner is running a 10 km race. It takes her 17.5 minutes to reach the 2.5 km mark. At that rate, how long will it take her to run the whole race?

Lesson 5-14

Solving Problems with Rational Numbers

NAME _______________ DATE _______________ PERIOD _______________

Learning Goal Let's use all four operations with signed numbers to solve problems.

Warm Up

14.1 Which One Doesn't Belong: Equations

Which equation doesn't belong?

(A.) $\frac{1}{2}x = -50$

(B.) $-60t = 30$

(C.) $x + 90 = -100$

(D.) $-0.01 = -0.001x$

Activity

14.2 Draining and Filling a Tank

A tank of water is being drained. Due to a problem, the sensor does not start working until some time into the draining process. The sensor starts its recording at time zero when there are 770 liters in the tank.

1. Given that the drain empties the tank at a constant rate of 14 liters per minute, complete the table:

Time after Sensor Starts (minutes)	Change in Water (liters)	Expression	Water in the Tank (liters)
0	0	770 + (0)(-14)	770
1	-14	770 + (1)(-14)	756
5	-70		
10			

2. Later, someone wants to use the data to find out how long the tank had been draining before the sensor started. Complete this table:

Time after Sensor Starts (minutes)	Change in Water (liters)	Expression	Water in the Tank (liters)
1	-14	770 + (1)(-14)	756
0	0	770 + (0)(-14)	770
-1	14	770 + (-1)(-14)	784
-2	28		
-3			
-4			
-5			

3. If the sensor started working 15 minutes into the tank draining, how much was in the tank to begin with?

Activity
14.3 Buying and Selling Power

A utility company charges $0.12 per kilowatt-hour for energy a customer uses. They give a credit of $0.025 for every kilowatt-hour of electricity a customer with a solar panel generates that they don't use themselves.

A customer has a charge of $82.04 and a credit of -$4.10 on this month's bill.

1. What is the amount due this month?

NAME ______________________ DATE ______________ PERIOD ______________

2. How many kilowatt-hours did they use?

3. How many kilowatt-hours did they generate that they didn't use themselves?

Are you ready for more?

1. Find the value of the expression without a calculator.

(2)(-30) + (-3)(-20) + (-6)(-10) − (2)(3)(10)

2. Write an expression that uses addition, subtraction, multiplication, and division and only negative numbers that has the same value.

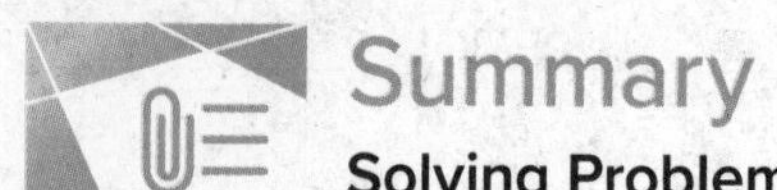

Summary

Solving Problems with Rational Numbers

We can apply the rules for arithmetic with rational numbers to solve problems.

In general, $a - b = a + -b$.

If $a - b = x$, then $x + b = a$.

We can add $-b$ to both sides of this second equation to get that $x = a + -b$.

Remember: the *distance* between two numbers is independent of the order, but the *difference* depends on the order.

And when multiplying or dividing:

- The sign of a positive number multiplied or divided by a negative number is always negative.
- The sign of a negative number multiplied or divided by a positive number is always negative.
- The sign of a negative number multiplied or divided by a negative number is always positive.

NAME ______________________ DATE ______________ PERIOD ______________

Practice

Solving Problems with Rational Numbers

1. A bank charges a service fee of $7.50 per month for a checking account. A bank account has $85.00. If no money is deposited or withdrawn except the service charge, how many months until the account balance is negative?

2. The table shows transactions in a checking account.

January	February	March	April
-38.50	250.00	-14.00	-86.80
126.30	-135.20	99.90	-570.00
429.40	35.50	-82.70	100.00
-265.00	-62.30	-1.50	-280.10

   a. Find the total of the transactions for each month.

   b. Find the mean total for the four months.

3. A large aquarium of water is being filled with a hose. Due to a problem, the sensor does not start working until some time into the filling process. The sensor starts its recording at the time zero minutes. The sensor initially detects the tank has 225 liters of water in it.

   a. The hose fills the aquarium at a constant rate of 15 liters per minute. What will the sensor read at the time 5 minutes?

b. Later, someone wants to use the data to find the amount of water at times before the sensor started. What should the sensor have read at the time -7 minutes?

4. A furniture store pays a wholesale price for a mattress. Then, the store marks up the retail price to 150% of the wholesale price. Later, they put the mattress on sale for 50% off the retail price. A customer just bought the mattress on sale and paid $1,200. **(Lesson 4-11)**

a. What was the retail price of the mattress, before the discount?

b. What was the wholesale price, before the markup?

5. Respond to each of the following. **(Lesson 4-10)**

a. A restaurant bill is $21. You leave a 15% tip. How much do you pay including the tip?

b. Which of the following represents the amount a customer pays including the tip of 15% if the bill was b dollars? Select **all** that apply.

(A.) $0.15b$

(B.) $15b$

(C.) $b + 0.15b$

(D.) $1.15b$

(E.) $1.015b$

(F.) $b + \frac{15}{100}b$

(G.) $b + 0.15$

Lesson 5-15

Solving Equations with Rational Numbers

NAME ______________________ DATE ______________ PERIOD ____________

Learning Goal Let's solve equations that include negative values.

Warm Up

15.1 Number Talk: Opposites and Reciprocals

The variables a through h all represent *different* numbers. Mentally find numbers that make each equation true.

$\frac{3}{5} \cdot \frac{5}{3} = a$ $\quad 7 \cdot b = 1$ $\quad c \cdot d = 1$

$-6 + 6 = e$ $\quad 11 + f = 0$ $\quad g + h = 0$

Activity

15.2 Match Solutions

Match each equation to its solution. Be prepared to explain your reasoning.

Equations	Solutions
1. $\frac{1}{2}x = -5$	$x = -4.5$
2. $-2x = -9$	$x = -\frac{1}{2}$
3. $-\frac{1}{2}x = \frac{1}{4}$	$x = -10$
4. $-2x = 7$	$x = 4.5$
5. $x + -2 = -6.5$	$x = 2\frac{1}{2}$
6. $-2 + x = \frac{1}{2}$	$x = -3.5$

Activity

15.3 Trip to the Mountains

The Hiking Club is on a trip to hike up a mountain.

1. The members increased their elevation 290 feet during their hike this morning. Now they are at an elevation of 450 feet.

   a. Explain how to find their elevation before the hike.

   b. Han says the equation $e + 290 = 450$ describes the situation. What does the variable e represent?

   c. Han says that he can rewrite his equation as $e = 450 + -290$ to solve for e. Compare Han's strategy to your strategy for finding the beginning elevation.

2. The temperature fell 4 degrees in the last hour. Now it is 21 degrees. Write and solve an equation to find the temperature it was 1 hour ago.

NAME ________________ DATE ________ PERIOD ________

3. There are 3 times as many students participating in the hiking trip this year than last year. There are 42 students on the trip this year.

a. Explain how to find the number of students that came on the hiking trip last year.

b. Mai says the equation $3s = 42$ describes the situation. What does the variable s represent?

c. Mai says that she can rewrite her equation as $s = \frac{1}{3} \cdot 42$ to solve for s. Compare Mai's strategy to your strategy for finding the number of students on last year's trip.

4. The cost of the hiking trip this year is $\frac{2}{3}$ of the cost of last year's trip. This year's trip cost \$32. Write and solve an equation to find the cost of last year's trip.

Are you ready for more?

A number line is shown below. The numbers 0 and 1 are marked on the line, as are two other rational numbers a and b.

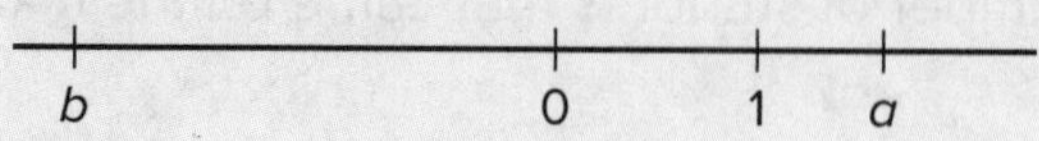

Decide which of the following numbers are positive and which are negative.

$a - 1$ $a - 2$ $-b$ $a + b$ $a - b$ $ab + 1$

Activity

15.4 Card Sort: Matching Inverses

Your teacher will give you a set of cards with numbers on them.

1. Match numbers with their additive inverses.

2. Next, match numbers with their multiplicative inverses.

3. What do you notice about the numbers and their inverses?

NAME ______________________________ DATE ______________ PERIOD ______________

Summary

Solving Equations with Rational Numbers

To solve the equation $x + 8 = -5$, we can add the opposite of 8, or -8, to each side:

$$\begin{aligned} x + 8 &= -5 \\ (x + 8) + -8 &= (-5) + -8 \\ x &= -13 \end{aligned}$$

Because adding the opposite of a number is the same as subtracting that number, we can also think of it as subtracting 8 from each side.

We can use the same approach for this equation:

$$\begin{aligned} -12 &= t + -\frac{2}{9} \\ (-12) + \frac{2}{9} &= \left(t + -\frac{2}{9}\right) + \frac{2}{9} \\ -11\frac{7}{9} &= t \end{aligned}$$

To solve the equation $8x = -5$, we can multiply each side by the reciprocal of 8, or $\frac{1}{8}$:

$$\begin{aligned} 8x &= -5 \\ \frac{1}{8}(8x) &= \frac{1}{8}(-5) \\ x &= -\frac{5}{8} \end{aligned}$$

Because multiplying by the reciprocal of a number is the same as dividing by that number, we can also think of it as dividing by 8. We can use the same approach for this equation:

$$\begin{aligned} -12 &= -\frac{2}{9}t \\ -\frac{9}{2}(-12) &= -\frac{9}{2}\left(-\frac{2}{9}t\right) \\ 54 &= t \end{aligned}$$

Glossary

variable

Practice

Solving Equations with Rational Numbers

1. Solve.

   a. $\frac{2}{5}t = 6$

   b. $-4.5 = a - 8$

   c. $\frac{1}{2} + p = -3$

   d. $12 = x \cdot 3$

   e. $-12 = -3y$

2. Match each equation to a step that will help solve the equation.

Equations

a. $5x = 0.4$

b. $\frac{x}{5} = 8$

c. $3 = \frac{-x}{5}$

d. $7 = -5x$

Steps

i. Multiply each side by 5.

ii. Multiply each side by -5.

iii. Multiply each side by $\frac{1}{5}$.

iv. Multiply each side by $\frac{-1}{5}$.

NAME ______________________________ DATE ______________ PERIOD ________

3. Evaluate each expression if x is $\frac{2}{5}$, y is -4, and z is -0.2. **(Lesson 5-13)**

a. $x + y$

b. $2x - z$

c. $x + y + z$

d. $y \cdot x$

4. Respond to each of the following.

a. Write an equation where a number is added to a variable, and a solution is -8.

b. Write an equation where a number is multiplied by a variable, and a solution is $\frac{-4}{5}$.

5. The markings on the number line are evenly spaced. Label the other markings on the number line. **(Lesson 5-8)**

-2.5   -1   0

6. In 2012, James Cameron descended to the bottom of Challenger Deep in the Marianas Trench, the deepest point in the ocean. The vessel he rode in was called DeepSea Challenger. Challenger Deep is 35,814 feet deep at its lowest point. **(Lesson 5-12)**

   a. DeepSea Challenger's descent was a change in depth of (-4) feet per second. We can use the equation $y = -4x$ to model this relationship, where y is the depth and x is the time in seconds that have passed. How many seconds does this model suggest it would take for DeepSea Challenger to reach the bottom?

   b. To end the mission, DeepSea Challenger made a one-hour ascent to the surface. How many seconds is this?

   c. The ascent can be modeled by a different proportional relationship $y = kx$. What is the value of k in this case?

Lesson 5-16

Representing Contexts with Equations

NAME ______________________ DATE ______________ PERIOD ____________

Learning Goal Let's write equations that represent situations.

Warm Up

16.1 Don't Solve It

Is the solution positive or negative?

1. $(-8.7)(1.4) = a$

2. $-8.7b = 1.4$

3. $-8.7 + c = -1.4$

4. $-8.7 - d = -1.4$

Activity

16.2 Warmer or Colder than Before?

For each situation:

- Find *two* equations that could represent the situation from the bank of equations. (Some equations will not be used.)
- Explain what the variable v represents in the situation.
- Determine the value of the variable that makes the equation true and explain your reasoning.

Bank of Equations:

$-3v = 9$	$v = -16 + 6$	$v = \frac{1}{3} \cdot (-6)$	$v + 12 = 4$
$-4 \cdot 3 = v$	$v = 4 + (-12)$	$v = -16 - (6)$	$v = 9 + 3$
$-4 \cdot -3 = v$	$-3v = -6$	$-6 + v = -16$	$-4 = \frac{1}{3}v$
$v = -\frac{1}{3} \cdot 9$	$v = -\frac{1}{3} \cdot (-6)$	$v = 4 + 12$	$4 = 3v$

1. Between 6 a.m. and noon, the temperature rose 12 degrees Fahrenheit to 4 degrees Fahrenheit.

2. At midnight the temperature was -6 degrees. By 4 a.m. the temperature had fallen to -16 degrees.

3. The temperature is 0 degrees at midnight and dropping 3 degrees per hour. The temperature is -6 degrees at a certain time.

4. The temperature is 0 degrees at midnight and dropping 3 degrees per hour. The temperature is 9 degrees at a certain time.

5. The temperature at 9 p.m. is one third the temperature at midnight.

NAME ______________________ DATE ____________ PERIOD ____________

Activity
16.3 Animals Changing Altitudes

1. Match each situation with a diagram on the next page.

   a. A penguin is standing 3 feet above sea level and then dives down 10 feet. What is its depth?

   b. A dolphin is swimming 3 feet below sea level and then jumps up 10 feet. What is its height at the top of the jump?

   c. A sea turtle is swimming 3 feet below sea level and then dives down 10 feet. What is its depth?

   d. An eagle is flying 10 feet above sea level and then dives down to 3 feet above sea level. What was its change in altitude?

   e. A pelican is flying 10 feet above sea level and then dives down reaching 3 feet below sea level. What was its change in altitude?

   f. A shark is swimming 10 feet below sea level and then swims up reaching 3 feet below sea level. What was its change in depth?

2. Next, write an equation to represent each animal's situation and answer the question. Be prepared to explain your reasoning.

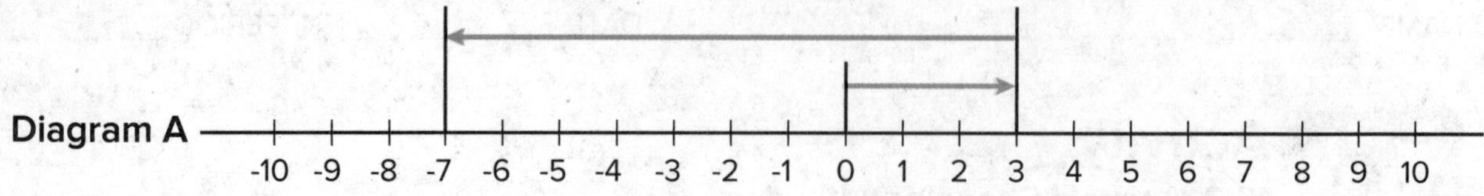
Diagram A
-10
-9
-8
-7
-6
-5
-4
-3
-2
-1
0
1
2
3
4
5
6
7
8
9
10

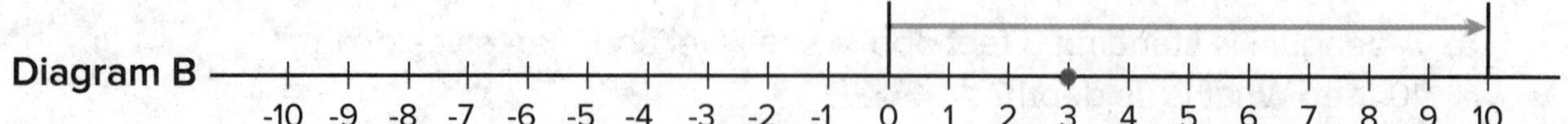
Diagram B
-10
-9
-8
-7
-6
-5
-4
-3
-2
-1
0
1
2
3
4
5
6
7
8
9
10

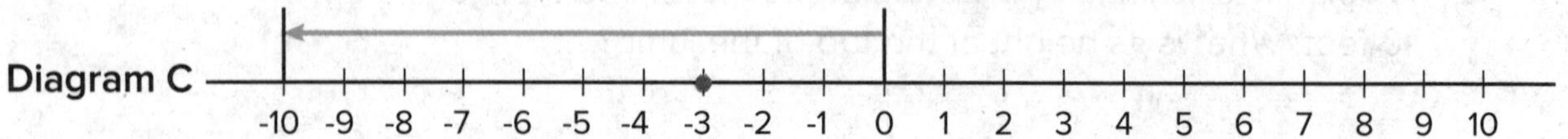
Diagram C
-10
-9
-8
-7
-6
-5
-4
-3
-2
-1
0
1
2
3
4
5
6
7
8
9
10

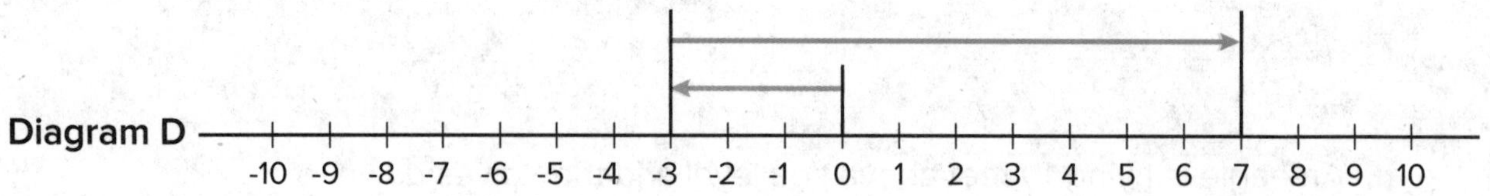
Diagram D
-10
-9
-8
-7
-6
-5
-4
-3
-2
-1
0
1
2
3
4
5
6
7
8
9
10

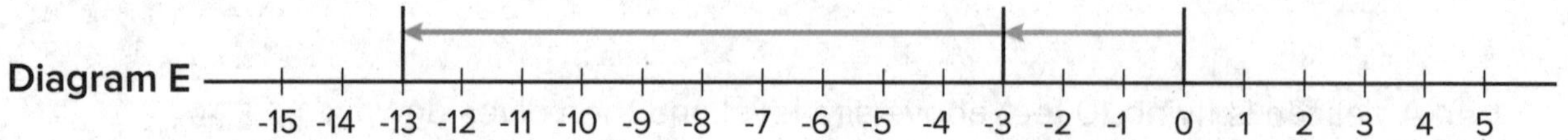
Diagram E
-15
-14
-13
-12
-11
-10
-9
-8
-7
-6
-5
-4
-3
-2
-1
0
1
2
3
4
5

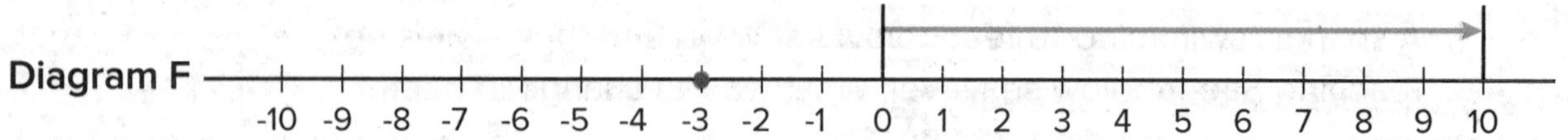
Diagram F
-10
-9
-8
-7
-6
-5
-4
-3
-2
-1
0
1
2
3
4
5
6
7
8
9
10

NAME ______________________ DATE ____________ PERIOD ____________

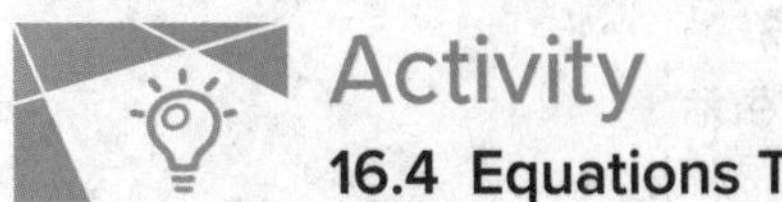

16.4 Equations Tell a Story

Your teacher will assign your group *one* of these situations. Create a visual display about your situation that includes:

- an equation that represents your situation
- what your variable and each term in the equation represent
- how the operations in the equation represent the relationships in the story
- how you use inverses to solve for the unknown quantity
- the solution to your equation

1. As a $7\frac{1}{4}$ inch candle burns down, its height decreases $\frac{3}{4}$ inch each hour. How many hours does it take for the candle to burn completely?

2. On Monday $\frac{1}{9}$ of the enrolled students in a school were absent. There were 4,512 students present. How many students are enrolled at the school?

3. A hiker begins at sea level and descends 25 feet every minute. How long will it take to get to an elevation of -750 feet?

4. Jada practices the violin for the same amount of time every day. On Tuesday she practices for 35 minutes. How much does Jada practice in a week?

5. The temperature has been dropping $2\frac{1}{2}$ degrees every hour and the current temperature is -15° F. How many hours ago was the temperature 0° F?

6. The population of a school increased by 12%, and now the population is 476. What was the population before the increase?

7. During a 5% off sale, Diego pays $74.10 for a new hockey stick. What was the original price?

8. A store buys sweaters for $8 and sells them for $26. How many sweaters does the store need to sell to make a profit of $990?

Are you ready for more?

Diego and Elena are 2 miles apart and begin walking towards each other. Diego walks at a rate of 3.7 miles per hour and Elena walks 4.3 miles per hour. While they are walking, Elena's dog runs back and forth between the two of them, at a rate of 6 miles per hour. Assuming the dog does not lose any time in turning around, how far has the dog run by the time Diego and Elena reach each other?

Summary

Representing Contexts with Equations

We can use variables and equations involving signed numbers to represent a story or answer questions about a situation.

For example, if the temperature is -3°C and then falls to -17°C, we can let x represent the temperature change and write the equation $-3 + x = -17$.

We can solve the equation by adding 3 to each side.
Since $-17 + 3 = -14$, the change is -14°C.

Here is another example. If a starfish is descending by $\frac{3}{2}$ feet every hour then we can solve $-\frac{3}{2}h = -6$ to find out how many hours h it takes the starfish to go down 6 feet.

We can solve this equation by multiplying each side by $-\frac{2}{3}$.
Since $-\frac{2}{3} \cdot -6 = 4$, we know it will take the starfish 4 hours to descend 6 feet.

NAME ______________________ DATE ______________ PERIOD ________

Practice

Representing Contexts with Equations

1. Match each situation to one of the equations.

Situations

a. A whale was diving at a rate of 2 meters per second. How long will it take for the whale to get from the surface of the ocean to an elevation of -12 meters at that rate?

b. A swimmer dove below the surface of the ocean. After 2 minutes, she was 12 meters below the surface. At what rate was she diving?

c. The temperature was -12 degrees Celsius and rose to 2 degrees Celsius. What was the change in temperature?

d. The temperature was 2 degrees Celsius and fell to -12 degrees Celsius. What was the change in temperature?

Equations

$-12 + x = 2$

$2 + x = -12$

$-2x = -12$

$2x = -12$

2. Starting at noon, the temperature dropped steadily at a rate of 0.8 degrees Celsius every hour.

For each of these situations, write and solve an equation and describe what your variable represents.

a. How many hours did it take for the temperature to decrease by 4.4 degrees Celsius?

b. If the temperature after the 4.4 degree drop was -2.5 degrees Celsius, what was the temperature at noon?

3. Kiran mixes $\frac{3}{4}$ cup of raisins, 1 cup peanuts, and $\frac{1}{2}$ cup of chocolate chips to make trail mix. How much of each ingredient would he need to make 10 cups of trail mix? Explain your reasoning. **(Lesson 4-3)**

4. Find the value of each expression. **(Lesson 5-6)**

   a. 12 + -10

   b. -5 − 6

   c. -42 + 17

   d. 35 − (-8)

   e. $-4\frac{1}{2} + 3$

5. The markings on the number line are evenly spaced. Label the other markings on the number line. **(Lesson 5-8)**

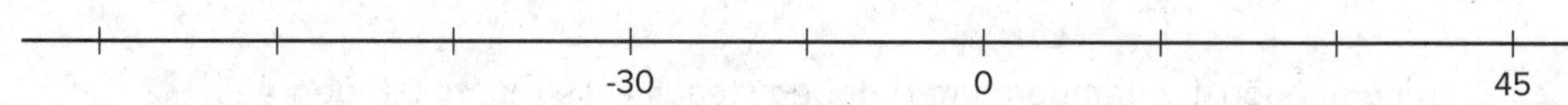

6. Kiran drinks 6.4 oz of milk each morning. How many days does it take him to finish a 32 oz container of milk?

   a. Write and solve an equation for the situation.

   b. What does the variable represent?

Lesson 5-17

The Stock Market

NAME ______________________ DATE ______________ PERIOD ______________

Learning Goal Let's learn about the stock market.

Warm Up

17.1 Revisiting Interest and Depreciation

1. Lin deposited $300 in a savings account that has a 2% interest rate per year. How much is in her account after 1 year? After 2 years?

2. Diego wants to sell his bicycle. It cost $150 when he bought it but has depreciated by 15%. How much should he sell it for?

Activity

17.2 Gains and Losses

1. Here is some information from the stock market in September 2016. Complete the table.

Company	Value on Day 1 (dollars)	Value on Day 2 (dollars)	Change in Value (dollars)	Change in Value as a Percentage of Day 1 Value
Mobile Tech Company	107.95	111.77	3.82	3.54
Electrical Appliance Company		114.03	2.43	2.18
Oil Corporation	26.14	25.14		-3.83
Department Store Company	7.38	7.17		
Jewelry Company		70.30		2.27

2. Which company's change in dollars had the largest magnitude?

3. Which company's change in percentage had the largest magnitude?

NAME ______________________ DATE ____________ PERIOD ____________

Activity
17.3 What Is a Stock Portfolio?

A person who wants to make money by investing in the stock market usually buys a portfolio, or a collection of different stocks. That way, if one of the stocks decreases in value, they won't lose all of their money at once.

1. Here is an example of someone's stock portfolio. Complete the table to show the total value of each investment.

Company	Price (dollars)	Number of Shares	Total Value (dollars)
Technology Company	107.75	98	
Airline Company	133.54	27	
Film Company	95.95	135	
Sports Clothing Company	58.96	100	

2. Here is the same portfolio the next year. Complete the table to show the new total value of each investment.

Company	Old Price (dollars)	Price Change	New Price (dollars)	Number of Shares	Total Value (dollars)
Technology Company	107.75	+2.43%		98	
Airline Company	133.54	-7.67%		27	
Film Company	95.95		87.58	135	
Sports Clothing Company	58.96	-5.56%		100	

3. Did the entire portfolio increase or decrease in value over the year?

Activity
17.4 Your Own Stock Portfolio

Your teacher will give you a list of stocks.

1. Select a combination of stocks with a total value close to, but no more than, $100.

2. Using the new list, how did the total value of your selected stocks change?

Learning Targets

Lesson	Learning Target(s)
5-1 Interpreting Negative Numbers	• I can compare rational numbers. • I can use rational numbers to describe temperature and elevation.
5-2 Changing Temperatures	• I can use a number line to add positive and negative numbers.
5-3 Changing Elevation	• I understand how to add positive and negative numbers in general.
5-4 Money and Debts	• I understand what positive and negative numbers mean in a situation involving money.

(continued on the next page)

(continued from the previous page)

Lesson	Learning Target(s)
5-5 Representing Subtraction	• I can explain the relationship between addition and subtraction of rational numbers. • I can use a number line to subtract positive and negative numbers.
5-6 Subtracting Rational Numbers	• I can find the difference between two rational numbers. • I understand how to subtract positive and negative numbers in general.
5-7 Adding and Subtracting to Solve Problems	• I can solve problems that involve adding and subtracting rational numbers.
5-8 Position, Speed, and Direction	• I can multiply a positive number with a negative number. • I can use rational numbers to represent speed and direction.

Lesson	Learning Target(s)
5-9 Multiplying Rational Numbers	• I can explain what it means when time is represented with a negative number in a situation about speed and direction. • I can multiply two negative numbers.
5-10 Multiply!	• I can solve problems that involve multiplying rational numbers.
5-11 Dividing Rational Numbers	• I can divide rational numbers.
5-12 Negative Rates	• I can solve problems that involve multiplying and dividing rational numbers. • I can solve problems that involve negative rates.
5-13 Expressions with Rational Numbers	• I can add, subtract, multiply, and divide rational numbers. • I can evaluate expressions that involve rational numbers.

(continued on the next page)

(continued from the previous page)

Lesson	Learning Target(s)
5-14 Solving Problems with Rational Numbers	• I can represent situations with expressions that include rational numbers. • I can solve problems using the four operations with rational numbers.
5-15 Solving Equations with Rational Numbers	• I can solve equations that include rational numbers and have rational solutions.
5-16 Representing Contexts with Equations	• I can explain what the solution to an equation means for the situation. • I can write and solve equations to represent situations that involve rational numbers.
5-17 The Stock Market	• I can solve problems about the stock market using rational numbers and percentages.

Notes:

Glossary

A

absolute value The absolute value of a number is its distance from 0 on the number line.

The absolute value of -7 is 7, because it is 7 units away from 0. The absolute value of 5 is 5, because it is 5 units away from 0.

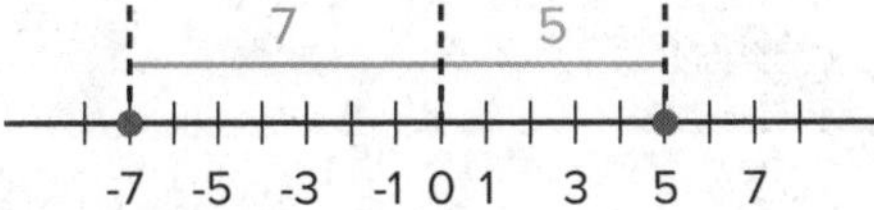

adjacent angles Adjacent angles share a side and a vertex.

In this diagram, angle *ABC* is adjacent to angle *DBC*.

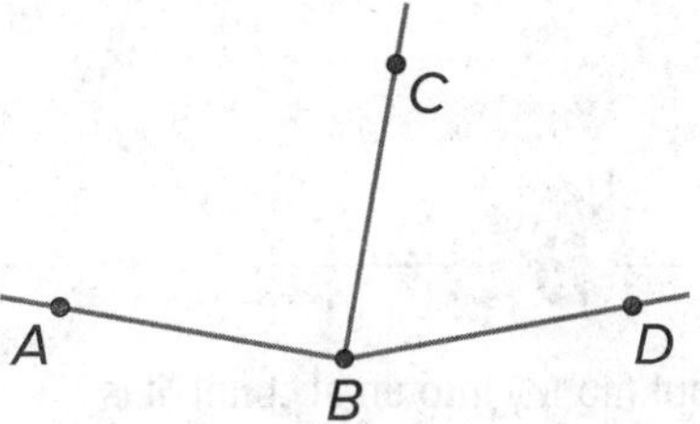

area Area is the number of square units that cover a two-dimensional region, without any gaps or overlaps.

For example, the area of region A is 8 square units. The area of the shaded region of B is $\frac{1}{2}$ square unit.

Region A

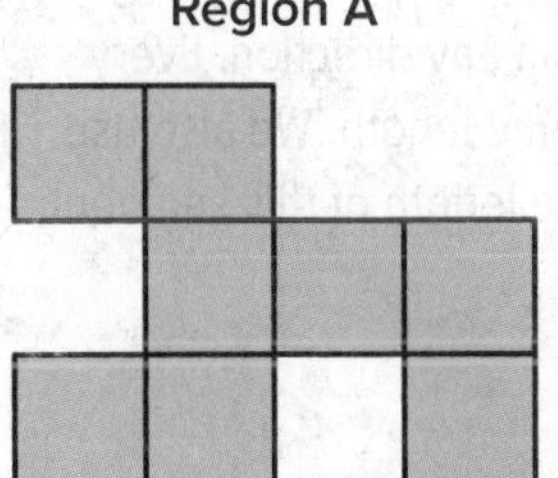

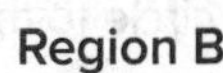

Region B

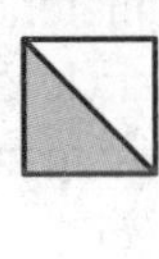

area of a circle If the radius of a circle is r units, then the area of the circle is πr^2 square units.

For example, a circle has radius 3 inches. Its area is $\pi 3^2$ square inches, or 9π square inches, which is approximately 28.3 square inches.

B

base (of a prism or pyramid) The word *base* can also refer to a face of a polyhedron. A prism has two identical bases that are parallel. A pyramid has one base. A prism or pyramid is named for the shape of its base.

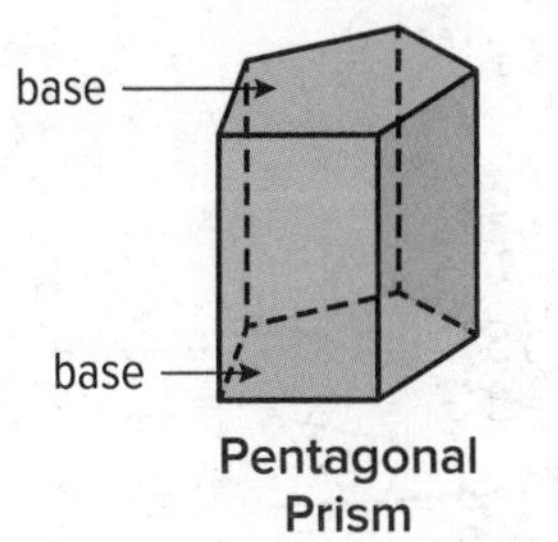

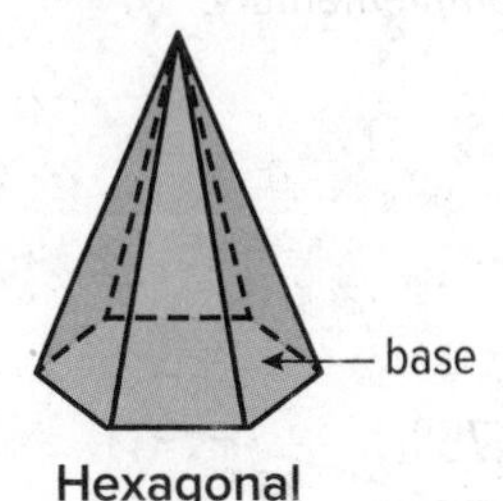

C

chance experiment A chance experiment is something you can do over and over again, and you don't know what will happen each time.

For example, each time you spin the spinner, it could land on red, yellow, blue, or green.

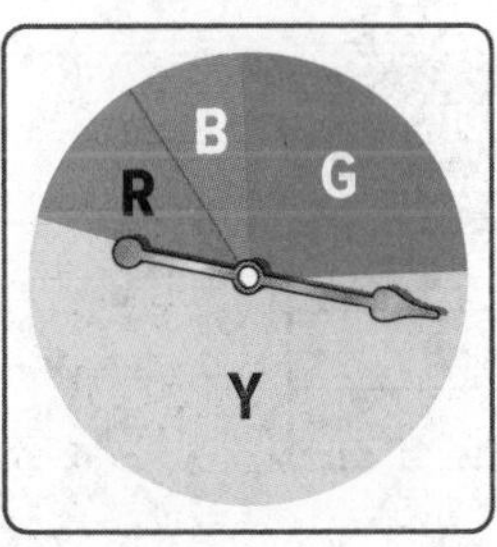

circle A circle is made out of all the points that are the same distance from a given point.

For example, every point on this circle is 5 cm away from point *A*, which is the center of the circle.

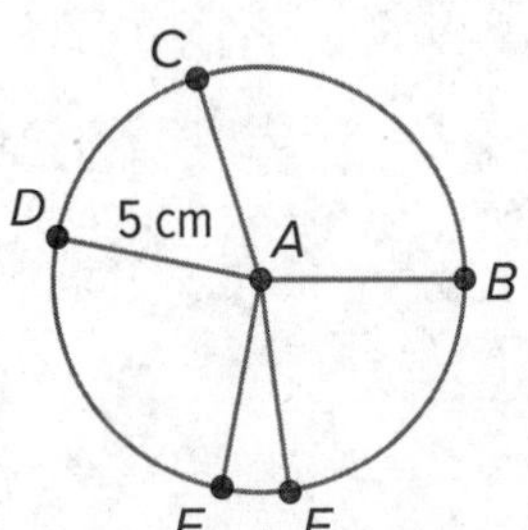

circumference The circumference of a circle is the distance around the circle. If you imagine the circle as a piece of string, it is the length of the string. If the circle has radius r then the circumference is $2\pi r$.

The circumference of a circle of radius 3 is $2 \cdot \pi \cdot 3$, which is 6π, or about 18.85.

complementary Complementary angles have measures that add up to 90 degrees.

For example, a 15° angle and a 75° angle are complementary.

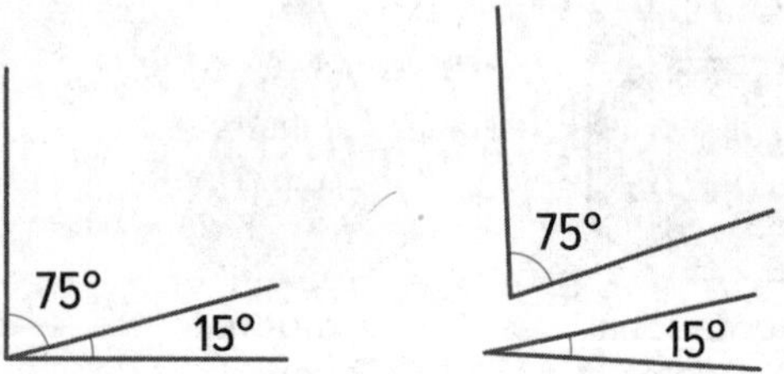

constant of proportionality In a proportional relationship, the values for one quantity are each multiplied by the same number to get the values for the other quantity. This number is called the constant of proportionality.

In this example, the constant of proportionality is 3, because $2 \cdot 3 = 6$, $3 \cdot 3 = 9$, and $5 \cdot 3 = 15$. This means that there are 3 apples for every 1 orange in the fruit salad.

Number of Oranges	Number of Apples
2	6
3	9
5	15

coordinate plane The coordinate plane is a system for telling where points are. For example, point R is located at (3, 2) on the coordinate plane, because it is three units to the right and two units up.

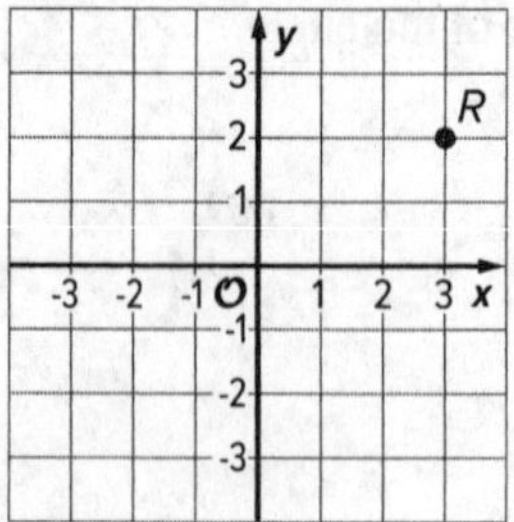

corresponding When part of an original figure matches up with part of a copy, we call them corresponding parts. These could be points, segments, angles, or distances.

For example, point B in the first triangle corresponds to point E in the second triangle. Segment AC corresponds to segment DF.

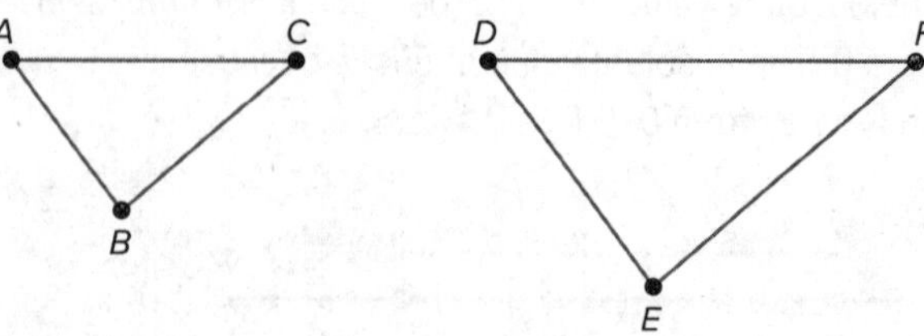

cross section A cross section is the new face you see when you slice through a three-dimensional figure.

For example, if you slice a rectangular pyramid parallel to the base, you get a smaller rectangle as the cross section.

D

deposit When you put money into an account, it is called a *deposit*.

For example, a person added \$60 to their bank account. Before the deposit, they had \$435. After the deposit, they had \$495, because $435 + 60 = 495$.

diameter A diameter is a line segment that goes from one edge of a circle to the other and passes through the center. A diameter can go in any direction. Every diameter of the circle is the same length. We also use the word *diameter* to mean the length of this segment.

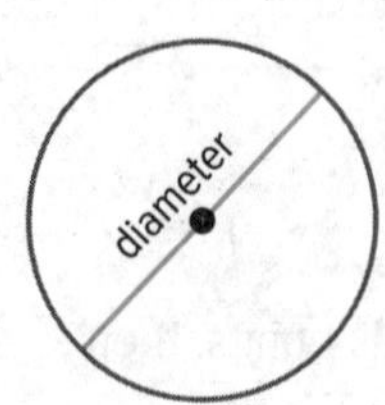

E

equivalent expressions Equivalent expressions are always equal to each other. If the expressions have variables, they are equal whenever the same value is used for the variable in each expression.

For example, $3x + 4x$ is equivalent to $5x + 2x$. No matter what value we use for x, these expressions are always equal. When x is 3, both expressions equal 21. When x is 10, both expressions equal 70.

equivalent ratios Two ratios are equivalent if you can multiply each of the numbers in the first ratio by the same factor to get the numbers in the second ratio. For example, 8 : 6 is equivalent to 4 : 3, because $8 \cdot \frac{1}{2} = 4$ and $6 \cdot \frac{1}{2} = 3$.

A recipe for lemonade says to use 8 cups of water and 6 lemons. If we use 4 cups of water and 3 lemons, it will make half as much lemonade. Both recipes taste the same, because 8 : 6 and 4 : 3 are equivalent ratios.

Cups of Water	Number of Lemons
8	6
4	3

event An event is a set of one or more outcomes in a chance experiment. For example, if we roll a number cube, there are six possible outcomes.

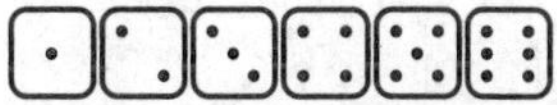

Examples of events are "rolling a number less than 3," "rolling an even number," or "rolling a 5."

expand To expand an expression, we use the distributive property to rewrite a product as a sum. The new expression is equivalent to the original expression.

For example, we can expand the expression $5(4x + 7)$ to get the equivalent expression $20x + 35$.

F

factor (an expression) To factor an expression, we use the distributive property to rewrite a sum as a product. The new expression is equivalent to the original expression.

For example, we can factor the expression $20x + 35$ to get the equivalent expression $5(4x + 7)$.

I

interquartile range (IQR) The interquartile range is one way to measure how spread out a data set is. We sometimes call this the IQR. To find the interquartile range we subtract the first quartile from the third quartile.

For example, the IQR of this data set is 20 because $50 - 30 = 20$.

22	29	30	31	32	43	44	45	50	50	59
		Q1			Q2			Q3		

L

long division Long division is a way to show the steps for dividing numbers in decimal form. It finds the quotient one digit at a time, from left to right.

For example, here is the long division for $57 \div 4$.

$$\begin{array}{r} 14.25 \\ 4\overline{)57.00} \\ -4 \\ \hline 17 \\ -16 \\ \hline 10 \\ -8 \\ \hline 20 \\ -20 \\ \hline 0 \end{array}$$

mean The mean is one way to measure the center of a data set. We can think of it as a balance point. For example, for the data set 7, 9, 12, 13, 14, the mean is 11.

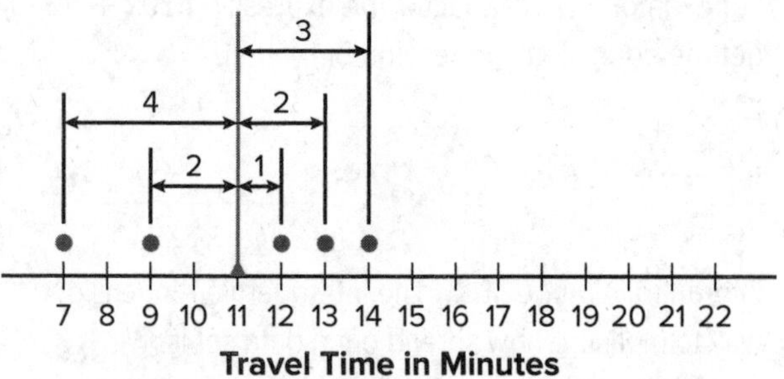

To find the mean, add up all the numbers in the data set. Then, divide by how many numbers there are. $7 + 9 + 12 + 13 + 14 = 55$ and $55 \div 5 = 11$.

mean absolute deviation (MAD) The mean absolute deviation is one way to measure how spread out a data set is. Sometimes we call this the MAD. For example, for the data set 7, 9, 12, 13, 14, the MAD is 2.4. This tells us that these travel times are typically 2.4 minutes away from the mean, which is 11.

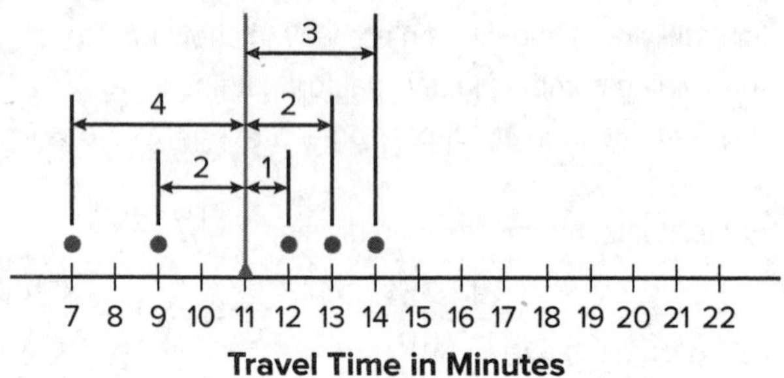

To find the MAD, add up the distance between each data point and the mean. Then, divide by how many numbers there are. $4 + 2 + 1 + 2 + 3 = 12$ and $12 \div 5 = 2.4$.

measurement error Measurement error is the positive difference between a measured amount and the actual amount.

For example, Diego measures a line segment and gets 5.3 cm. The actual length of the segment is really 5.32 cm. The measurement error is 0.02 cm, because $5.32 - 5.3 = 0.02$.

median The median is one way to measure the center of a data set. It is the middle number when the data set is listed in order.

For the data set 7, 9, 12, 13, 14, the median is 12.

For the data set 3, 5, 6, 8, 11, 12, there are two numbers in the middle. The median is the average of these two numbers. $6 + 8 = 14$ and $14 \div 2 = 7$.

negative number A negative number is a number that is less than zero. On a horizontal number line, negative numbers are usually shown to the left of 0.

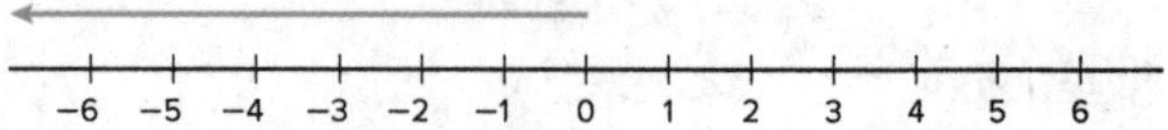

origin The origin is the point (0, 0) in the coordinate plane. This is where the horizontal axis and the vertical axis cross.

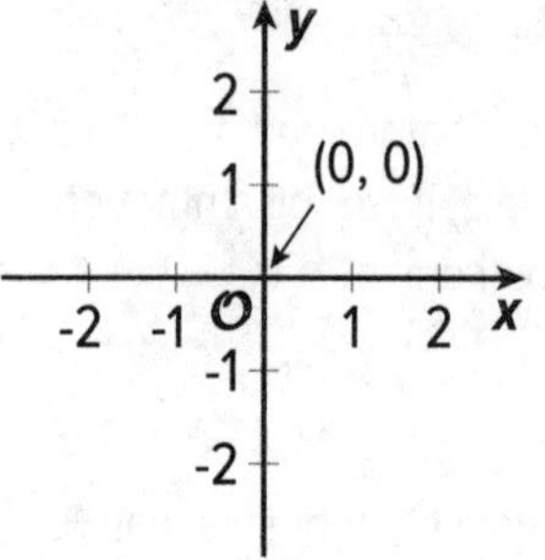

outcome An outcome of a chance experiment is one of the things that can happen when you do the experiment. For example, the possible outcomes of tossing a coin are heads and tails.

percent error Percent error is a way to describe error, expressed as a percentage of the actual amount.

For example, a box is supposed to have 150 folders in it. Clare counts only 147 folders in the box. This is an error of 3 folders. The percent error is 2%, because 3 is 2% of 150.

percentage A percentage is a rate per 100.

For example, a fish tank can hold 36 liters. Right now there is 27 liters of water in the tank. The percentage of the tank that is full is 75%.

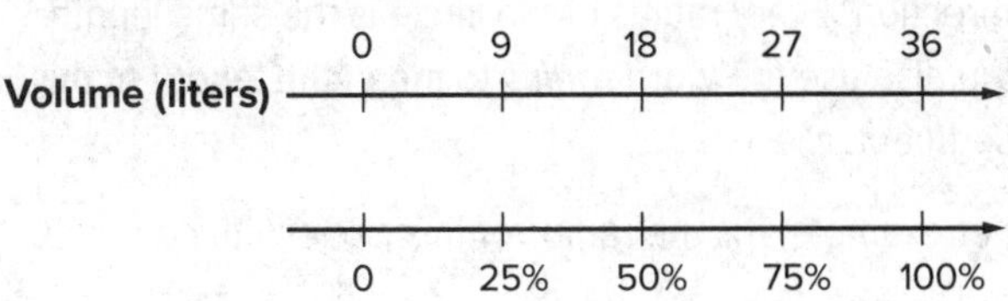

percentage decrease A percentage decrease tells how much a quantity went down, expressed as a percentage of the starting amount.

For example, a store had 64 hats in stock on Friday. They had 48 hats left on Saturday. The amount went down by 16.

This was a 25% decrease, because 16 is 25% of 64.

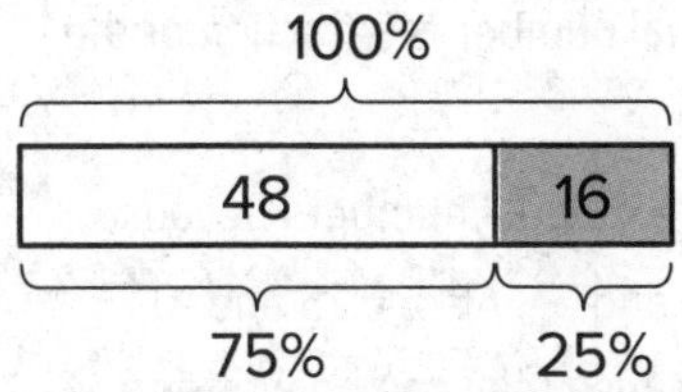

percentage increase A percentage increase tells how much a quantity went up, expressed as a percentage of the starting amount.

For example, Elena had $50 in the bank on Monday. She had $56 on Tuesday. The amount went up by $6.

This was a 12% increase, because 6 is 12% of 50.

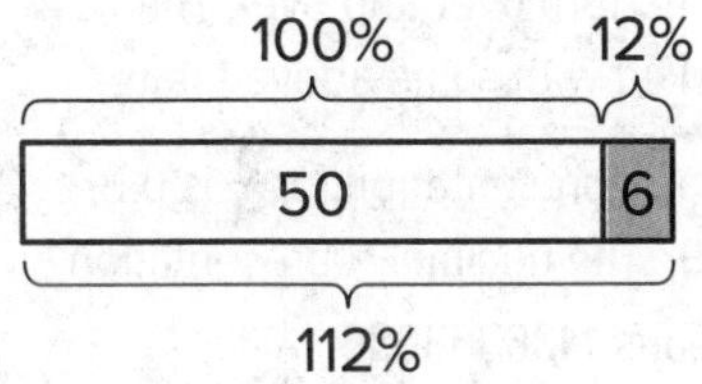

pi (π) There is a proportional relationship between the diameter and circumference of any circle. The constant of proportionality is pi. The symbol for pi is π.

We can represent this relationship with the equation $C = \pi d$, where C represents the circumference and d represents the diameter.

Some approximations for π are $\frac{22}{7}$, 3.14, and 3.14159.

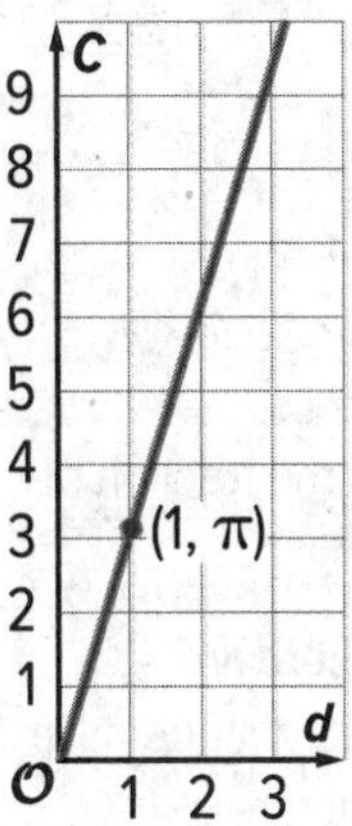

population A population is a set of people or things that we want to study.

For example, if we want to study the heights of people on different sports teams, the population would be all the people on the teams.

positive number A positive number is a number that is greater than zero. On a horizontal number line, positive numbers are usually shown to the right of 0.

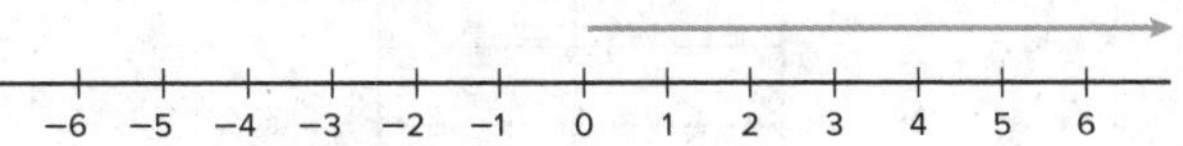

prism A prism is a type of polyhedron that has two bases that are identical copies of each other. The bases are connected by rectangles or parallelograms.

Here are some drawings of prisms.

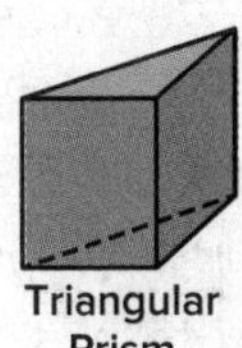
Triangular Prism

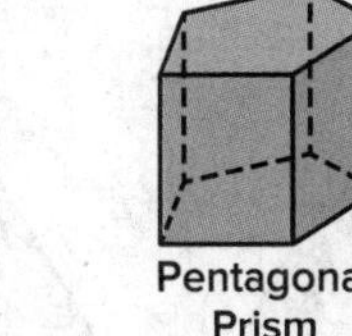
Pentagonal Prism

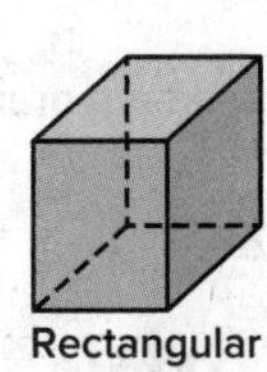
Rectangular Prism

probability The probability of an event is a number that tells how likely it is to happen. A probability of 1 means the event will always happen. A probability of 0 means the event will never happen.

For example, the probability of selecting a moon block at random from this bag is $\frac{4}{5}$.

proportion A proportion of a data set is the fraction of the data in a given category.

For example, a class has 20 students. There are 2 left-handed students and 18 right-handed students in the class. The proportion of students who are left-handed is $\frac{2}{20}$, or 0.1.

proportional relationship In a proportional relationship, the values for one quantity are each multiplied by the same number to get the values for the other quantity. For example, in this table every value of p is equal to 4 times the value of s on the same row. We can write this relationship as $p = 4s$. This equation shows that s is proportional to p.

s	p
2	8
3	12
5	20
10	40

pyramid A pyramid is a type of polyhedron that has one base. All the other faces are triangles, and they all meet at a single vertex.

Here are some drawings of pyramids.

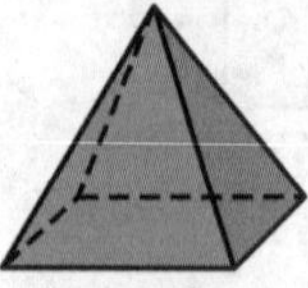

Rectangular Pyramid

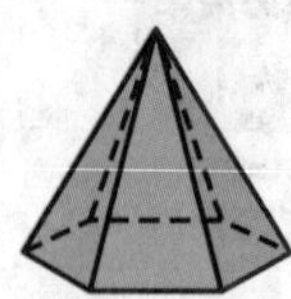

Hexagonal Pyramid

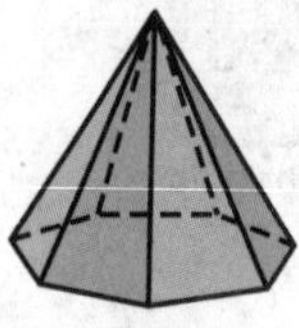

Heptagonal Pyramid

R

radius A radius is a line segment that goes from the center to the edge of a circle. A radius can go in any direction. Every radius of the circle is the same length. We also use the word *radius* to mean the length of this segment.

For example, r is the radius of this circle with center O.

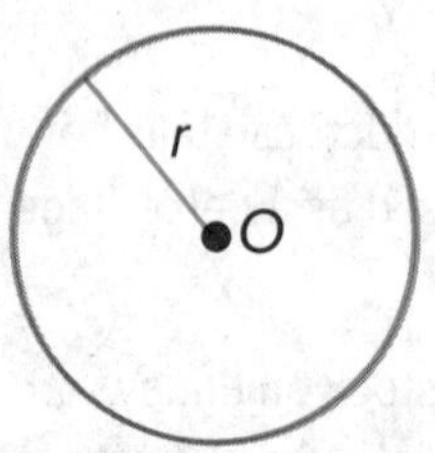

random Outcomes of a chance experiment are random if they are all equally likely to happen.

rational number A rational number is a fraction or the opposite of a fraction.

For example, 8 and -8 are rational numbers because they can be written as $\frac{8}{1}$ and $-\frac{8}{1}$. Also, 0.75 and -0.75 are rational numbers because they can be written as $\frac{75}{100}$ and $-\frac{75}{100}$.

reciprocal Dividing 1 by a number gives the reciprocal of that number. For example, the reciprocal of 12 is $\frac{1}{12}$, and the reciprocal of $\frac{2}{5}$ is $\frac{5}{2}$.

repeating decimal A repeating decimal has digits that keep going in the same pattern over and over. The repeating digits are marked with a line above them.

For example, the decimal representation for $\frac{1}{3}$ is $0.\overline{3}$, which means 0.3333333...The decimal representation for $\frac{25}{22}$ is $1.1\overline{36}$ which means 1.136363636...

representative A sample is representative of a population if its distribution resembles the population's distribution in center, shape, and spread.

For example, this dot plot represents a population.

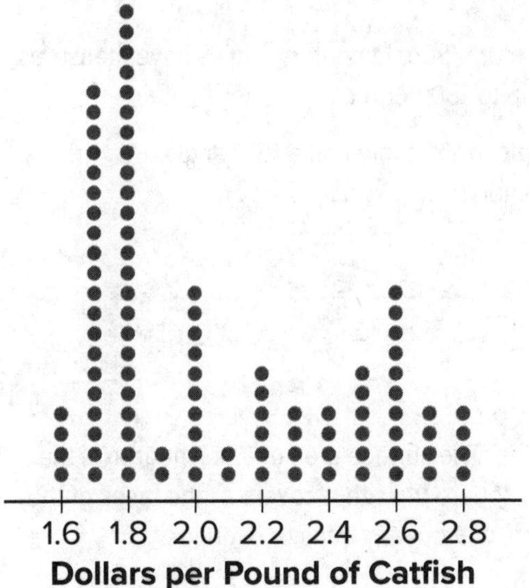

This dot plot shows a sample that is representative of the population.

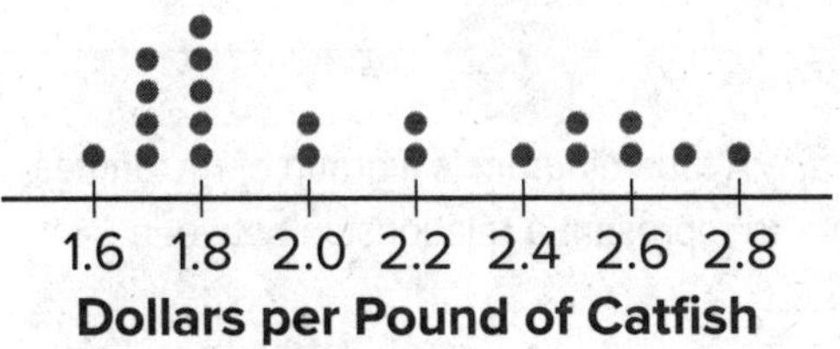

right angle A right angle is half of a straight angle. It measures 90 degrees.

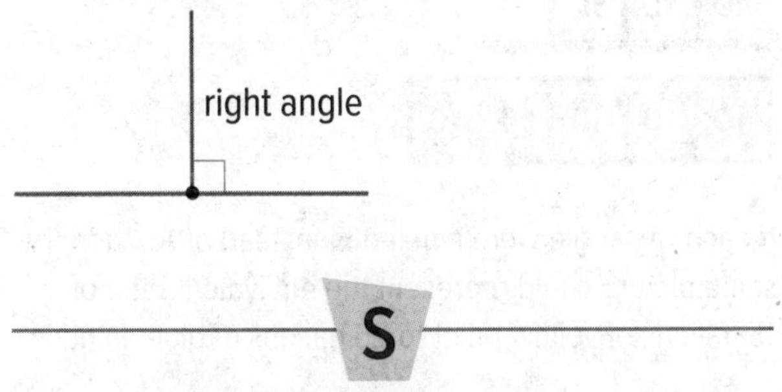

S

sample A sample is part of a population. For example, a population could be all the seventh grade students at one school. One sample of that population is all the seventh grade students who are in band.

sample space The sample space is the list of every possible outcome for a chance experiment.

For example, the sample space for tossing two coins is:

heads-heads	tails-heads
heads-tails	tails-tails

scale A scale tells how the measurements in a scale drawing represent the actual measurements of the object.

For example, the scale on this floor plan tells us that 1 inch on the drawing represents 8 feet in the actual room. This means that 2 inches would represent 16 feet, and $\frac{1}{2}$ inch would represent 4 feet.

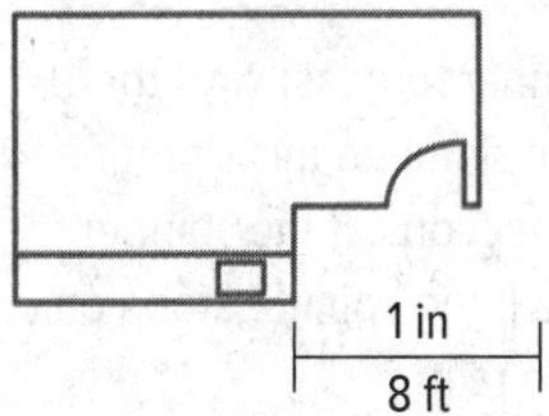

scale drawing A scale drawing represents an actual place or object. All the measurements in the drawing correspond to the measurements of the actual object by the same scale.

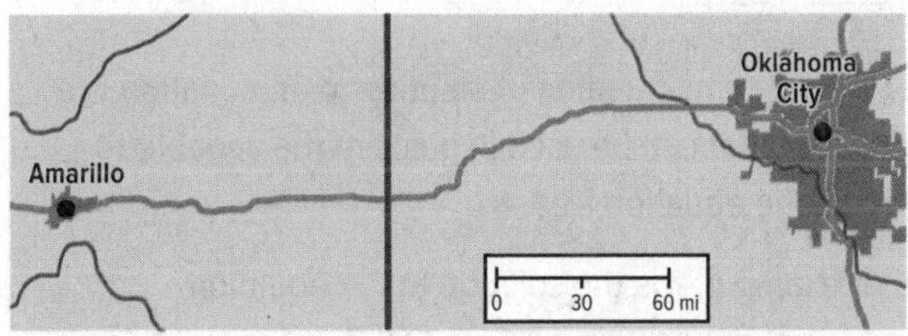

scale factor To create a scaled copy, we multiply all the lengths in the original figure by the same number. This number is called the scale factor.

In this example, the scale factor is 1.5, because $4 \cdot (1.5) = 6$, $5 \cdot (1.5) = 7.5$, and $6 \cdot (1.5) = 9$.

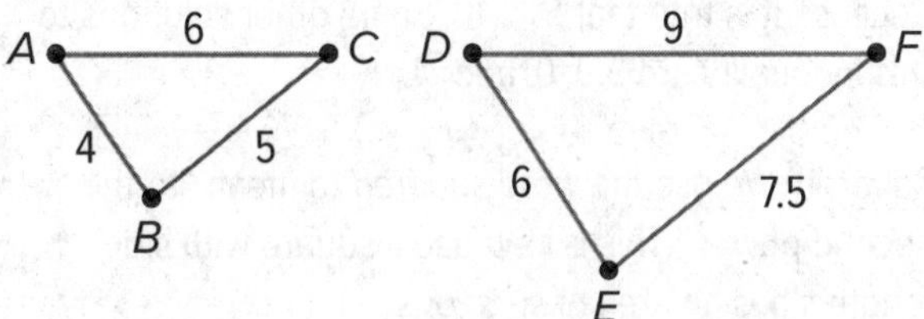

scaled copy A scaled copy is a copy of a figure where every length in the original figure is multiplied by the same number.

For example, triangle *DEF* is a scaled copy of triangle *ABC*. Each side length on triangle *ABC* was multiplied by 1.5 to get the corresponding side length on triangle *DEF*.

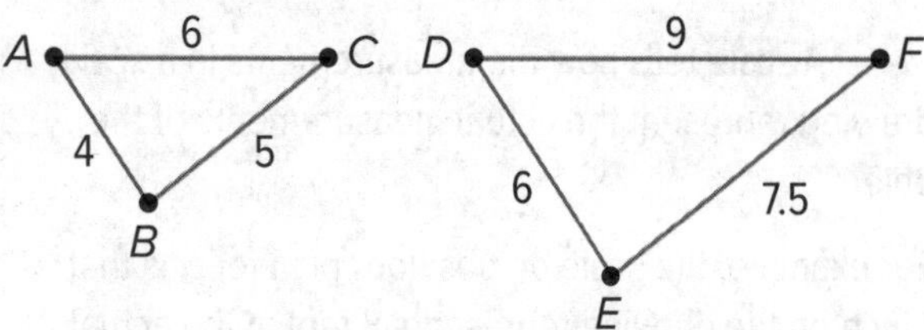

simulation A simulation is an experiment that is used to estimate the probability of a real-world event.

For example, suppose the weather forecast says there is a 25% chance of rain. We can simulate this situation with a spinner with four equal sections. If the spinner stops on red, it represents rain. If the spinner stops on any other color, it represents no rain.

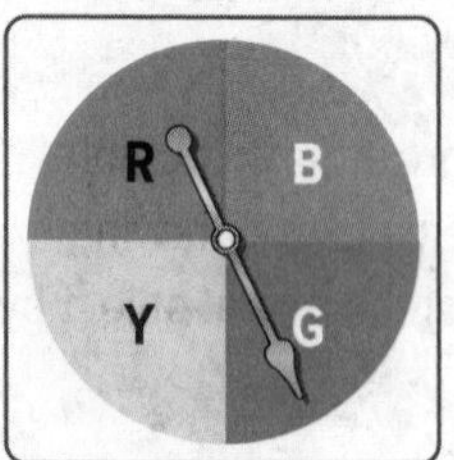

solution to an equation A solution to an equation is a number that can be used in place of the variable to make the equation true.

For example, 7 is the solution to the equation $m + 1 = 8$, because it is true that $7 + 1 = 8$. The solution to $m + 1 = 8$ is not 9, because $9 + 1 \neq 8$.

solution to an inequality A solution to an inequality is a number that can be used in place of the variable to make the inequality true.

For example, 5 is a solution to the inequality $c < 10$, because it is true that $5 < 10$. Some other solutions to this inequality are 9.9, 0, and -4.

squared We use the word *squared* to mean "to the second power." This is because a square with side length s has an area of $s \cdot s$, or s^2.

straight angle A straight angle is an angle that forms a straight line. It measures 180 degrees.

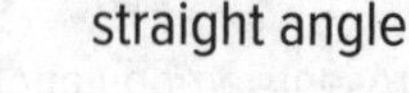

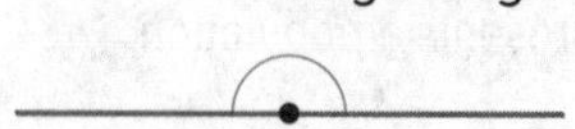

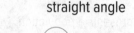

supplementary Supplementary angles have measures that add up to 180 degrees.

For example, a 15° angle and a 165° angle are supplementary.

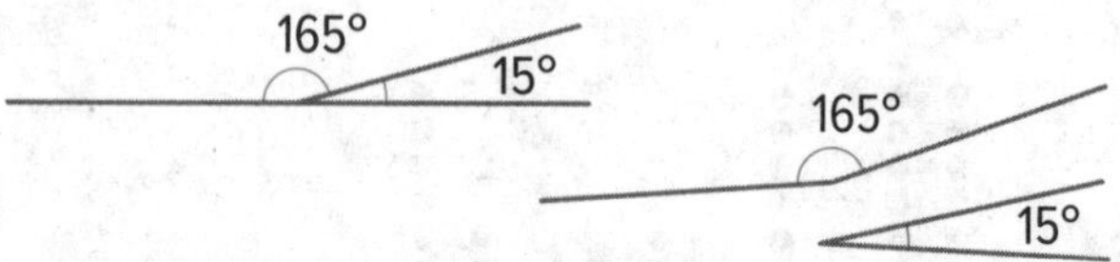

surface area The surface area of a polyhedron is the number of square units that covers all the faces of the polyhedron, without any gaps or overlaps.

For example, if the faces of a cube each have an area of 9 cm^2, then the surface area of the cube is $6 \cdot 9$, or 54 cm^2.

tape diagram A tape diagram is a group of rectangles put together to represent a relationship between quantities.

For example, this tape diagram shows a ratio of 30 gallons of yellow paint to 50 gallons of blue paint.

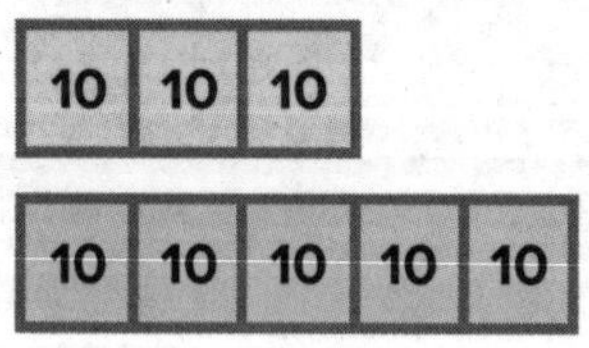

If each rectangle were labeled 5, instead of 10, then the same picture could represent the equivalent ratio of 15 gallons of yellow paint to 25 gallons of blue paint.

term A term is a part of an expression. It can be a single number, a variable, or a number and a variable that are multiplied together. For example, the expression $5x + 18$ has two terms. The first term is $5x$ and the second term is 18.

U

unit rate A unit rate is a rate per 1.

For example, 12 people share 2 pies equally. One unit rate is 6 people per pie, because $12 \div 2 = 6$. The other unit rate is $\frac{1}{6}$ of a pie per person, because $2 \div 12 = \frac{1}{6}$.

V

variable A variable is a letter that represents a number. You can choose different numbers for the value of the variable.

For example, in the expression $10 - x$, the variable is x. If the value of x is 3, then $10 - x = 7$, because $10 - 3 = 7$. If the value of x is 6, then $10 - x = 4$, because $10 - 6 = 4$.

vertical angles Vertical angles are opposite angles that share the same vertex. They are formed by a pair of intersecting lines. Their angle measures are equal.

For example, angles *AEC* and *DEB* are vertical angles. If angle *AEC* measures 120°, then angle *DEB* must also measure 120°.

Angles *AED* and *BEC* are another pair of vertical angles.

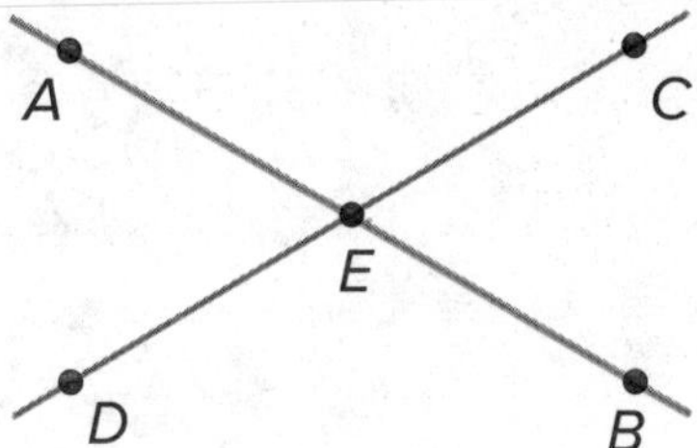

volume Volume is the number of cubic units that fill a three-dimensional region, without any gaps or overlaps.

For example, the volume of this rectangular prism is 60 units^3, because it is composed of 3 layers that are each 20 units^3.

W

withdrawal When you take money out of an account, it is called a *withdrawal*.

For example, a person removed \$25 from their bank account. Before the withdrawal, they had \$350. After the withdrawal, they had \$325, because $350 - 25 = 325$.

Index

V

W